Lecture Notes in Mathematics 1699

Editors:
A. Dold, Heidelberg
F. Takens, Groningen
B. Teissier, Paris

T0255596

Springer
Berlin
Heidelberg
New York
Barcelona
Hong Kong
London
Milan
Paris
Singapore
Tokyo

Atsushi Inoue

Tomita-Takesaki Theory in Algebras of Unbounded Operators

Springer

Author

Atsushi Inoue
Department of Applied Mathematics
Fukuoka University
Fukuoka 814-0180, Japan
e-mail: sm010888@ssat.fukuoka-u.ac.jp

Cataloging-in-Publication Data applied for

Die Deutsche Bibliothek - CIP-Einheitsaufnahme

Inoue, Atsushi:
Tomitak-Takesaki theory in algebras of unbounded operators / Atsushi Inoue. - Berlin ; Heidelberg ; New York ; Barcelona ; Budapest ; Hong Kong ; London ; Milan ; Paris ; Santa Clara ; Singapore ; Tokyo : Springer, 1998
(Lecture notes in mathematics ; 1699)
ISBN 3-540-65194-2

Mathematics Subject Classification (1991): 47D40, 47D25, 46L60, 46N50

ISSN 0075-8434
ISBN 3-540-65194-2 Springer-Verlag Berlin Heidelberg New York

This work is subject to copyright. All rights are reserved, whether the whole or part of the material is concerned, specifically the rights of translation, reprinting, re-use of illustrations, recitation, broadcasting, reproduction on microfilms or in any other way, and storage in data banks. Duplication of this publication or parts thereof is permitted only under the provisions of the German Copyright Law of September 9, 1965, in its current version, and permission for use must always be obtained from Springer-Verlag. Violations are liable for prosecution under the German Copyright Law.

© Springer-Verlag Berlin Heidelberg 1998
Printed in Germany

The use of general descriptive names, registered names, trademarks, etc. in this publication does not imply, even in the absence of a specific statement, that such names are exempt from the relevant protective laws and regulations and therefore free for general use.

Typesetting: Camera-ready TEX output by the author
SPIN: 10650140 41/3143-543210 - Printed on acid-free paper

Preface

Main part of this note is a summary of my studies during several years of the Tomita-Takesaki theory in O*-algebra. In 1995 I began this work for the preparation of the seminars during the summer of 1996 at the Mathematics Institute of Leipzig university. I wish to thank Professor K. D. Kürsten and K. Schmüdgen for their warm hospitality, for their interest in this work and for their encouragement. Further, I wish to thank them for many helpful discussions and suggestions when they visited the Department of our university in 1996, 1997. I also acknowlege Professor J. P. Antoine (Louvain Catholique University), Van Daele (Leuven Katholieke University), W. Karwowski (Wroclaw University), G. Epifanio and C. Trapani (Palermo University) and A. Arai and M. Kishimoto (Hokkaido University) for giving me opportunities of the seminars and lectures about this work for their colleagues and graduate students of their Mathematical Departments and for many invaluable discussions and many helpful suggestions. I should like to thank Professor H. Kurose and Dr. Ogi for their encouragement and many helpful conversations. It remains for me to express my gratitude to M. Takakura for typing this mannuscript in TeX. This work was supported in part by Japan Society for the Promotion Science and Japan Private School Promotion Foundation.

August 1998 *Atsushi Inoue*

Contents

Introduction

This note is devoted to a study of the O*-approach of the Tomita-Takesaki theory (T-T theory) for von Neumann algebras. As well-known, the T-T theory plays an important rule for the structure of von Neumann algebras and the physical applications. An extension of the T-T theory to $*$-algebras of closable operators called O*-algebras has been given for the Wightman quantum field theory and for quantum mechanics, and we have tried to develop systematically the T-T theory in O*-algebras for studies of the structure of O*-algebras and the physical applications for several years (Inoue [4, 5, 9, 10, 14, 17 -19], Antoine-Inoue-Ogi-Trapani [1], Inoue-Karwowski [1], Inoue-Karwowski-Ogi [1], Inoue-Kürsten [1]). The main purpose of this note is to summarize and develop these studies.

Let $\mathcal{D}$ be a dense subspace in a Hilbert space $\mathcal{H}$ and denote by $\mathcal{L}^{\dagger}(\mathcal{D})$ the set of all linear operators X from $\mathcal{D}$ to $\mathcal{D}$ such that $\mathcal{D}(X^*) \supset \mathcal{D}$ and $X^*\mathcal{D} \subset \mathcal{D}$. Then $\mathcal{L}^{\dagger}(\mathcal{D})$ is a $*$-algebra under the usual operations and the involution $X \to X^{\dagger} \equiv X^*\lceil\mathcal{D}$. A $*$-subalgebra of $\mathcal{L}^{\dagger}(\mathcal{D})$ is an *O*-algebra* on $\mathcal{D}$ in $\mathcal{H}$. O*-algebras were considered for the first time in 1962, independently by Borchers [1] and Uhlmann [1], in the Wightman formulation of quantum field theory. A systematic study was undertaken only at the begining of 1970, first by Powers [1] and Lassner [1], then by many mathematicians, from the pure mathematical situations (operator theory, topological $*$-algebras, representations of Lie algebras etc.) and the physical applications (Wightman quantum field theory, unbounded CCR-algebras, quantum groups etc.). Powers defined and studied the notions of closedness, self-adjointness and integrability of O*-algebras in analogy with the notions of closedness and self-adjointness of a single operator, respectively, and investigated the weak commutant which plays a fundamental role in the general theory. The *weak commutant* $\mathcal{M}'_{\mathrm{w}}$ of an O*-algebra $\mathcal{M}$ on $\mathcal{D}$ in $\mathcal{H}$ is defined by $\mathcal{M}'_{\mathrm{w}} = \{C \in \mathcal{B}(\mathcal{H}); (CX\xi|\eta) = (C\xi|X^{\dagger}\eta), {}^{\forall}X \in \mathcal{M}, {}^{\forall}\xi, \eta \in \mathcal{D}\}$, and it is weakly closed $*$-invariant subspace of the $*$-algebra $\mathcal{B}(\mathcal{H})$ of all bounded linear operators on $\mathcal{H}$, however is not an algebra in general. The self-adjointness of $\mathcal{M}$ implies $\mathcal{M}'_{\mathrm{w}}\mathcal{D} \subset \mathcal{D}$ and it implies that $\mathcal{M}'_{\mathrm{w}}$ is a von Neumann algebra. A survey of the theory of O*-algebras may be found in the recent monograph of Schmüdgen [21]. In Chapter I the general theory of O*-algebras is introduced.

In Chapter II the notion of standard generalized vectors which makes possible to develop the T-T theory in O*-algebras is defined and studied. We started such a study in case of O*-algebras with cyclic vectors. But, an O*-algebra is not spatially isomorphic with a direct sum of O*-algebras possessing cyclic vectors in general, in other words only a very special subfamily of O*-algebras has representations with cyclic vectors. On the other hand, the concept of cyclic vectors proved to be very useful for studies of O*-algebras. These facts suggest that a generalization of the notion of cyclic vectors would provide a useful tool for investigations of wider class of O*-algebras. Here we pursue this idea. We explain how to define the notion of generalized vectors. Let $\mathcal{M}$ be an O*-algebra on $\mathcal{D}$ in $\mathcal{H}$ and $\xi_0 \in \mathcal{H}$. If $\xi_0 \notin \mathcal{D}$ then every operator in $\mathcal{M}$ doesn't act on ξ_0. How do we treat with ξ_0? Three ways are considered: (i) Rigged Hilbert space : $\mathcal{D}[t_{\mathcal{M}}] \subset \mathcal{H} \subset \mathcal{D}^{\dagger}[t_{\mathcal{M}}]$. As usual, ξ_0 is regarded as an element of the topological conjugate dual $\mathcal{D}^{\dagger}[t_{\mathcal{M}}]$ of the locally convex space $\mathcal{D}[t_{\mathcal{M}}]$ equipped with the graph topology $t_{\mathcal{M}}$ on $\mathcal{D}$. (ii) Generalized vectors: A *generalized vector* λ for $\mathcal{M}$ is a linear map of a left ideal $\mathcal{D}(\lambda)$ of $\mathcal{M}$ into $\mathcal{D}$ such that $\lambda(AX) = A\lambda(X)$ for each $A \in \mathcal{M}$ and $X \in \mathcal{D}(\lambda)$. The generalized vector λ_{ξ_0} for $\mathcal{M}$ by ξ_0 is defined by $\mathcal{D}(\lambda_{\xi_0}) = \{X \in \mathcal{M}; \xi_0 \in \mathcal{D}(X^{\dagger *}) \text{ and } X^{\dagger *}\xi_0 \in \mathcal{D}\}$ and $\lambda_{\xi_0}(X) = X^{\dagger *}\xi_0$ for $X \in \mathcal{D}(\lambda_{\xi_0})$. This is the reason we call such a map λ generalized vector. (iii) Quasi-weights. A *quasi-weight* φ is a map of the algebraic positive cone $\mathcal{P}(\mathfrak{N}_{\varphi}) \equiv \{\sum_k X_k^{\dagger} X_k; X_k \in \mathfrak{N}_{\varphi}\}$ of a left ideal $\mathfrak{N}_{\varphi}$ of $\mathcal{M}$ into $\mathbb{R}_+$ such that $\varphi(A+B) = \varphi(A) + \varphi(B)$ and $\varphi(\lambda A) = \lambda\varphi(A)$ for each $A, B \in \mathcal{P}(\mathfrak{N}_{\varphi})$ and $\lambda \geq 0$. The quasi-weight ω_{ξ_0} by ξ_0 is defined by $\mathfrak{N}_{\omega_{\xi_0}} = \mathcal{D}(\lambda_{\xi_0})$ and $\omega_{\xi_0}(\sum_k X_k^{\dagger} X_k) = \sum_k \|X_k^{\dagger *}\xi_0\|^2$ for $\sum_k X_k^{\dagger} X_k \in \mathcal{P}(\mathfrak{N}_{\omega_{\xi_0}})$. Let $\mathcal{M}$ be a closed O*-algebra on $\mathcal{D}$ in $\mathcal{H}$ such that $\mathcal{M}'_{\mathrm{w}}\mathcal{D} \subset \mathcal{D}$ and λ be a generalized vector for $\mathcal{M}$ satisfying $(\mathrm{S})_1$ $\lambda((\mathcal{D}(\lambda)^{\dagger} \cap \mathcal{D}(\lambda))^2)$ is total in $\mathcal{H}$. The commutant λ^{C} of λ is defined by $\mathcal{D}(\lambda^{\mathrm{C}}) = \{K \in \mathcal{M}'_{\mathrm{w}}; {}^{\exists}\xi_K \in \mathcal{D} \text{ s.t. } K\lambda(X) = X\xi_K, {}^{\forall}X \in \mathcal{D}(\lambda)\}$ and $\lambda^{\mathrm{C}}(K) = \xi_K$ for $K \in \mathcal{D}(\lambda^{\mathrm{C}})$, and it is a generalized vector for the von Neumann algebra $\mathcal{M}'_{\mathrm{w}}$. Suppose that $(\mathrm{S})_2$ $\lambda^{\mathrm{C}}((\mathcal{D}(\lambda^{\mathrm{C}})^* \cap \mathcal{D}(\lambda^{\mathrm{C}}))^2)$ is total in $\mathcal{H}$. Then the commutant λ^{CC} of λ^{C} is similarly defined and it is a generalized vector for the von Neumann algebra $(\mathcal{M}'_{\mathrm{w}})'$. The maps $\lambda(X) \to \lambda(X^{\dagger})$, $X \in \mathcal{D}(\lambda)^{\dagger} \cap \mathcal{D}(\lambda)$ and $\lambda^{\mathrm{CC}}(A) \to \lambda^{\mathrm{CC}}(A^*), A \in \mathcal{D}(\lambda^{\mathrm{CC}})^* \cap \mathcal{D}(\lambda^{\mathrm{CC}})$ are closable conjugate linear maps whose closures are denoted by S_{λ} and $S_{\lambda^{\mathrm{CC}}}$, respectively. Let $S_{\lambda} = J_{\lambda}\Delta_{\lambda}^{1/2}$ and $S_{\lambda^{\mathrm{CC}}} = J_{\lambda^{\mathrm{CC}}}\Delta_{\lambda^{\mathrm{CC}}}^{1/2}$ be polar decompositions of S_{λ} and $S_{\lambda^{\mathrm{CC}}}$, respectively. By the Tomita fundamental theorem, $J_{\lambda^{\mathrm{CC}}}(\mathcal{M}'_{\mathrm{w}})'J_{\lambda^{\mathrm{CC}}} = \mathcal{M}'_{\mathrm{w}}$ and $\Delta_{\lambda^{\mathrm{CC}}}^{it}(\mathcal{M}'_{\mathrm{w}})'\Delta_{\lambda^{\mathrm{CC}}}^{-it} = (\mathcal{M}'_{\mathrm{w}})', {}^{\forall}t \in \mathbb{R}$. But, we don't know how $\{\Delta_{\lambda^{\mathrm{CC}}}^{it}\}_{t\in\mathbb{R}}$ acts on the O*-algebra $\mathcal{M}$ in general, and so we define a generalized vector which has the best condition. A generalized vector λ for $\mathcal{M}$ is *standard* if $(\mathrm{S})_1$, $(\mathrm{S})_2$ and the following conditions $(\mathrm{S})_3$ $\sim$ $(\mathrm{S})_5$ hold: $(\mathrm{S})_3$ $\Delta_{\lambda^{\mathrm{CC}}}^{it}\mathcal{D} \subset \mathcal{D}, {}^{\forall}t \in \mathbb{R}$. $(\mathrm{S})_4$ $\Delta_{\lambda^{\mathrm{CC}}}^{it}\mathcal{M}\Delta_{\lambda^{\mathrm{CC}}}^{-it} = \mathcal{M}, {}^{\forall}t \in \mathbb{R}$. $(\mathrm{S})_5$ $\Delta_{\lambda^{\mathrm{CC}}}^{it}(\mathcal{D}(\lambda)^{\dagger} \cap \mathcal{D}(\lambda))\Delta_{\lambda^{\mathrm{CC}}}^{-it} = \mathcal{D}(\lambda)^{\dagger} \cap \mathcal{D}(\lambda), {}^{\forall}t \in \mathbb{R}$. Suppose λ is a standard generalized vector for $\mathcal{M}$. Then $S_\lambda = S_{\lambda cc}, t \in \mathbb{R} \to \sigma_t^\lambda(X) \equiv \Delta_\lambda^{it} X \Delta_\lambda^{-it} (X \in \mathcal{M})$ is a one-parameter group of $*$-automorphisms of $\mathcal{M}$ and the quasi-weight φ_λ generated by λ satisfies the KMS-condition with respect to $\{\sigma_t^\lambda\}_{t\in\mathbb{R}}$. To apply the unbounded T-T theory to more examples we weaken the above conditions $(S)_3 \sim (S)_5$ and define the notion of modular generalized vectors. A generalized vector λ for $\mathcal{M}$ is *modular* if it satisfies the conditions $(S)_1$, $(S)_2$ and the following condition (M): There exists a dense subspace $\mathcal{E}$ of $\mathcal{D}[t_\mathcal{M}]$ such that $(M)_1$ $\mathcal{M}\mathcal{E} \subset \mathcal{E}$ and $(M)_2$ $\Delta_{\lambda cc}^{it}\mathcal{E} \subset \mathcal{E}$ for all $t \in \mathbb{R}$. We need the notion of generalized von Neumann algebras which is an unbounded generalization of von Neumann algebras. An O*-algebra $\mathcal{M}$ is a *generalized von Neumann algebra* if $\mathcal{M}'_{\mathrm{w}}\mathcal{D} \subset \mathcal{D}$ and $\mathcal{M} = (\mathcal{M}'_{\mathrm{w}})'_{\mathrm{c}} \equiv \{X \in \mathcal{L}^\dagger(\mathcal{D}); CX \subset XC, {}^\forall C \in \mathcal{M}'_{\mathrm{w}}\}$. The other unbounded generalizations of von Neumann algebras were considered by Dixon [2], Araki-Jurzak [1] and Schmüdgen [19]. Suppose λ is a modular generalized vector for $\mathcal{M}$. Then there exists the largest subdomain $\mathcal{D}_\lambda^M$ of $\mathcal{D}$ satifying the conditions $(M)_1$ and $(M)_2$, and λ can be extended to a standard generalized vector λ_s for the generalized von Neumann algebra $(\mathcal{M}\lceil\mathcal{D}_\lambda^M)''_{\mathrm{wc}}$. The standardness and the modularity of generalized vectors in the following special cases are investigated: A. Standard systems with vectors. B. Standard tracial generalized vectors. C. Standard systems for semifinite O*-algebras. D. Standard generalized vectors in the Hilbert space of all Hilbert-Schmidt operators. E. Standard systems constructed by von Neumann algebras with standard generalized vectors. The Connes cocycle theorem and the Pedersen-Takesaki Radon-Nikodym theorem for von Neumann algebras are generalized to standard generalized vectors for generalized von Neumann algebras.

In Chapter III the notion of standard weights is defined and studied. Weights on O*-algebras (that is, linear functionals that take positive, but not finite valued) have naturally appeared in the studies of the unbounded T-T theory and the quantum physics. The algebraic positive cone $\mathcal{P}(\mathcal{M})$ and the operational positive cone $\mathcal{M}_+$ are defined and the corresponding weights are defined. The GNS-construction of a weight φ is important for studies of O*-algebras like for positive linear functionals on O*-algebras and for weights on von Neumann algebras. In the bounded case $\mathfrak{N}_\varphi^0 \equiv \{X \in \mathcal{M}; \varphi(X^\dagger X) < \infty\}$ is a left ideal of $\mathcal{M}$, but it is not necessarily a left ideal of $\mathcal{M}$ in the unbounded case. For example, the condition $\varphi(I) < \infty$ does not necessarily imply $\varphi(X^\dagger X) < \infty$ for all $X \in \mathcal{M}$. So, using the left ideal $\mathfrak{N}_\varphi \equiv \{X \in \mathcal{M}; \varphi((AX)^\dagger(AX)) < \infty, {}^\forall A \in \mathcal{M}\}$ of $\mathcal{M}$, the GNS-representation π_φ and the vector representation λ_φ are constructed. We give two important examples of weights. For any $\xi \in \mathcal{D}$ the positive linear functional ω_ξ on $\mathcal{M}$ is defined by $\omega_\xi(X) = (X\xi|\xi), X \in \mathcal{M}$, but if $\xi \in \mathcal{H} \setminus \mathcal{D}$ then the definition of the above ω_ξ is impossible. So, we need to study the quasi-weight ω_ξ defined by ξ as stated at the begining of this section. Another important example is a regular (quasi-)weight. A (quasi-)weight φ is *regular* if $\varphi = \sup_\alpha f_\alpha$ for some net $\{f_\alpha\}$ of positive linear functionals on $\mathcal{M}$. An important class in regular (quasi-)weights which is possible to develop the T-T theory in O*-algebras is defined and studied. A faithful (quasi-)weight φ on $\mathcal{P}(\mathcal{M})$ is *standard* if the generalized vector Λ_φ defined by $\Lambda_\varphi(\pi_\varphi(X)) = \lambda_\varphi(X)$, $X \in \mathfrak{N}_\varphi$ is standard. Suppose φ is standard, then the modular automorphism group $\{\sigma_t^\varphi\}_{t\in\mathbb{R}}$ of $\mathfrak{N}_\varphi^\dagger \cap \mathfrak{N}_\varphi$ is defined and φ is a $\{\sigma_t^\varphi\}$-KMS (quasi-)weight. To generalize the Connes cocycle theorem for (quasi-)weights on O*-algebras, some difficult problems arise. Let φ and ψ be standard (quasi-)weights on $\mathcal{P}(\mathcal{M})$. In case of von Neumann algebras π_φ and π_ψ are unitarily equivalent, but in case of O*-algebras they are not necessarily unitarily equivalent. So, the unitary equivalence of π_φ and π_ψ is first characterized, and in this case the Connes cocycle theorem for standard (quasi-)weights on generalized von Neumann algebras is generalized. The Radon-Nikodym theorem for (quasi-)weights on O*-algebras is also studied.

In Chapter IV the T-T theory in O*-algebras studied in Chapter II, III is applied to quantum statistical mechanics and the Wightman quantum field theory. The quantum moment problem for states on an O*-algebra is first studied. Many important examples of states f in quantum physics are trace functionals, that is, they are of the form $f(X) = tr\overline{TX}$ with a certain trace operator T. It is important to consider the quantum moment problem (QMP): Under what conditions is every strongly positive linear functional on an O*-algebra a trace functional? This was studied by Sherman [1], Woronowicz [1,2], Lassner-Timmermann [1] and Schmüdgen [2,4,21] etc. Main results of Schmüdgen are here introduced. QMP is also closely related to unbounded T-T theory. In fact, if f is a trace functional on $\mathcal{M}$, then $f(X^\dagger X) = tr(X\Omega)^*(X\Omega)$, $X \in \mathcal{M}$ for some positive Hilbert-Schmidt operator Ω, and so π_f is unitarily equivalent to a subrepresentation of the $*$-representation π of $\mathcal{M}$ on the Hilbert space $\mathcal{H} \otimes \overline{\mathcal{H}}$ of all Hilbert-Schmidt operators on $\mathcal{H}$ defined by $\pi(A)X\Omega = AX\Omega$ for $A, X \in \mathcal{M}$. As stated in special cases D in Chapter II, such a representation π is useful for the T-T theory in O*-algebras. Hence, as QMP for weights, it is important to consider when a weight φ on $\mathcal{M}_+$ is represented as $\varphi(X^\dagger X) = tr(X^{\dagger *}\Omega)^* \overline{X^{\dagger *}\Omega}$, $X \in \mathfrak{N}_\varphi^0$ for some positive self-adjoint operator Ω. This problem was recently considered by Inoue-Kürsten [1] and is here introduced.

Standard generalized vectors in unbounded CCR-algebras are studied. Let $\mathcal{A}$ be the canonical algebra of one degree of freedom, that is, a $*$-algebra generated by identity 1 and hermitian elements p and q satisfying the Heisenberg commutation relation: $[p, q] = -i1$. The Schrödinger representation π_0 of $\mathcal{A}$ is defined by $(\pi_0(p)\xi)(t) = -i\frac{d}{dt}\xi(t)$ and $(\pi_0(q)\xi)(t) = t\xi(t)$, $\xi \in \mathcal{S}(\mathbb{R})$. Von Neumann [1], Dixmier [1] and Powers [1] considered when a self-adjoint representation π of $\mathcal{A}$ is unitarily equivalent to a direct sum $\bigoplus_{\alpha\in I_0} \pi_\alpha$ of $*$-representations π_α which are unitarily equivalent to the Schrödinger representation π_0. Such a representation is called a Weyl representation of the cardinal I_0. The Powers results are here introduced. Furthermore, it is shown that a Weyl representation of countable cardinal is unitarily equivalent to the self-adjoint representation $\pi_\otimes$ of $\mathcal{A}$ defined by $\mathcal{D}(\pi_\otimes) = \mathcal{S}(\mathbb{R}) \otimes \overline{L^2(\mathbb{R})} \equiv \{T \in L^2(\mathbb{R}) \otimes \overline{L^2(\mathbb{R})}; TL^2(\mathbb{R}) \subset \mathcal{S}(\mathbb{R})\}$ and $\pi_\otimes(a)T = \pi_0(a)T$ for $a \in \mathcal{A}$ and $T \in \mathcal{D}(\pi_\otimes)$, and $\Omega_\beta \equiv \sum_{n=0}^{\infty} e^{-n\beta/2} f_n \otimes \overline{f_n}$ ($\beta > 0$) is a standard vector for $\pi_\otimes(\mathcal{A})$, where $\{f_n\}_{n=0,1,\cdots}$ is an ONB in $L^2(\mathbb{R})$ consisting of the normalized Hermite functions. Let $\mathcal{M}$ be an O*-algebra on $\mathcal{S}(\mathbb{R})$ generated by $\pi_0(\mathcal{A})$ and $f_0 \otimes \overline{f_0}$ and π a self-adjoint representation of $\mathcal{M}$ on $\mathcal{S}(\mathbb{R}) \otimes \overline{L^2(\mathbb{R})}$ defined by $\pi(X)T = XT$ for $X \in \mathcal{M}$ and $T \in \mathcal{S}(\mathbb{R}) \otimes \overline{L^2(\mathbb{R})}$. Then the positive self-adjoint operator $\Omega_{\{\alpha_n\}} \equiv \sum_{n=0}^{\infty} \alpha_n f_n \otimes \overline{f_n}$ ($\alpha_n > 0, n = 0, 1, \cdots$) defines a modular generalized vector $\lambda_{\Omega_{\{\alpha_n\}}}$ for the self-adjoint O*-algebra $\pi(\mathcal{M})$, and $\lambda_{\Omega_{\{\alpha_n\}}}$ is standard if and only if $\alpha_n = e^{\beta n}, n \in \mathbb{N} \cup \{0\}$ for some $\beta \in \mathbb{R}$. Standard generalized vectors and modular generalized vectors in an interacting Boson model and the BCS-Bogoluvov model are given.

Standard generalized vectors in the Wightman quantum field theory are studied. The general theory of quantum fields has been developed along two main lines: One is based on the Wightman axioms and makes use of unbounded field operators, and the other is the theory of local nets of bounded observables initiated by Haag-Kastler [1] and Araki [1]. The passage from a Wightman field to a local net of von Neumann algebras is here characterized by the existence of standard systems from the right wedge-region in Minkowski space.

1. Fundamentals of O*-algebras

In this chapter we state about the basic definitions and properties of O*-algebras without the proofs except for Section 1.9. In Section 1.1 the notion of closedness (self-adjointness, integrability) of an O*-algebra is defined and studied in analogy with the notion of a closed (self-adjoint) operator.

In Section 1.2 the relation between the self-adjointness of an O*-algebra $\mathcal{M}$ and the weak commutant $\mathcal{M}'_{\mathrm{w}}$ are investigated. In Section 1.3 invariant subspaces for O*-algebras are studied. For a closed O*-algebra $\mathcal{M}$ on $\mathcal{D}$ in $\mathcal{H}$, there are some pathologies between a subspace $\mathfrak{M}$ of $\mathcal{D}$ which is $\mathcal{M}$-invariant (i.e. $\mathcal{M}\mathfrak{M} \subset \mathfrak{M}$) and the projection $P_{\overline{\mathfrak{M}}}$ onto the closure $\overline{\mathfrak{M}}$ of $\mathfrak{M}$ with respect to the Hilbert space norm. For any $\mathcal{M}$-invariant subspace $\mathfrak{M}$ of $\mathcal{D}$ the closure $\mathcal{M}_{\mathfrak{M}}$ of the O*-algebra $\mathcal{M}\lceil\mathfrak{M} \equiv \{X\lceil\mathfrak{M}; X \in \mathcal{M}\}$ is a closed O*-algebra on the closure $\overline{\mathfrak{M}}^{t_{\mathcal{M}}}$ of $\mathfrak{M}$ with respect to the graph topology $t_{\mathcal{M}}$ in $\overline{\mathfrak{M}}$. But, the projection $P_{\overline{\mathfrak{M}}}$ of $\mathcal{H}$ onto $\overline{\mathfrak{M}}$ does not belong to $\mathcal{M}'_{\mathrm{w}}$ in general. When $P_{\overline{\mathfrak{M}}} \in \mathcal{M}'_{\mathrm{w}}$, another closed O*-algebra $\mathcal{M}_{P_{\overline{\mathfrak{M}}}}$ on $P_{\overline{\mathfrak{M}}}\mathcal{D}$ in $\overline{\mathfrak{M}}$ can be defined, and $\mathcal{M}_{P_{\overline{\mathfrak{M}}}}$ is an extension of $\mathcal{M}_{\mathfrak{M}}$ (denote by $\mathcal{M}_{\mathfrak{M}} \prec \mathcal{M}_{P_{\overline{\mathfrak{M}}}}$), but $\mathcal{M}_{\mathfrak{M}} \neq \mathcal{M}_{P_{\overline{\mathfrak{M}}}}$ in general. In this section the relation between $\mathcal{M}_{\mathfrak{M}}$ and $\mathcal{M}_{P_{\overline{\mathfrak{M}}}}$, and the self-adjointness of these two O*-algebras are investigated in detail. The different notions of cyclic vectors and strongly cyclic vectors for O*-algebras are defined and investigated. In Section 1.4 induced extensions of O*-algebras are introduced. Let $\mathcal{M}$ be a closed O*-algebra on $\mathcal{D}$ in $\mathcal{H}$. Suppose $\mathcal{M}'_{\mathrm{w}}\mathcal{D} \subset \mathcal{D}$. Then $\mathcal{M}'_{\mathrm{w}}$ is a von Neumann algebra and $\overline{X}$ is affiliated with the von Neumann algebra $(\mathcal{M}'_{\mathrm{w}})'$ for each $X \in \mathcal{M}$, so that it is possible to make use of the von Neumann algebra $(\mathcal{M}'_{\mathrm{w}})'$ for studies of O*-algebras. Thus the condition $\mathcal{M}'_{\mathrm{w}}\mathcal{D} \subset \mathcal{D}$ is useful for studies of O*-algebras. Even if $\mathcal{M}'_{\mathrm{w}}$ is a von Neumann algebra, the condition $\mathcal{M}'_{\mathrm{w}}\mathcal{D} \subset \mathcal{D}$ does not necessarily hold. In this section it is shown that if $\mathcal{M}'_{\mathrm{w}}$ is a von Neumann algebra then there exists a closed O*-algebra $\mathcal{N}$ on a dense subspace $\mathcal{E}$ in $\mathcal{H}$ such that $\mathcal{M} \prec \mathcal{N}, \mathcal{N}'_{\mathrm{w}} = \mathcal{M}'_{\mathrm{w}}$ and $\mathcal{N}'_{\mathrm{w}}\mathcal{E} \subset \mathcal{E}$. In Section 1.5 the relation between a commutative O*-algebra $\mathcal{M}$ and the von Neumann algebra $(\mathcal{M}'_{\mathrm{w}})'$ is investigated. The commutativity of an O*-algebra $\mathcal{M}$ doesn't necessarily imply the commutativity of the von Neumann algebra $(\mathcal{M}'_{\mathrm{w}})'$. There exists a commutative self-adjoint O*-algebra $\mathcal{P}(A, B)$ generated by commutative essentially self-adjoint operators A and B such that $(\mathcal{P}(A, B)'_{\mathrm{w}})'$ is a purely infinite von Neumann algebra. It is shown that $\mathcal{M}$ is integrable if and only if

$(\mathcal{M}'_{\mathrm{w}})'$ is commutative. In Section 1.6 several topologies on O*-algebras are introduced. The (quasi-)uniform topology, ρ-topology and λ-topology which are generalizations of the operator-norm topology in case of bounded case are introduced. They are different in general in case of unbounded case. Further, the other topologies called weak, σ-weak, strong, strong*, σ-strong and σ-strong* are introduced. The relations among these topologies are investigated. In Section 1.7 the notions of extended W*-algebras and generalized von Neumann algebras which are unbounded generalizations of von Neumann algebras are introduced. In Section 1.8 the basic properties of $*$-representations of $*$-algebras are noted. In Section 1.9 strongly positive linear functionals on O*-algebras are studied. Many important examples of states in quantum physics are trace functionals, that is, they are of the form $f(X) = \mathrm{tr}TX$ with some positive trace class operator T. Here trace functionals are studied for the preparation of quantum moment problem studied in Chapter IV.

1.1 O*-algebras

Let $\mathcal{D}$ be a dense subspace in a Hilbert space $\mathcal{H}$ with inner product $(\ |\ )$. By $\mathcal{L}(\mathcal{D})$ (resp. $\mathcal{L}_c(\mathcal{D})$) we denote the set of all linear operators (resp. closable linear operators) from $\mathcal{D}$ to $\mathcal{D}$, and put

$$\mathcal{L}^{\dagger}(\mathcal{D}) = \{X \in \mathcal{L}(\mathcal{D}); \mathcal{D}(X^*) \supset \mathcal{D} \text{ and } X^*\mathcal{D} \subset \mathcal{D}\}.$$

Then $\mathcal{L}^{\dagger}(\mathcal{D}) \subset \mathcal{L}_c(\mathcal{D}) \subset \mathcal{L}(\mathcal{D})$ and $\mathcal{L}(\mathcal{D})$ is an algebra with the usual operations $X + Y, \alpha X$ and XY, and $\mathcal{L}^{\dagger}(\mathcal{D})$ is a $*$-algebra with the involution $X \to X^{\dagger} \equiv X^*\lceil\mathcal{D}$. We remark that if there exists an element X of $\mathcal{L}^{\dagger}(\mathcal{D})$ which is closed, then $\mathcal{D} = \mathcal{H}$ and hence $\mathcal{L}^{\dagger}(\mathcal{D})$ equals the algebra $\mathcal{B}(\mathcal{H})$ of all bounded linear operators on $\mathcal{H}$ (Lemma 2.2 in Lassner [1]).

Definition 1.1.1. A subalgebra of $\mathcal{L}(\mathcal{D})$ contained in $\mathcal{L}_c(\mathcal{D})$ is said to be an *O-algebra* on $\mathcal{D}$ in $\mathcal{H}$, and a $*$-subalgebra of $\mathcal{L}^{\dagger}(\mathcal{D})$ is said to be an *O*-algebra* on $\mathcal{D}$ in $\mathcal{H}$.

We first define the notion of a closed O-algebra in analogy with the notion of a closed operator.

Definition 1.1.2. Let $\mathcal{M}_1$ and $\mathcal{M}_2$ be O-algebras on $\mathcal{D}_1$ and $\mathcal{D}_2$ in $\mathcal{H}$, respectively. We say that $\mathcal{M}_2$ is an *extension* of $\mathcal{M}_1$ if $\mathcal{D}_1 \subset \mathcal{D}_2$ and there exists a bijection ι of $\mathcal{M}_1$ onto $\mathcal{M}_2$ such that $\iota(X)\lceil\mathcal{D}_1 = X$ for each $X \in \mathcal{M}_1$, and denoted by $\mathcal{M}_1 \prec \mathcal{M}_2$.

Let $\mathcal{M}$ be an O-algebra on $\mathcal{D}$ in $\mathcal{H}$. We define a natural graph topology on $\mathcal{D}$. This topology is a locally convex topology defined by a family $\{\|\ \|_X; X \in$

$\mathcal{M}\}$ of seminorms: $\|\xi\|_X \equiv \|\xi\| + \|X\xi\|$, $\xi \in \mathcal{D}$, and it is called the *induced* (or *graph*) topology on $\mathcal{D}$ and denoted by $t_{\mathcal{M}}$.

Definition 1.1.3. An O-algebra $\mathcal{M}$ on $\mathcal{D}$ is *closed* if the locally convex space $\mathcal{D}[t_{\mathcal{M}}]$ is complete.

Let $\mathcal{M}$ be an O-algebra on $\mathcal{D}$ in $\mathcal{H}$. We denote by $\widetilde{\mathcal{D}}(\mathcal{M})$ the completion of the locally convex space $\mathcal{D}[t_{\mathcal{M}}]$. Then it is clear that

$$\widetilde{\mathcal{D}}(\mathcal{M}) \subset \widehat{\mathcal{D}}(\mathcal{M}) \equiv \bigcap_{X \in \mathcal{M}} \mathcal{D}(\overline{X}).$$

For the closure of an O-algebra we have the following

Theorem 1.1.4. (1) Let $\mathcal{M}$ be an O-algebra on $\mathcal{D}$ in $\mathcal{H}$. We put

$$\widetilde{X} = \overline{X} \lceil \widetilde{\mathcal{D}}(\mathcal{M}), \quad X \in \mathcal{M},$$
$$\widetilde{\mathcal{M}} = \{\widetilde{X}; X \in \mathcal{M}\}.$$

Then $\widetilde{\mathcal{M}}$ is a closed O-algebra on $\widetilde{\mathcal{D}}(\mathcal{M})$ in $\mathcal{H}$ which is the smallest closed extension of $\mathcal{M}$.

(2) Let $\mathcal{M}$ be an O*-algebra on $\mathcal{D}$ in $\mathcal{H}$. Then $\widetilde{\mathcal{D}}(\mathcal{M}) = \widehat{\mathcal{D}}(\mathcal{M})$ and $\widetilde{\mathcal{M}}$ is a closed O*-algebra on $\widetilde{\mathcal{D}}(\mathcal{M})$ in $\mathcal{H}$ which is the smallest closed extension of $\mathcal{M}$.

Proof. See Lemma 2.5 in Powers [1].

The above $\widetilde{\mathcal{M}}$ is called the *closure* of $\mathcal{M}$.

Example 1.1.5. Let $\mathcal{H} = L^2(\mathbb{R})$ and $\mathcal{D}$ the space $C_0^\infty(\mathbb{R})$ of all infinitely differentialbe functions with compact supports. We define O*-algebras $\mathcal{M}_1$ and $\mathcal{M}_2$ on $\mathcal{D}$ as follows:

$$\mathcal{M}_1 = \{\sum_{k=0}^{m} \sum_{l=0}^{n} \alpha_{kl}\, t^k \left(\frac{d}{dt}\right)^l ; \alpha_{k,l} \in \mathbb{C},\ n, m \in \mathbb{N} \cup \{0\}\},$$

$$\mathcal{M}_2 = \{\sum_{k=0}^{n} f_k(t) \left(\frac{d}{dt}\right)^k ; f_k \in C^\infty(\mathbb{R}),\ n \in \mathbb{N} \cup \{0\}\}.$$

Then $\widetilde{\mathcal{D}}(\mathcal{M}_1)$ equals the Schwartz space $\mathcal{S}(\mathbb{R})$ of all infinitely differentiable rapidly decreasing functions. In fact, $\widetilde{\mathcal{M}_1}$ is the O*-algebra of the Schrödinger representation of the $*$-algebra generated by identity *1* and two self-adjoint elements p, q satisfying the Heisenberg commutation relation: $pq - qp = -i1$. This will often appear in this note. The O*-algebra $\mathcal{M}_2$ on $\mathcal{D}$ is closed.

We next define the notion of self-adjointness of O*-algebras. Let $\mathcal{M}$ be an O*-algebra on $\mathcal{D}$ in $\mathcal{H}$. We put

$$\begin{cases} \mathcal{D}^*(\mathcal{M}) = \bigcap_{X \in \mathcal{M}} \mathcal{D}(X^*), \\ \mathcal{M}^* = \{\iota^*(X) \equiv X^{\dagger *} \lceil \mathcal{D}^*(\mathcal{M}); X \in \mathcal{M}\}, \end{cases}$$

$$\begin{cases} \mathcal{D}^{**}(\mathcal{M}) = \bigcap_{X \in \mathcal{M}} \mathcal{D}((X^* \lceil \mathcal{D}^*(\mathcal{M}))^*), \\ \mathcal{M}^{**} = \{\iota^{**}(X) \equiv (X^* \lceil \mathcal{D}^*(\mathcal{M}))^* \lceil \mathcal{D}^{**}(\mathcal{M}); X \in \mathcal{M}\}. \end{cases}$$

Then we have the following

Proposition 1.1.6. Let $\mathcal{M}$ be an O*-algebra on $\mathcal{D}$ in $\mathcal{H}$. Then $\mathcal{M}^*$ is a closed O-algebra on $\mathcal{D}^*(\mathcal{M})$ and $\mathcal{M}^{**}$ is a closed O*-algebra on $\mathcal{D}^{**}(\mathcal{M})$ such that $\mathcal{M} \prec \widetilde{\mathcal{M}} \prec \mathcal{M}^{**} \prec \mathcal{M}^*$. These O-algebras $\mathcal{M}, \widetilde{\mathcal{M}}, \mathcal{M}^{**}$ and $\mathcal{M}^*$ don't coincide in general.

Proof. See Lemma 4.1 in Powers [1] and Proposition 8.1.12 in Schmüdgen [21].

Definition 1.1.7. An O*-algebra $\mathcal{M}$ on $\mathcal{D}$ in $\mathcal{H}$ is said to be *algebraically self-adjoint* (resp. *essentially self-adjoint* , *self-adjoint*) if $\mathcal{M}^{**} = \mathcal{M}^*$ (resp. $\widetilde{\mathcal{M}} = \mathcal{M}^*, \mathcal{M} = \mathcal{M}^*$). A closed O*-algebra $\mathcal{M}$ is said to be *integrable* (or *standard*) if $X^* = \overline{X^\dagger}$ for each $X \in \mathcal{M}$.

For integrable O*-algebras we have the following

Proposition 1.1.8. Let $\mathcal{M}$ be a closed O*-algebra on $\mathcal{D}$ in $\mathcal{H}$. The following statements are equivalent:

(i) $\mathcal{M}$ is integrable.

(ii) $\overline{X}$ is a self-adjoint operator for each $X \in \mathcal{M}_h \equiv \{X \in \mathcal{M}; X = X^\dagger\}$.

If this is true, then $\mathcal{M}$ is self-adjoint and $\overline{\mathcal{M}} \equiv \{\overline{X}; X \in \mathcal{M}\}$ is a $*$-algebra of closed operators in $\mathcal{H}$ equipped with the strong sum $\overline{X} \boldsymbol{+} \overline{Y} \equiv \overline{\overline{X} + \overline{Y}} = \overline{X+Y}$, the strong scalar multiplication $\alpha {\cdot} \overline{X} = \begin{matrix} \alpha\overline{X}, & \alpha \neq 0 \\ 0 & \alpha = 0 \end{matrix} = \overline{\alpha X}$, the strong product $\overline{X} {\cdot} \overline{Y} \equiv \overline{\overline{X}\,\overline{Y}} = \overline{XY}$ and the involution $X \to X^* = \overline{X^\dagger}$.

Proof. See Theorem 7.1 in Powers [1] and Theorem 2.3 in Inoue [1].

Example 1.1.9. Let μ be a regular Borel measure on $\mathbb{R}^n$ and $\mathcal{H} = L^2(\mathbb{R}^n, \mu)$. We put

$$\mathcal{D} = \{f \in \mathcal{H}; \int |p(x)f(x)|^2 d\mu(x) < \infty \text{ for all polynomials } p\},$$

and for any polynomial p define an operator $\pi(p)$ on $\mathcal{D}$ by

$$(\pi(p)f)(x) = p(x)f(x), f \in \mathcal{D}.$$

Then $\mathcal{M} \equiv \{\pi(p); p \text{ is a polynomial}\}$ is an integrable O*-algebra on $\mathcal{D}$.

Example 1.1.10. We define an essentially self-adjoint operator H_1 and symmetric operators ${}_nH_1$ in $L^2[0,1]$ by

$$\begin{cases} \mathcal{D}_1 = \{f \in C^\infty[0,1]; f^{(n)}(0) = f^{(n)}(1),\ n = 0,1,2,\cdots\} \\ H_1 = -i\dfrac{d}{dt}\lceil \mathcal{D}_1 \end{cases}$$

$$\begin{cases} {}_n\mathcal{D}_1 = \{f \in \mathcal{D}_1; f^{(k)}(0) = f^{(k)}(1) = 0,\ k = n, n+1, \cdots\} \\ {}_nH_1 = -i\dfrac{d}{dt}\lceil {}_n\mathcal{D}_1, n \in \mathbb{N} \cup \{0\}. \end{cases}$$

Let $\mathcal{M}_1$ and ${}_n\mathcal{M}_1$ be the closed O*-algebras $\mathcal{P}(H_1)$ and $\mathcal{P}({}_nH_1)$ on $\mathcal{D}_1$ and ${}_n\mathcal{D}_1$ generated by H_1 and ${}_nH_1$, respectively. Then we have

$$\begin{aligned} {}_0\mathcal{M}_1 &= {}_0\mathcal{M}_1^{**} \\ &\precneqq {}_1\mathcal{M}_1 \precneqq \cdots \precneqq {}_n\mathcal{M}_1 \precneqq \cdots \precneqq {}_n\mathcal{M}_1^{**} = {}_n\mathcal{M}_1^* = \mathcal{M}_1 = \mathcal{M}_1^* \precneqq {}_0\mathcal{M}_1^*, n \in \mathbb{N} \end{aligned}$$

and hence $\mathcal{M}_1$ is self-adjoint and ${}_n\mathcal{M}_1$ is algebraically self-adjoint but not self-adjoint for each $n \in \mathbb{N}$, and ${}_0\mathcal{M}_1$ is not even algebraically self-adjoint.

Example 1.1.11. The O*-algebra $\mathcal{M}_1$ in Example 1.1.5 is essentially self-adjoint. In fact, it is well-known that $N \equiv \left(t^2 - \dfrac{d^2}{dt^2}\right)\lceil \mathcal{S}(\mathbb{R})$ is an essentially self-adjoint operator in $\widetilde{\mathcal{M}_1}$ and $\mathcal{S}(\mathbb{R}) = \mathcal{D}^\infty(\overline{N}) \equiv \bigcap_{k=1}^\infty \mathcal{D}(\overline{N}^k)$, which implies

$$\mathcal{D}^*(\mathcal{M}_1) \equiv \bigcap_{k=1}^\infty \mathcal{D}(X^*) \subset \bigcap_{k=1}^\infty \mathcal{D}(\overline{N}^k) = \mathcal{S}(\mathbb{R}) = \widetilde{\mathcal{D}}(\mathcal{M}_1).$$

Hence $\mathcal{M}_1$ is essentially self-adjoint. The O*-algebra $\mathcal{M}_2$ in Example 1.1.5 is not self-adjoint. In fact, suppose $\mathcal{M}_2$ is self-adjoint. Then $\mathcal{M}_2 \prec \widetilde{\mathcal{M}_1}$ and $\widetilde{\mathcal{M}_1}$ is self-adjoint, it follows that $\widetilde{\mathcal{M}_1} = \mathcal{M}_2$, which contradicts $C_0^\infty(\mathbb{R}) \subsetneqq \mathcal{S}(\mathbb{R})$. Hence $\mathcal{M}_2$ is not self-adjoint.

In analogy with the notion of a closed (symmetric, self-adjoint) operator, we have defined the notion of a closed O-algebra (O*-algebra, self-adjoint O*-algebra). We here arrange those relations on the following diagram:

Single operator T		Operator algebra $\mathcal{M}$
T : closable		$\mathcal{M}$: O-algebra
T : symmetric $\Updownarrow$ $T \subset T^*$	$\longleftrightarrow$	$\mathcal{M}$: O*-algebra $\Updownarrow$ $X \subset X^*, \quad {}^\forall X^\dagger = X \in \mathcal{M}$
T : closed $\Updownarrow$ $\mathcal{D}(T)$ is complete w.r.t. the graph norm $\| \ \|_T$ $\Updownarrow$ $\mathcal{D}(T) = \mathcal{D}(\overline{T})$	$\longleftrightarrow$ $\longleftrightarrow$	$\mathcal{M}$: closed $\Updownarrow$ $\mathcal{D}$ is complete w.r.t. the graph topology $t_{\mathcal{M}}$ $\Updownarrow$ $\mathcal{D} = \bigcap_{X \in \mathcal{M}} \mathcal{D}(\overline{X})$
T : self-adjoint $\Updownarrow$ $\mathcal{D}(T^*) = \mathcal{D}(T)$ $\Updownarrow$ $T^* = T$	 $\longleftrightarrow$ $\longleftrightarrow$	$\mathcal{M}$: self-adjoint $\Updownarrow$ $\mathcal{D}^*(\mathcal{M}) \equiv \bigcap_{X \in \mathcal{M}} \mathcal{D}(X^*) = \mathcal{D}$ $\Uparrow \ \not\Downarrow$ $\mathcal{M}$: integrable $\Updownarrow$ $X^* = \overline{X}, \quad {}^\forall X^\dagger = X \in \mathcal{M}.$

1.2 Weak commutant

In this section we define a weak commutant of an O*-algebra which plays an important rule for the studies of O*-algebras.

Definition 1.2.1. Let $\mathcal{M}$ be an O*-algebra on $\mathcal{D}$ in $\mathcal{H}$. The weak commutant $\mathcal{M}'_{\mathrm{w}}$ of $\mathcal{M}$ is defined by

$$\mathcal{M}'_{\mathrm{w}} = \{C \in \mathcal{B}(\mathcal{H}); (CX\xi|\eta) = (C\xi|X^\dagger\eta) \text{ for all } X \in \mathcal{M} \text{ and } \xi, \eta \in \mathcal{D}\}.$$

Proposition 1.2.2. $\mathcal{M}'_{\mathrm{w}}$ is a weak-operator closed $*$-invariant subspace in $\mathcal{B}(\mathcal{H})$ such that $\widetilde{\mathcal{M}}'_{\mathrm{w}} = (\mathcal{M}^{**})'_{\mathrm{w}} = \mathcal{M}'_{\mathrm{w}}$.

Proof. See Proposition 7.2.9 in Schmüdgen [21] and Lemma 2.3 in Inoue-Takesue [1].

In general, $\mathcal{M}'_{\mathrm{w}}$ is not an algebra as Proposition 1.2.4 shows later. So, we define the strong commutants $\mathcal{M}'_{\mathrm{s}}$ and $\mathcal{M}'_{\mathrm{ss}}$ of $\mathcal{M}$ as follows:

$$\mathcal{M}'_{\mathrm{s}} = \{C \in \mathcal{M}'_{\mathrm{w}} \ ; \ C\mathcal{D} \subset \mathcal{D}\},$$
$$\mathcal{M}'_{\mathrm{ss}} = \{C \in \mathcal{M}'_{\mathrm{w}} \ ; \ C\mathcal{D} \subset \mathcal{D} \text{ and } C^*\mathcal{D} \subset \mathcal{D}\}.$$

It is easy to show that $\mathcal{M}'_s$ is a subalgebra of $\mathcal{B}(\mathcal{H})$ contained in $\widetilde{\mathcal{M}'_s}$ but, $\mathcal{M}'_s$ is not $*$-invariant in general. If $\mathcal{M}$ is closed, then $\mathcal{M}'_s$ is weak-operator closed and $\mathcal{M}'_{ss}$ is a von Neumann algebra on $\mathcal{H}$. We have the following

Proposition 1.2.3. Let $\mathcal{M}$ be a closed O*-algebra on $\mathcal{D}$ in $\mathcal{H}$. Consider the following statements:
(i) $\mathcal{M}$ is self-adjoint.
(ii) $\mathcal{M}'_w = \mathcal{M}'_s$.
(ii)′ $\overline{X}$ is affiliated with the von Neumann algebra $(\mathcal{M}'_w)'$ for each $X \in \mathcal{M}$.
(iii) $\mathcal{M}'_w = (\mathcal{M}^{**})'_s$.
(iii)′ $\iota^{**}(X)$ is affiliated with $(\mathcal{M}'_w)'$ for each $X \in \mathcal{M}$.
(iv) $\mathcal{M}'_w$ is a von Neumann algebra.
Then the following implications hold:

$$\text{(i)} \Rightarrow \begin{matrix}\text{(ii)}\\ \Updownarrow\\ \text{(ii)}'\end{matrix} \Rightarrow \begin{matrix}\text{(iii)}\\ \Updownarrow\\ \text{(iii)}'\end{matrix} \Rightarrow \text{(iv)}.$$

Any converse implications don't hold in general.

Proof. See Lemma 4.5 and Lemma 4.6 in Powers [1].

We investigate the relations of the statements in Proposition 1.2.3 in case of the O*-algebra generated by one symmetric operator.

Lemma 1.2.4. Let $\mathcal{D}$ be a dense subspace in a Hilbert space $\mathcal{H}$ and $\mathcal{P}(A)$ the O*-algebra on $\mathcal{D}$ generated by $A^\dagger = A$ in $\mathcal{L}^\dagger(\mathcal{D})$. Let U_A be the partial isometry on $\mathcal{H}$ defined by the Cayley transformation $(\overline{A} - iI)(\overline{A} + iI)^{-1}$ of the closed symmetric operator $\overline{A}$. Then $U_A^n \in \mathcal{P}(A)'_w$ for each $n \in \mathbb{N}$.

Proof. See Lemma 3.2 in Powers [1] and Example 8.5.1 in Schmüdgen [21].

Proposition 1.2.5. Let $\mathcal{D}$ be a dense subspace in a Hilbert space $\mathcal{H}$ and $\mathcal{P}(A)$ an O*-algebra on $\mathcal{D}$ generated by $A^\dagger = A \in \mathcal{L}^\dagger(\mathcal{D})$. Then the following statements are equivalent:
(i) A is essentially self-adjoint.
(ii) $\mathcal{P}(A)'_w$ is a von Neumann algebra.
(iii) $\mathcal{P}(A)$ is algebraically self-adjoint.

Proof. See Lemma 3.2 in Powers [1] and Theorem 2.4 in Inoue-Takesue [1].

Proposition 1.2.6. Let $\mathcal{D}$ and A be in Proposition 1.2.4. Suppose $\mathcal{P}(A)$ is closed. Then the following statements are equivalent:
(i) $\mathcal{P}(A)$ is integrable.

(ii) $\mathcal{P}(A)$ is self-adjoint.
(iii) $\mathcal{P}(A)'_{\mathrm{w}} = \mathcal{P}(A)'_{\mathrm{s}}$.
(iv) A is essentially self-adjoint and $\mathcal{D} = \bigcap_{n=1}^{\infty} \mathcal{D}(\overline{A}^n)$.
(v) A^n is essentially self-adjoint for $n = 1, 2, \cdots$.

Proof. See Theorem 2.1 in Inoue-Takesue [1].

Examaple 1.2.7. Let $\mathcal{D}_1, H_1, \mathcal{M}_1$ and ${}_n\mathcal{D}_1, {}_nH_1, {}_n\mathcal{M}_1 (n \in \mathbb{N} \cup \{0\})$ be in Example 1.1.10. Then the following results hold:

(1) Let T be a closed operator in $L^2[0,1]$ defined by

$$\mathcal{D}(T) = \{f \in C[0,1]; f(t) - f(0) = \int_0^t g(s)ds \text{ for some } g \in L^2[0,1]\},$$

$$Tf = -ig, \quad f \in \mathcal{D}(T) \quad (f(t) - f(0) = \int_0^t g(s)ds).$$

Then H_1 and ${}_nH_1$ $(n \in \mathbb{N})$ are essentially self-adjoint operators in $L^2[0,1]$ such that $\overline{{}_nH_1} = H_1$, and ${}_0H_1$ is a symmetric operator in $L^2[0,1]$ such that $\mathcal{D}(\overline{{}_0H_1}) = \{f \in \mathcal{D}(T); f(1) = f(0) = 0\}$ and $\overline{{}_0H_1} \subset T$.

(2) $\mathcal{M}_1 = \mathcal{P}(H_1)$ and ${}_n\mathcal{M}_1 = \mathcal{P}({}_nH_1)$, and so $\mathcal{P}(H_1)$ is self-adjoint, $\mathcal{P}({}_nH_1)$ is algebraically self-adjoint but not self-adjoint for each $n \in \mathbb{N}$, and $\mathcal{P}({}_0H_1)$ is not even algebraically self-adjoint.

(3) $\mathcal{D}_1 = \bigcap_{k=1}^{\infty} \mathcal{D}(\overline{H_1}^k)$,

$${}_0\mathcal{D}_1 = \bigcap_{k=1}^{\infty} \mathcal{D}(\overline{{}_0H_1}^k) \subset \mathcal{D}^*(\mathcal{P}({}_0H_1)) = C^\infty[0,1],$$

$${}_n\mathcal{D}_1 \subsetneqq \bigcap_{k=1}^{\infty} \mathcal{D}(\overline{{}_nH_1}^k) = \mathcal{D}_1, \quad n \in \mathbb{N}.$$

Remark 1.2.8. (1) $\mathcal{P}(A)'_{\mathrm{w}}$ is not a von Neumann algebra if A is not essentially self-adjoint.

(2) The algebraic self-adjointness of $\mathcal{P}(A)$ does not necessarily imply the self-adjointness of $\mathcal{P}(A)$ (see $\mathcal{P}({}_nH_1), n \in \mathbb{N}$ in Example 1.2.7), and so the implication (iii) $\Rightarrow$ (ii) in Proposition 1.2.3 does not hold in general. In particular, even if $\mathcal{M}'_{\mathrm{w}}$ is a von Neumann algebra, $\mathcal{M}'_{\mathrm{w}} \neq \mathcal{M}'_{\mathrm{s}}$ in general.

(3) The condition $\mathcal{D} = \bigcap_{n=1}^{\infty} \mathcal{D}(\overline{A}^n)$ does not always imply that A is essentially self-adjoint (see $\mathcal{P}({}_0H_1)$ in Example 1.2.7). The essential self-adjointness of A does not necessarily imply the condition $\mathcal{D} = \bigcap_{n=1}^{\infty} \mathcal{D}(\overline{A}^n)$. If so, then $\mathcal{P}(A)$ is always self-adjoint by Proposition 1.2.6, which contradicts the above (2).

(4) The essential self-adjointness of A does not necessarily imply that of A^n for $n = 1, 2, \cdots$ (Proposition 1.2.6 and the above (2)).

(5) The implication (ii) $\Rightarrow$ (i) in Proposition 1.2.3 does not necessarily hold as the following example shows: Let $n \in \mathbb{N}$. Let $\mathcal{M}$ be a closed O*-algebra on ${}_n\mathcal{D}_1$ generated by ${}_nH_1$ and $\{\xi \otimes \overline{\eta}; \xi, \eta \in {}_n\mathcal{D}_1\}$, where $(\xi \otimes \overline{\eta})\zeta = (\zeta|\eta)\xi, \zeta \in {}_n\mathcal{D}_1$. As seen in Example 1.2.7, ${}_nH_1$ is essentially self-adjoint and $\mathcal{P}({}_nH_1)$ is not self-adjoint, and so $\mathcal{M}$ is not self-adjoint. It is clear that $\mathcal{M}'_{\rm w} = \mathcal{M}'_{\rm s} = \mathbb{C}I$.

1.3 Invariant subspaces for O*-algebras

Let $\mathcal{M}$ be a closed O*-algebra on $\mathcal{D}$ in $\mathcal{H}$. We denote by $\overline{\mathcal{K}}$ the closure of a subset $\mathcal{K}$ of $\mathcal{H}$ with respect to the Hilbert space norm and denote by $\overline{\mathfrak{M}}^{t_\mathcal{M}}$ the closure of a subset $\mathfrak{M}$ of $\mathcal{D}$ with respect to the graph topology $t_\mathcal{M}$. We first define an induction of $\mathcal{M}$. For any projection E in $\mathcal{M}'_{\rm w}$ we put

$$X_E E\xi = EX\xi, \quad X \in \mathcal{M}, \xi \in \mathcal{D};$$
$$\mathcal{M}_E = \{X_E; X \in \mathcal{M}\}.$$

Then $\mathcal{M}_E$ is an O*-algebra on $E\mathcal{D}$ in $E\mathcal{H}$ and is called an *induction* of $\mathcal{M}$. For inductions of O*-algebras we have the following

Proposition 1.3.1. Let $\mathcal{M}$ be a closed O*-algebra on $\mathcal{D}$ in $\mathcal{H}$ and E a projection in $\mathcal{M}'_{\rm w}$. The following statements hold:

(1) If $\mathcal{M}'_{\rm w}$ is a von Neumann algebra, then $(\mathcal{M}_E)'_{\rm w}$ is a von Neumann algebra which coincides with the reduction $(\mathcal{M}'_{\rm w})_E$ of the von Neumann algebra $\mathcal{M}'_{\rm w}$.

(2) $E\mathcal{D} \subset \mathcal{D}^*(\mathcal{M}_E) \subset E\mathcal{D}^*(\mathcal{M})$.

(3) Suppose $E\mathcal{D} \subset \mathcal{D}$. Then the following statements hold:

$(3)_1$ $\mathcal{M}_E$ is closed.

$(3)_2$ $\mathcal{D}^*(\mathcal{M}_E) = E\mathcal{D}^*(\mathcal{M})$.

$(3)_3$ $\mathcal{M}_E$ is self-adjoint if and only if $E\mathcal{D} = E\mathcal{D}^*(\mathcal{M})$.

Proof. See Proposition 3.1 in Ikeda-Inoue [1].

Let $\mathfrak{M}$ be any $\mathcal{M}$-invariant subspace of $\mathcal{D}$. The closure $\mathcal{M}_\mathfrak{M}$ of the O*-algebra $\mathcal{M}\lceil\mathfrak{M} \equiv \{X\lceil\mathfrak{M}; X \in \mathcal{M}\}$ is a closed O*-algebra on $\overline{\mathfrak{M}}^{t_\mathcal{M}}$ in $\overline{\mathfrak{M}}$. The following questions for the projection $P_{\overline{\mathfrak{M}}}$ and the O*-algebra $\mathcal{M}_\mathfrak{M}$ arise:

Q_1. Does the projection $P_{\overline{\mathfrak{M}}}$ belong to $\mathcal{M}'_{\rm w}$?

Q_2. Suppose $P_{\overline{\mathfrak{M}}} \in \mathcal{M}'_{\rm w}$. Does the closed O*-algebras $\mathcal{M}_{P_{\overline{\mathfrak{M}}}}$ and $\mathcal{M}_\mathfrak{M}$ coincide?

The following example shows that Q_1 is not affirmative in general.

Example 1.3.2. Let $\mathcal{H} = L^2(\mathbb{R})$ and

$$\mathcal{D} = \{f \in C^\infty(\mathbb{R}); \frac{d^n}{dt^n} f \in \mathcal{H} \text{ for } n = 0, 1, \cdots\},$$

$$A = -i\frac{d}{dt}\lceil \mathcal{D}.$$

Then $\mathcal{M} \equiv \mathcal{P}(A)$ is a self-adjoint O*-algebra on $\mathcal{D}$ in $\mathcal{H}$. We put

$$\mathfrak{M} = \{f \in C^\infty(\mathbb{R}); f(t) = 0 \text{ for each } t \in \mathbb{R} \setminus [0,1]\}.$$

Then $\mathfrak{M}$ is a $\mathcal{M}$-invariant subspace of $\mathcal{D}$ and $P_{\overline{\mathfrak{M}}}$ is given by

$$\left(P_{\overline{\mathfrak{M}}} f\right)(t) = \begin{matrix} f(t), & t \in [0,1] \\ 0, & t \notin [0,1] \end{matrix}$$

for all $f \in \mathcal{H}$. By the simple calculation we have $P_{\overline{\mathfrak{M}}} \notin \mathcal{M}'_{\mathrm{w}}$.

We next have the following results for the relations between $\mathcal{M}_{\mathfrak{M}}$ and $\mathcal{M}_{P_{\overline{\mathfrak{M}}}}$:

Theorem 1.3.3. Let $\mathcal{M}$ be a closed O*-algebra on $\mathcal{D}$ in $\mathcal{H}$ and $\mathfrak{M}$ a $\mathcal{M}$-invariant subspace of $\mathcal{D}$. Consider the following statements:
(i) $\mathfrak{M}$ is essentially self-adjoint , that is, $\mathcal{M}_{\mathfrak{M}}$ is self-adjoint.
(ii) $P_{\overline{\mathfrak{M}}}\mathcal{D}^*(\mathcal{M}) = \overline{\mathfrak{M}}^{t_\mathcal{M}}$.
(iii) $P_{\overline{\mathfrak{M}}} \in \mathcal{M}'_{\mathrm{w}}$ and $\mathcal{M}_{P_{\overline{\mathfrak{M}}}} = \mathcal{M}_{\mathfrak{M}}$.
(iv) $P_{\overline{\mathfrak{M}}}\mathcal{D} \subset \mathcal{D}$.
(v) $P_{\overline{\mathfrak{M}}} \in \mathcal{M}'_{\mathrm{w}}$.
Then the following implications hold:

$$\begin{matrix} \text{(i)} & & & & \\ \Updownarrow & \Rightarrow & \text{(iii)} & \Rightarrow \text{(iv)} & \Rightarrow \text{(v)}. \\ \text{(ii)} & & & & \end{matrix}$$

In particular, if $\mathcal{M}$ is self-adjoint, then the above statements (i), (ii) and (iii) are equivalent and the above statements (iv) and (v) are equivalent. Any converse implications don't hold in general.

Proof. See Theorem 4.7 in Powers [1] and Theorem 3.3 in Ikeda-Inoue [1].

Example 1.3.4. Let $\mathcal{D}$ be a dense subspace in a Hilbert space $\mathcal{H}$ and ξ any non-zero element of $\mathcal{D}$. Then $\mathcal{L}^\dagger(\mathcal{D})\xi = \mathcal{D}$, and so $P_{\overline{\mathcal{L}^\dagger(\mathcal{D})\xi}} = I$ and $\mathcal{L}^\dagger(\mathcal{D})_{P_{\overline{\mathcal{L}^\dagger(\mathcal{D})\xi}}} = \mathcal{L}^\dagger(\mathcal{D})_{\mathcal{L}^\dagger(\mathcal{D})\xi} = \mathcal{L}^\dagger(\mathcal{D})$. This shows that the implication (iii) $\Rightarrow$ (i) in Theorem 1.3.3 does not hold in general.

Example 1.3.5. Let

$$\mathcal{D} = \{f \in C_0^\infty[0,2]; f^{(n)}(0) = f^{(n)}(1) = f^{(n)}(2), n = 0, 1, \cdots\},$$
$$A = -i\frac{d}{dt}\lceil\mathcal{D}.$$

Then $\mathcal{M} \equiv \mathcal{P}(A)$ is a closed O*-algebra on $\mathcal{D}$ in $L^2[0,2]$. We put

$$\mathfrak{M} = \{f \in C^\infty[0,2]; f^{(n)}(t) = f^{(n)}(t) = 0 \text{ for each } t \in [1,2] \text{ and } n = 0, 1, \cdots\}.$$

The following statements hold:

(1) $\mathfrak{M}$ is a $\mathcal{M}$-invariant subspace of $\mathcal{D}$ and

$$(P_{\overline{\mathfrak{M}}}f)(t) = \begin{cases} f(t), & 0 \leq t \leq 1 \\ 0, & 1 < t \leq 2 \end{cases}$$

for each $f \in L^2[0,2]$. Hence we have $P_{\overline{\mathfrak{M}}} \in \mathcal{M}'_{\mathrm{w}}$ but $P_{\overline{\mathfrak{M}}}\mathcal{D} \not\subset \mathcal{D}$. This shows that the implication (v) $\Rightarrow$ (iv) in Theorem 1.3.3 does not hold in general.

(2) $\mathcal{M}_{\mathfrak{M}} \neq \mathcal{M}_{P_{\overline{\mathfrak{M}}}}$. This follows since $\mathcal{M}_{\mathfrak{M}}$ and $\mathcal{M}_{P_{\overline{\mathfrak{M}}}}$ are unitarily equivalent to the non-selfadjoint O*-algebra ${}_0\mathcal{M}_1$ and the self-adjiont O*-algebra $\mathcal{M}_1$ in Example 1.1.10, respectively.

We define the notions of cyclic vectors for O*-algebras.

Definition 1.3.6. Let $\mathcal{M}$ be a closed O*-algebra on $\mathcal{D}$ in $\mathcal{H}$. A vector ξ_0 in $\mathcal{D}$ is said to be *ultra-cyclic* (resp. *strongly cyclic* , *cyclic*) if $\mathcal{M}\xi_0 = \mathcal{D}$ (resp. $\overline{\mathcal{M}\xi_0}^{t_\mathcal{M}} = \mathcal{D}, \overline{\mathcal{M}\xi_0} = \mathcal{H}$).

As seen later in Example 1.3.11, two notions of cyclic vectors and strongly cyclic vectors are different. It is shown later that the two notions coincides in case of integrable commutative O*-algebras with the metrizable graph topology (see Theorem 1.5.5).

Proposition 1.3.7. Let $\mathcal{M}$ be a closed O*-algebra on $\mathcal{D}$ in $\mathcal{H}$. Consider the following statements:

(i) Every non-zero vector in $\mathcal{D}$ is strongly cyclic.

(ii) Suppose $\mathfrak{M}$ is any $\mathcal{M}$-invariant subspace of $\mathcal{D}$. Then $\mathfrak{M} = \{0\}$ or $\overline{\mathfrak{M}}^{t_\mathcal{M}} = \mathcal{D}$.

(iii) Every non-zero vector in $\mathcal{D}$ is cyclic.

(iv) $\mathcal{M}'_{\mathrm{ss}} \equiv \{C \in \mathcal{M}'_{\mathrm{s}}; C^*\mathcal{D} \subset \mathcal{D}\} = \mathbb{C}I$.

(v) Supose $\mathfrak{M}$ is any essentially self-adjoint $\mathcal{M}$-invariant subspace of $\mathcal{D}$. Then $\mathfrak{M} = \{0\}$ or $\overline{\mathfrak{M}}^{t_\mathcal{M}} = \mathcal{D}$.

Then the following implications hold:

$$\begin{array}{c} \text{(i)} \\ \Updownarrow \\ \text{(ii)} \end{array} \Rightarrow \text{(iii)} \Rightarrow \text{(iv)} \Rightarrow \text{(v)}.$$

In particular, if $\mathcal{M}$ is self-adjoint, then the statements (iv) and (v) are equivalent to $\mathcal{M}'_{\rm w} = \mathbb{C}I$.

Proof. See Theorem 5 in Gudder-Scruggs[1].

We give some concrete examples for (strongly) cyclic vectors:

Example 1.3.8. Let $\mathcal{D}$ be a dense subspace in a Hilbert space $\mathcal{H}$. Every non-zero vector ξ in $\mathcal{D}$ is ultra-cyclic for $\mathcal{L}^\dagger(\mathcal{D})$.

Example 1.3.9. Let $\mathcal{M}_1$ be in Example 1.1.5 and $\mathcal{M} = \widetilde{\mathcal{M}_1}$, that is, $\mathcal{M}$ is a self-adjoint O*-algebra on $\mathcal{S}(\mathbb{R})$ in $L^2(\mathbb{R})$ generated by the momentum operator P and the position operator Q on $\mathcal{S}(\mathbb{R})$ defined as follows:

$$(Pf)(t) = -if'(t) \text{ and } (Qf)(t) = tf(t), \quad f \in \mathcal{S}(\mathbb{R}).$$

Then the following statements hold:

(1) $\mathcal{M}'_{\rm w} = \mathbb{C}I$. In fact, take an arbitrary $C \in \mathcal{M}'_{\rm w}$. Since C commutes the spectral projections of $\overline{P}$ and $\overline{Q}$, it follows that C commutes with the unitary operators $U(s) \equiv e^{is\overline{P}}$ and $V(s) = e^{is\overline{Q}}$ for all $s \in \mathbb{R}$. Here, $(U(s)f)(t) = e^{ist}f(t)$ and $(V(s)f)(t) = f(s+t)$ for $f \in \mathcal{S}(\mathbb{R})$ and $s \in \mathbb{R}$. It is well known that $\{U(s), V(s); s \in \mathbb{R}\}' = \mathbb{C}I$ (see von Neumann [1]), which implies $C = \alpha I$ for some $\alpha \in \mathbb{C}$.

(2) A vector $\xi_0(t) \equiv e^{-\frac{t^2}{2}}$ in $\mathcal{S}(\mathbb{R})$ is strongly cyclic for $\mathcal{M}$.

(3) Let ξ be any non-zero vector in $C_0^\infty(\mathbb{R})$. Then, $0 \lneqq P_{\overline{\mathcal{M}\xi}} \lneqq I$, and so ξ is not cyclic for $\mathcal{M}$. By (1) we have $P_{\overline{\mathcal{M}\xi}} \notin \mathcal{M}'_{\rm w}$. This shows that the implication (iv) $\Rightarrow$ (iii) in Proposition 1.3.7 does not hold in general.

Example 1.3.10. Let $\mathcal{M}_2$ be a closed O*-algebra on $C_0^\infty(\mathbb{R})$ given in Example 1.1.5. Since $\mathcal{M}_2 \subset \mathcal{M}_1$, it follows from Example 1.3.9, (1) that $(\mathcal{M}_2)'_{\rm w} = \mathbb{C}I$, which implies that every vector in $C_0^\infty(\mathbb{R})$ is not cyclic for $\mathcal{M}_2$.

Next example shows the difference between cyclic vectors and strong cyclic vectors.

Example 1.3.11. Let $\mathcal{M}$ be a closed O*-algebra on $\mathcal{D}$ in $\mathcal{H}$, $\mathcal{E}$ a dense subspace of $\mathcal{H}$ contained in $\mathcal{D}$ and $\mathcal{N}$ a closed O*-algebra on $\mathcal{D}$ generated by $\mathcal{M}$ and $\{(\xi \otimes \overline{\eta})\lceil\mathcal{D}; \xi, \eta \in \mathcal{E}\}$. Then every non-zero element of $\mathcal{E}$ is cyclic for $\mathcal{N}$, but $\mathcal{N}$ has no strongly cyclic vector. The concrete example of a cyclic but not strongly cyclic vector for an O*-algebra has been given in Example 8.3.18 in Schmüdgen [21].

1.4 Induced extensions

Let $\mathcal{M}$ be a closed O*-algebra on $\mathcal{D}$ in $\mathcal{H}$. Let $\mathcal{C}$ be a $*$-invariant subset of $\mathcal{M}'_{\rm w}$ containing I. We put

$$\widehat{\mathcal{D}} = \text{ linear span of } \mathcal{C}\mathcal{D},$$
$$\widehat{X}(\sum_k K_k \xi_k) = \sum_k K_k X \xi_k, \quad X \in \mathcal{M}, \quad \sum_k K_k \xi_k \in \widehat{\mathcal{D}},$$
$$\widehat{\mathcal{M}} = \{\widehat{X}; X \in \mathcal{M}\}.$$

Then it is easily shown that $\widehat{\mathcal{M}}$ is an O-algebra on $\widehat{\mathcal{D}}$ such that $\mathcal{M} \prec \widehat{\mathcal{M}} \prec \mathcal{e}_{\mathcal{C}}(\mathcal{M}) \prec \mathcal{M}^*$, where $\mathcal{e}_{\mathcal{C}}(\mathcal{M})$ is the closure of $\widehat{\mathcal{M}}$. We call this O-algebra $\mathcal{e}_{\mathcal{C}}(\mathcal{M})$ the *induced extension* of $\mathcal{M}$ determined by $\mathcal{C}$. The induced extension $\mathcal{e}_{\mathcal{C}}(\mathcal{M})$ is not an O*-algebra in general. We consider when $\mathcal{e}_{\mathcal{C}}(\mathcal{M})$ is an O*-algebra.

Proposition 1.4.1. Let $\mathcal{M}$ be a closed O*-algebra on $\mathcal{D}$ in $\mathcal{H}$ and $\mathcal{C}$ a $*$-invariant subset of $\mathcal{M}'_{\rm w}$ containing I. The induced extension $\mathcal{e}_{\mathcal{C}}(\mathcal{M})$ is an O*-algebra if and only if $\mathcal{C}^2 \subset \mathcal{M}'_{\rm w}$. In particular, $\mathcal{e}_{\mathcal{M}'_{\rm w}}(\mathcal{M})$ is an O*-algebra if and only if $\mathcal{M}'_{\rm w}$ is a von Neumann algebra.

Proof. See Proposition 2.1 in Inoue-Kurose-Ôta [1].

We next investigate whether the domain $\mathcal{D}(\mathcal{e}_{\mathcal{C}}(\mathcal{M}))$ is $\mathcal{C}$-invariant.

Theorem 1.4.2. Let $\mathcal{M}$ be a closed O*-algebra on $\mathcal{D}$ in $\mathcal{H}$. Suppose $\mathcal{C}$ is a $*$-invariant subset of $\mathcal{M}'_{\rm w}$ such that $I \in \mathcal{C}$ and $\mathcal{C}^2 \subset \mathcal{M}'_{\rm w}$. Then the following statements are equivalent:

(i) $\mathcal{D}(\mathcal{e}_{\mathcal{C}}(\mathcal{M}))$ is $\mathcal{C}$-invariant.

(ii) $\mathcal{C}'' \subset \mathcal{M}'_{\rm w}$ and $\mathcal{e}_{\mathcal{C}}(\mathcal{M}) = \mathcal{e}_{\mathcal{C}^2}(\mathcal{M}) = \cdots = \mathcal{e}_{\mathcal{C}''}(\mathcal{M})$.

If this is true, then $\mathcal{C}'' \subset e_{\mathcal{C}}(\mathcal{M})'_{\rm w}$. In particular, if $\mathcal{M}'_{\rm w}$ is a von Neumann algebra, then $\mathcal{e}_{\mathcal{M}'_{\rm w}}(\mathcal{M})$ is a closed O*-algebra such that $\mathcal{M} \prec \mathcal{e}_{\mathcal{M}'_{\rm w}}(\mathcal{M})$, $\mathcal{e}_{\mathcal{M}'_{\rm w}}(\mathcal{M})'_{\rm w} = \mathcal{M}'_{\rm w}$ and $\mathcal{M}'_{\rm w}\mathcal{D}(\mathcal{e}_{\mathcal{M}'_{\rm w}}(\mathcal{M})) \subset \mathcal{D}(\mathcal{e}_{\mathcal{M}'_{\rm w}}(\mathcal{M}))$.

Proof. See Theorem 3.1 and Corollary 3.2 in Inoue-Kurose-Ôta [1].

Example 1.4.3. Let $\mathcal{D}$ be a dense subspace in $\mathcal{H}$ and $A = A^\dagger \in \mathcal{L}^\dagger(\mathcal{D})$. We denote by $\mathcal{M}$ the closure of the O*-algebra $\mathcal{P}(A)$. Let $U \equiv U_A$ be the partial isometry defined by the Cayley transformation of $\overline{A}$. By Lemma 1.2.4, we have $U^n \in \mathcal{M}'_{\rm w}$ for each $n \in \mathbb{N}$. We have that $\mathcal{e}_{\{I,U,U^*\}}(\mathcal{M})$ is an O*-algebra if and only if A is essentially self-adjoint. In fact, suppose $\mathcal{e}_{\{I,U,U^*\}}(\mathcal{M})$ is an O*-algebra. By Proposition 1.4.1 we have $U^*U = UU^* = I$, and hence $\overline{A}$ is self-adjoint. The converse follows from Lemma 1.2.4 and Proposition 1.4.1.

1.5 Integrability of commutative O*-algebras

For the integrability of commutative O*-algebras we have the following

Theorem 1.5.1. Let $\mathcal{M}$ be a closed commutative O*-algebra on $\mathcal{D}$ in $\mathcal{H}$. The following statements are equivalent:

(i) $\mathcal{M}$ is integrable.

(ii) $\mathcal{M}$ is self-adjoint and $(\mathcal{M}'_{\mathrm{w}})'$ is a commutative von Neumann algebra.

(iii) $\mathcal{M}'_{\mathrm{w}} = \mathcal{M}'_{\mathrm{s}}$ and $(\mathcal{M}'_{\mathrm{w}})'$ is a commutative von Neumann algebra.

(iv) The von Neumann algebra $(\mathcal{M}'_{\mathrm{ss}})'$ is commutative.

(v) There is a commutative von Neumann algebra $\mathcal{A}$ such that $\overline{X}$ is affiliated with $\mathcal{A}$ for each $X \in \mathcal{M}$.

(vi) There is a commutative von Neumann algebra $\mathcal{A}$ such that $\overline{X}$ is affiliated with $\mathcal{A}$ for each $X \in \mathcal{M}_h$.

Proof. See Theorem 7.1 in Powers [1] and Theorem 9.1.7 in Schmüdgen [21].

For the integrable extension of commutative O*-algebras we have the following

Proposition 1.5.2. Let $\mathcal{M}$ be a commutative O*-algebra on $\mathcal{D}$ in $\mathcal{H}$. The following statements are equivalent:

(i) There exists an integrable O*-algebra $\mathcal{M}_1$ acting in the same Hilbert space $\mathcal{H}$ as $\mathcal{M}$ such that $\mathcal{M} \prec \mathcal{M}_1$.

(ii) There exists a von Neumann algebra $\mathcal{C}$ on $\mathcal{H}$ contained in $\mathcal{M}'_{\mathrm{w}}$ such that $\mathcal{C}'$ is commutative.

If this is true, then the induced extension $e_{\mathcal{C}}(\mathcal{M})$ determined by $\mathcal{C}$ can be taken for $\mathcal{M}_1$.

Proof. See Proposition 9.1.12 in Schmüdgen [21].

Proposition 1.5.3. Let $\mathcal{M}$ be a closed commutative O*-algebra on $\mathcal{D}$ in $\mathcal{H}$. Suppose there is a subset $\mathcal{N}$ of $\mathcal{M}_h$ such that $\mathcal{N}$ generates $\mathcal{M}$, and $\overline{B_1}$ and $\overline{B_2}$ are strongly commuting self-adjoint operators for each $B_1, B_2 \in \mathcal{N}$. Then $\mathcal{M}^*$ is an integrable commutative O*-algebra, and $\mathcal{D}^*(\mathcal{M}) = \bigcap_{B\in\mathcal{N}} \bigcap_{n\in\mathbb{N}} \mathcal{D}(\overline{B}^n)$. Further, we have that $\mathcal{M}^* = e_{\mathcal{M}'_{\mathrm{w}}}(\mathcal{M})$ and $(\mathcal{M}^*)'_{\mathrm{w}} = \bigcap_{B\in\mathcal{N}} \{\overline{B}\}'$.

Proof. See Theorem 9.1.3 in Schmüdgen [21].

Theorem 1.5.4. Let $\mathcal{M}$ be a closed commutative O*-algebra on $\mathcal{D}$ in $\mathcal{H}$ and $\mathcal{N}$ a subset of $\mathcal{M}_h$ such that $\mathcal{N}$ generates $\mathcal{M}$. The following statements are equivalent:

(i) $\mathcal{M}$ is integrable.

(ii) $\mathcal{M}$ is self-adjoint and $(B_1 + iB_2)^* = \overline{B_1 - iB_2}$ for each $B_1, B_2 \in \mathcal{N}$.

(iii) Every element B of $\mathcal{N}$ is essentially self-adjoint, $\mathcal{D} = \bigcap_{B \in \mathcal{N}} \bigcap_{n \in \mathbb{N}} \mathcal{D}(\overline{B}^n)$ and $(B_1 + iB_2)^* = \overline{B_1 - iB_2}$ for each $B_1, B_2 \in \mathcal{N}$.

(iv) $\overline{B_1}$ and $\overline{B_2}$ are strongly commuting self-adjoint operators for each $B_1, B_2 \in \mathcal{N}$ and $\mathcal{M}'_{\mathrm{w}}\mathcal{D} \subset \mathcal{D}$.

Proof. This follows from Proposition 1.5.3 and Corollary 9.1.14 in Schmüdgen [21].

For cyclic vectors for an integrable commutative O*-algebra we have the following

Theorem 1.5.5. Let $\mathcal{M}$ be an integrable commutative O*-algebra on $\mathcal{D}$ in $\mathcal{H}$ such that the graph topology $t_{\mathcal{M}}$ is metrizable. Then the following statements are equivalent:

(i) $\mathcal{M}$ has a strongly cyclic vector.

(ii) $\mathcal{M}$ has a cyclic vector.

(iii) The von Neumann algebra $(\mathcal{M}'_{\mathrm{w}})'$ has a cyclic vector.

Proof. See Theorem 9.2.13 in Schmüdgen [21].

The main assertion in Theorem 1.5.5 (the implication (iii) $\Rightarrow$ (i)) is no longer true in general if the graph topology is not metrizable as seen in next example.

Example 1.5.6. We denote by $\mathcal{D}$ the set of all functions f in $L^2(\mathbb{R})$ with compact support, and define $M_g f = gf$ for $g \in C(\mathbb{R})$ and $f \in \mathcal{D}$. Then $M_{C(\mathbb{R})} \equiv \{\mathcal{M}_f; f \in C(\mathbb{R})\}$ is an integrable O*-algebra on $\mathcal{D}$ in $L^2(\mathbb{R})$ which has no cyclic vector, and $((\mathcal{M}_{C(\mathbb{R})})'_{\mathrm{w}})' = M_{L^\infty(\mathbb{R})}$. The vector $\xi_0(t) \equiv e^{-t^2}$, $t \in \mathbb{R}$ is cyclic for the commutative von Neumann algebra $M_{L^\infty(\mathbb{R})}$.

For the polynomial algebra $\mathcal{P}(A, B)$ generated by commutative symmetric operators A and B in $\mathcal{L}^\dagger(\mathcal{D})$ we have the following

Corollary 1.5.7. The following statements are equivalent:

(i) $\mathcal{P}(A, B)$ is integrable.

(ii) $\mathcal{P}(A, B)$ is self-adjoint and there exists a normal operator C which is an extension of $A + iB$.

(iii) A and B are essentially self-adjoint, $\mathcal{D} = \bigcap_{n=1}^{\infty} (\mathcal{D}(\overline{A}^n) \cap \mathcal{D}(\overline{B}^n))$ and there exists a normal operator C which is an extension of $A + iB$.

(iv) $\overline{A}$ and $\overline{B}$ are strongly commuting self-adjoint operators and $\mathcal{P}(A,B)$ is a closed O*-algebra satisfying $\mathcal{P}(A,B)'_{\mathrm{w}}\mathcal{D} \subset \mathcal{D}$.

We next study self-adjoint extensions of $\mathcal{P}(A,B)$. Suppose that A and B are essentially self-adjoint. We put

$$\mathcal{D}^\infty(\overline{A},\overline{B}) = \{\xi \in \bigcap_{n,m\in\mathbb{N}} \mathcal{D}(\overline{A}^n\overline{B}^m) \cap \mathcal{D}(\overline{B}^m\overline{A}^n); \overline{A}^n\overline{B}^m\xi = \overline{B}^m\overline{A}^n\xi \quad \text{for all } n,m \in \mathbb{N}\},$$

$$A_\infty = \overline{A}\lceil\mathcal{D}_\infty(\overline{A},\overline{B}), \quad B_\infty = \overline{B}\lceil\mathcal{D}_\infty(\overline{A},\overline{B}).$$

Then it is easily shown that $A_\infty, B_\infty \in \mathcal{L}^\dagger(\mathcal{D}^\infty(\overline{A},\overline{B}))$, and $A \subset A_\infty \subset \overline{A}$ and $B \subset B_\infty \subset \overline{B}$. For the polynomial algebra $\mathcal{P}(A_\infty, B_\infty)$ we have the following

Proposition 1.5.8. Let A and B be commuting essentially self-adjoint operators in $\mathcal{L}^\dagger(\mathcal{D})$. The following statements hold:

(1) $\mathcal{P}(A_\infty, B_\infty) = \mathcal{P}(A,B)^*$, and so they are self-adjoint O*-algebras on $\mathcal{D}^\infty(\overline{A},\overline{B})$ and $\mathcal{P}(A,B)'_{\mathrm{w}}$ is a von Neumann algebra.

(2) Suppose $\mathcal{P}(A,B)$ is self-adjoint. Then $\mathcal{D} = \mathcal{D}^\infty(\overline{A},\overline{B})$ and $\mathcal{P}(A,B) = \mathcal{P}(A_\infty, B_\infty)$.

(3)Suppose $\overline{A}$ and $\overline{B}$ are strongly commuting. Then $\mathcal{D}^*(\mathcal{P}(A,B)) = \mathcal{D}_\infty(\overline{A},\overline{B}) = \bigcap_{n=1}^{\infty}(\mathcal{D}(\overline{A}^n) \cap \mathcal{D}(\overline{B}^n))$, $\mathcal{P}(A,B)^* = \mathcal{P}(A_\infty, B_\infty) = e_{\mathcal{P}(A,B)'_{\mathrm{w}}}(\mathcal{P}(A,B))$ and they are integrable.

Proof. See Proposition 9.3.13 in Schmüdgen [21].

We showed in Proposition 1.5.8, (1) that if A and B are essentially self-adjoint, then $\mathcal{P}(A,B)^*$ is self-adjoint. But the converse of this result doesn't necessarily hold as seen in next example.

Example 1.5.9. Let A be an essentially self-adjoint operator in $\mathcal{H}$ such that A^2 and $A+A^2$ are both not essentially self-adjoint. Since $\overline{A}$ is self-adjoint, $\mathcal{P}(A)^*$ is integrable by Proposition 1.5.3. Further, since $\mathcal{P}(A^2, A+A^2) = \mathcal{P}(A)$, it follows that $\mathcal{P}(A^2, A+A^2)^*$ is integrable though the operators $\overline{A^2}$ and $\overline{A+A^2}$ are both not self-adjoint.

We finally introduce the Schmüdgen construction of non-integrable self-adjoint commutative O*-algebras.

Theorem 1.5.10. Suppose $\mathcal{A}$ is a properly infinite von Neumann algebra on a separable Hilbert space $\mathcal{H}$. Then there exists a self-adjoint O*-algebra $\mathcal{P}(A,B)$ on $\mathcal{D}$ in $\mathcal{H}$ generated by commuting essentially self-adjoint operators A and B in $\mathcal{L}^\dagger(\mathcal{D})$ such that $(\mathcal{P}(A,B)'_{\mathrm{w}})' = \mathcal{A}$, and A^n and B^n are essentially self-adjoint operators for each $n \in \mathbb{N}$.

Proof. See Theorem 9.4.1 in Schmüdgen [21].

1.6 Topologies of O*-algebras

In this section we introduce several topologies of an O*-algebra. Let $\mathcal{M}$ be a closed O*-algebra on $\mathcal{D}$ in $\mathcal{H}$.

A. Weak, strong and strong* topologies

Let $\mathcal{L}^\dagger(\mathcal{D},\mathcal{H})$ be the set of all linear operators X from $\mathcal{D}$ to $\mathcal{H}$ such that $\mathcal{D}(X^*) \supset \mathcal{D}$. Then $\mathcal{L}^\dagger(\mathcal{D},\mathcal{H})$ is a $*$-preserving vector space equipped with the usual operators $X+Y, \alpha X$ and the involution $X^\dagger \equiv X^*\lceil\mathcal{D}$. For each $\xi,\eta \in \mathcal{D}$ we put

$$\begin{aligned} p_{\xi,\eta}(X) &= |(X\xi|\eta)|, \\ p_\xi(X) &= \|X\xi\|, \\ p_\xi^*(X) &= p_\xi(X) + p_\xi(X^\dagger), \quad X \in \mathcal{L}^\dagger(\mathcal{D},\mathcal{H}). \end{aligned}$$

The topology on $\mathcal{L}^\dagger(\mathcal{D},\mathcal{H})$ defined by the family $\{p_{\xi,\eta}(\cdot);\xi,\eta \in \mathcal{D}\}$ (resp. $\{p_\xi(\cdot);\xi \in \mathcal{D}\}$, $\{p_\xi^*(\cdot);\xi \in \mathcal{D}\}$) of the seminorms is called the *weak* (resp. *strong* , *strong**) topology on $\mathcal{L}^\dagger(\mathcal{D},\mathcal{H})$ and denoted by τ_{w} (resp. $\tau_{\mathrm{s}}, \tau_{\mathrm{s}}^*$). It is shown that the locally convex space $\mathcal{L}^\dagger(\mathcal{D},\mathcal{H})[\tau_{\mathrm{s}}^*]$ is complete. For each subset $\mathcal{N}$ of $\mathcal{L}^\dagger(\mathcal{D},\mathcal{H})$ we denoted by $\overline{\mathcal{N}}^{\tau_{\mathrm{w}}}$ (resp. $\overline{\mathcal{N}}^{\tau_{\mathrm{s}}}$, $\overline{\mathcal{N}}^{\tau_{\mathrm{s}}^*}$) the closure of $\mathcal{N}$ in $\mathcal{L}^\dagger(\mathcal{D},\mathcal{H})[\tau_{\mathrm{w}}]$ (resp. $\mathcal{L}^\dagger(\mathcal{D},\mathcal{H})[\tau_{\mathrm{s}}], \mathcal{L}^\dagger(\mathcal{D},\mathcal{H})[\tau_{\mathrm{s}}^*]$). The induced topology of the weak (resp. strong, strong*) topology on $\mathcal{M}$ is called the *weak* (resp. *strong, strong**) topology on $\mathcal{M}$. It is easily shown that $\mathcal{M}[\tau_{\mathrm{w}}]$ and $\mathcal{M}[\tau_{\mathrm{s}}^*]$ are locally convex $*$-algebras.

B. σ-weak, σ-strong and σ-strong* topologies

We put

$$\mathcal{D}^\infty(\mathcal{M}) = \{\{\xi_n\} \subset \mathcal{D}; \sum_{n=1}^{\infty}(\|\xi_n\|^2 + \|X\xi_n\|^2) < \infty \text{ for all } X \in \mathcal{M}\},$$

$$p_{\{\xi_n\},\{\eta_n\}}(X) = |\sum_{n=1}^{\infty}(X\xi_n|\eta_n)|,$$

$$p_{\{\xi_n\}}(X) = [\sum_{n=1}^{\infty}\|X\xi_n\|^2]^{1/2},$$

$$p^*_{\{\xi_n\}}(X) = p_{\{\xi_n\}}(X) + p_{\{\xi_n\}}(X^\dagger), \quad X \in \mathcal{M}$$

for $\{\xi_n\},\{\eta_n\} \in \mathcal{D}^\infty(\mathcal{M})$. The locally convex topology defined by the family $\{p_{\{\xi_n\},\{\eta_n\}}(\cdot)\ ; \{\xi_n\},\{\eta_n\} \in \mathcal{D}^\infty(\mathcal{M})\}$ (resp. $\{p_{\{\xi_n\}}(\cdot);\{\xi_n\} \in \mathcal{D}^\infty(\mathcal{M})\}$,

$\{p^*_{\{\xi_n\}}(\cdot); \{\xi_n\} \in \mathcal{D}^\infty(\mathcal{M})\}$) is called the $(\mathcal{M})$-σ-*weak* (resp. $(\mathcal{M})$-σ-*strong*, $(\mathcal{M})$-σ-*strong**) topology on $\mathcal{M}$ and denoted by $\tau^{\mathcal{M}}_{\sigma w}$ (resp. $\tau^{\mathcal{M}}_{\sigma s}$, $\tau^{*\mathcal{M}}_{\sigma s}$). It is easily seen that $\mathcal{M}[\tau^{\mathcal{M}}_{\sigma w}]$ and $\mathcal{M}[\tau^{*\mathcal{M}}_{\sigma s}]$ are locally convex $*$-algebras. In particular, the $(\mathcal{L}^\dagger(\mathcal{D}))$-$\sigma$-weak ($(\mathcal{L}^\dagger(\mathcal{D}))$-$\sigma$-strong, $(\mathcal{L}^\dagger(\mathcal{D}))$-$\sigma$-strong*) topology on $\mathcal{M}$ is simply called the σ-*weak* (resp σ-*strong*, σ-*strong**) topology and denoted by $\tau_{\sigma w}$ (resp. $\tau_{\sigma s}$, $\tau^*_{\sigma s}$). It is easily shown that $\mathcal{M}[\tau^{\mathcal{M}}_{\sigma w}], \mathcal{M}[\tau_{\sigma w}], \mathcal{M}[\tau^{*\ \mathcal{M}}_{\sigma s}]$ and $\mathcal{M}[\tau^*_{\sigma s}]$ are locally convex $*$-algebras, and the following relations among these topologies hold:

$$\begin{array}{ccccc} \tau_w & \prec & \tau_s & \prec & \tau_s^* \\ \curlywedge & & \curlywedge & & \curlywedge \\ \tau_{\sigma w} & \prec & \tau_{\sigma s} & \prec & \tau^*_{\sigma s} \\ \curlywedge & & \curlywedge & & \curlywedge \\ \tau^{\mathcal{M}}_{\sigma w} & \prec & \tau^{\mathcal{M}}_{\sigma s} & \prec & \tau^{*\ \mathcal{M}}_{\sigma s}, \end{array}$$

where the symbols $\tau_2 \succ \tau_1$ and $\overset{\tau_1}{\underset{\tau_2}{\curlywedge}}$ mean that the topology τ_2 is finer than the topology τ_1.

C. Uniform and quasi-uniform topologies

A subset $\mathfrak{M}$ of $\mathcal{D}$ is said to be $\mathcal{M}$-*bounded* if it is a bounded subset of the locally convex space $\mathcal{D}[t_{\mathcal{M}}]$. Let $\mathfrak{M}$ be a $\mathcal{M}$-bounded subset of $\mathcal{D}$. We put

$$\|X\|_{\mathfrak{M}} = \sup_{\xi,\eta \in \mathfrak{M}} |(X\xi|\eta)|, \quad X \in \mathcal{M}$$

and

$$P_{Y,\mathfrak{M}}(X) = \sup_{\xi \in \mathfrak{M}} \|YX\xi\|, \quad X \in \mathcal{M}$$

for $Y \in \mathcal{M}_I$, where $\mathcal{M}_I$ is an O*-algebra on $\mathcal{D}$ obtained by adjoining the identity operator I. The locally convex topology defined by the family $\{\|\cdot\|_{\mathfrak{M}}; \mathfrak{M}$ is $\mathcal{M}$-bounded $\}$ (resp. $\{P_{Y,\mathfrak{M}}(\cdot); Y \in \mathcal{M}_I, \mathfrak{M}$ is $\mathcal{M}$-bounded$\}$) is called the *uniform* (resp. *quasi-uniform*) topology on $\mathcal{M}$. Since $\mathcal{M}$ is closed, it follows from Corollary 2.3.11 in Schmüdgen [21] that $\mathfrak{M}$ is a bounded subset of $\mathcal{D}[t_{\mathcal{M}}]$ iff $\mathfrak{M}$ is a bounded subset of $\mathcal{D}[t_{\mathcal{L}^\dagger(\mathcal{D})}]$, and so the uniform (resp. quasi-uniform) topology on $\mathcal{M}$ equals the topology induced by the uniform (resp. quasi-uniform) topology on $\mathcal{L}^\dagger(\mathcal{D})$. Hence we denote by τ_u (resp. τ_{qu}) the uniform (resp. quasi-uniform) topology on $\mathcal{M}$.

Theorem 1.6.1. $\mathcal{M}[\tau_u]$ is a locally convex $*$-algebra and $\mathcal{M}[\tau_{qu}]$ is a locally convex algebra such that $\tau_u \prec \tau_{qu}$. The topologies τ_u and τ_{qu} equal if and only if the locally convex $*$-algebra $\mathcal{M}[\tau_u]$ has the jointly continous multiplication.

Proof. See Theorem 3.1 and Theorem 3.2 in Lassner [1].

Further studies of the uniform topology and the quasi-uniform topology have appeared in Inoue-Kuriyama-Ôta [1], Lassner [1] and Schmüdgen [21].

D. ρ-topology and λ-topology

For each $A \in (\mathcal{M}_I)_+$ and put

$$\rho_A(X) = \sup_{\xi \in \mathcal{D}} \frac{|(X\xi|\xi)|}{(A\xi|\xi)}, \quad X \in \mathcal{M},$$

where $\lambda/0 = \infty$ for $\lambda > 0$. This defines the normed space

$$\mathfrak{A}_A = \{X \in \mathcal{M}; \rho_A(X) < \infty\}$$

with the norm $\rho_A \equiv \rho_A \lceil \mathfrak{A}_A$. We note that $\bigcup_{A \in (\mathcal{M}_I)_+} \mathfrak{A}_A = \mathcal{M}$, further, the relation $0 \leq A \leq B$ implies that the injection $\iota_{A,B} : (\mathfrak{A}_A : \rho_A) \rightarrow (\mathfrak{A}_B : \rho_B)$ is a norm-decreasing map. The inductive limit topology for the normed spaces $\{(\mathfrak{A}_A : \rho_A); A \in (\mathcal{M}_I)_+\}$ is called the *ρ-topology* on $\mathcal{M}$ and denoted by τ_ρ.

Theorem 1.6.2. $\mathcal{M}[\tau_\rho]$ is a bornological locally convex $*$-algebra.

Proof. See Theorem 1 in Jurzak [1].

For each $A \in \mathcal{M}_I$ we put

$$\lambda_A(X) = \sup_{\xi \in \mathcal{D}} \frac{\|X\xi\|}{\|A\xi\|}, \quad X \in \mathcal{M},$$

where $\lambda/0 = \infty$ for $\lambda > 0$. This defines the normed space

$$\mathfrak{B}_A = \{X \in \mathcal{M}; \lambda_A(X) < \infty\}$$

with the norm $\lambda_A \equiv \lambda_A \lceil \mathfrak{B}_A$, and the spaces $\mathfrak{B}_A$ constitute a direct set. The inductive limit topology for the normed spaces $\{(\mathfrak{B}_A; \lambda_A); A \in \mathcal{M}_I\}$ is called the *λ-topology* on $\mathcal{M}$ and denoted by τ_λ. It is clear that $\tau_\rho \prec \tau_\lambda$, but $\mathcal{M}[\tau_\lambda]$ is not even a locally convex algebra, in general. Further studies of the ρ-topology and λ-topology have appeared in Arnal-Jurzak [1], Inoue-Kuriyama-Ôta [1] and Jurzak [1].

We here arrange the relations among all topologies defined above:

Theorem 1.6.3. The following diagram among the topologies holds:

$$\begin{array}{ccccccccc} \tau_{qu} & \succ & \tau_u & \succ & \tau_w & \prec & \tau_s & \prec & \tau_{qu} \\ \curlywedge & & \curlywedge & & \curlywedge & & \curlywedge & & \curlywedge \\ \tau_\lambda & \succ & \tau_\rho & & \tau_{\sigma w} & \prec & \tau_{\sigma s} & \prec & \tau_\lambda. \end{array}$$

1.7 Unbounded generalizations of von Neumann algebras

In this section we introduce extended EW*-algebras and generalized von Neumann algebras which are unbouded generalizations of von Neumann algebras.

A. Extended W*-algebras

An O*-algebra $\mathcal{M}$ on $\mathcal{D}$ in $\mathcal{H}$ is said to be *symmetric* if $(I + X^\dagger X)^{-1}$ exists and lies in $\mathcal{M}_b$ for all $X \in \mathcal{M}$, where $\mathcal{M}_b$ is the set of all bounded linear operators in $\mathcal{M}$.

Theorem 1.7.1. A closed symmetric O*-algebra is integrable.

Proof. See Theorem 2.3 in Inoue [1].

Definition 1.7.2. A closed O*-algebra $\mathcal{M}$ on $\mathcal{D}$ in $\mathcal{H}$ is said to be an extended W*-algebra (simply, an EW*-algebra) if it is symmetric and $\overline{\mathcal{M}_b} \equiv \{\overline{A}; A \in \mathcal{M}_b\}$ is a von Neumann algebra on $\mathcal{H}$.

B. Generalized von Neumann algebras

To define another unbounded generalization of von Neumann algebras, we first introduce unbounded commutants of an O*-algebra. Let $\mathcal{M}$ be a closed O*-algebra on $\mathcal{D}$ in $\mathcal{H}$. We define unbounded commutants and unbounded bicommutants of $\mathcal{M}$ as follows:

$$\begin{aligned}
\mathcal{M}'_\sigma &= \{S \in \mathcal{L}^\dagger(\mathcal{D},\mathcal{H}); (SX\xi|\eta) = (S\xi|X^\dagger\eta) \text{ for all } X \in \mathcal{M} \text{ and } \xi,\eta \in \mathcal{D}\},\\
\mathcal{M}'_c &= \{S \in \mathcal{L}^\dagger(\mathcal{D}); SX = XS \text{ for all } X \in \mathcal{M}\},\\
\mathcal{M}''_{\mathrm{wc}} &= \{X \in \mathcal{L}^\dagger(\mathcal{D}); (CX\xi|\eta) = (C\xi|X^\dagger\eta) \text{ for all } C \in \mathcal{M}'_{\mathrm{w}} \text{ and } \xi,\eta \in \mathcal{D}\},\\
\mathcal{M}''_{cc} &= \{X \in \mathcal{L}^\dagger(\mathcal{D}); XS = SX \text{ for all } S \in \mathcal{M}'_c\}.
\end{aligned}$$

Then we have the following

Proposition 1.7.3. (1) $\mathcal{M}'_\sigma$ is a subspace of $\mathcal{L}^\dagger(\mathcal{D},\mathcal{H})$ and $(\mathcal{M}'_\sigma)_b = \mathcal{M}'_{\mathrm{w}}\lceil\mathcal{D}$.

(2) $\mathcal{M}'_c$ is an O*-algebra on $\mathcal{D}$, which equals $\mathcal{M}'_\sigma \cap \mathcal{L}^\dagger(\mathcal{D})$.

(3) $\mathcal{M}''_{\mathrm{wc}}$ is a closed O*-algebra on $\mathcal{D}$ containing $\overline{\mathcal{M}}^{\tau_s^*} \cap \mathcal{L}^\dagger(\mathcal{D})$ and $(\mathcal{M}''_{\mathrm{wc}})'_{\mathrm{w}} = \mathcal{M}'_{\mathrm{w}}$.

(4) $\mathcal{M}''_{cc}$ is a closed O*-algebra on $\mathcal{D}$ containing $\overline{\mathcal{M}}^{\tau_{\mathrm{w}}} \cap \mathcal{L}^\dagger(\mathcal{D})$.

(5) Suppose $\mathcal{M}'_{\mathrm{w}}\mathcal{D} \subset \mathcal{D}$. Then

$$\mathcal{M}''_{\mathrm{wc}} = \{X \in \mathcal{L}^\dagger(\mathcal{D}); \overline{X} \text{ is affiliated with } (\mathcal{M}'_{\mathrm{w}})'\} \supset \mathcal{M}''_{cc} \supset \overline{\mathcal{M}}^{\tau_{\mathrm{w}}} \cap \mathcal{L}^\dagger(\mathcal{D}).$$

Proof. See Lemma 4.1 and Theorem 4.2 in Inoue [9].

Further studies of unbounded commutants of O*-algebras are treated in Epifanio-Trapani [1], Inoue [9], Inoue-Ueda-Yamauchi [1], Kasparek-Van Daele [1], Mathot [1] and Schmüdgen [17].

Definition 1.7.4. A closed O*-algebra $\mathcal{M}$ on $\mathcal{D}$ in $\mathcal{H}$ is said to be a generalized von Neumann algebra if $\mathcal{M}'_{\mathrm{w}}\mathcal{D} \subset \mathcal{D}$ and $\mathcal{M}''_{\mathrm{wc}} = \mathcal{M}$.

Proposition 1.7.5. Let $\mathcal{M}$ be a closed O*-algebra on $\mathcal{D}$ in $\mathcal{H}$ such that $\mathcal{M}'_{\mathrm{w}}\mathcal{D} \subset \mathcal{D}$. The following statements are equivalent:
(i) $\mathcal{M}$ is a generalized von Neumann algebra.
(ii) $\mathcal{M} = \{X \in \mathcal{L}^\dagger(\mathcal{D}); \overline{X}$ is affiliated with the von Neumann algebra $(\mathcal{M}'_{\mathrm{w}})'\}$.
(iii) $\mathcal{M} = \overline{(\mathcal{M}'_{\mathrm{w}})'\lceil\mathcal{D}}^{\tau_s^*} \cap \mathcal{L}^\dagger(\mathcal{D})$.

Proof. See Proposition 2.6 in Inoue-Ueda-Yamauchi [1].

1.8 $*$-representations of $*$-algebras

In this section we define $*$-representations of $*$-algebras and note their basic properties.

Definition 1.8.1. Let $\mathcal{A}$ be an algebra. A *representation* π of $\mathcal{A}$ on a Hilbert space $\mathcal{H}$ is a homomorphism of $\mathcal{A}$ onto an O-algebra on a dense subspace $\mathcal{D}(\pi)$ of $\mathcal{H}$ satisfying $\pi(1) = I$ whenever $\mathcal{A}$ has an identity 1. A representation π of a $*$-algebra $\mathcal{A}$ is said to be *hermitian* or a *$*$-representation* if $\pi(x^*) = \pi(x)^\dagger$ for all $x \in \mathcal{A}$.

Definition 1.8.2. Let π_1 and π_2 be representations of $\mathcal{A}$ on a Hilbert space $\mathcal{H}$. If $\pi_1(x) \subset \pi_2(x)$ for each $x \in \mathcal{A}$, that is, $\mathcal{D}(\pi_1) \subset \mathcal{D}(\pi_2)$ and $\pi_1(x)\xi = \pi_2(x)\xi$ for all $x \in \mathcal{A}$ and $\xi \in \mathcal{D}(\pi_1)$, then π_2 is said to be an *extension* of π_1 and denoted by $\pi_1 \subset \pi_2$.

Definition 1.8.3. Let π be a representation of $\mathcal{A}$ on a Hilbert space $\mathcal{H}$. We denote by t_π the graph topology $t_{\pi(\mathcal{A})}$ on $\mathcal{D}(\pi)$ with respect to the O-algebra $\pi(\mathcal{A})$. If $\mathcal{D}(\pi)[t_\pi]$ is complete, then π is said to be *closed* .

Let π be a representation of an algebra $\mathcal{A}$. We denote by $\widetilde{\mathcal{D}}(\pi)$ the completion of $\mathcal{D}(\pi)[t_\pi]$. Then we have

$$\widetilde{\mathcal{D}}(\pi) \subset \widehat{\mathcal{D}}(\pi) \equiv \bigcap_{x \in \mathcal{A}} \mathcal{D}(\overline{\pi(x)}).$$

Proposition 1.8.4. Let π be a representation of $\mathcal{A}$. We put

$$\tilde{\pi}(x) = \overline{\pi(x)}\lceil\tilde{\mathcal{D}}(\pi),$$
$$\hat{\pi}(x) = \overline{\pi(x)}\lceil\hat{\mathcal{D}}(\pi), \quad x \in \mathcal{A}.$$

Then $\tilde{\pi}$ is a closed representation of $\mathcal{A}$ which is the smallest closed extension of π and $\hat{\pi}$ is a closed representation of $\mathcal{A}$ satisfying $\pi \subset \tilde{\pi} \subset \hat{\pi}$. If π is a $*$-representation of a $*$-algebra $\mathcal{A}$, then $\tilde{\pi} = \hat{\pi}$.

Proof. See Lemma 2.6 in Powers [1].

Hereafter let $\mathcal{A}$ be a $*$-algebra. For a $*$-representation π of $\mathcal{A}$ we put

$$\begin{cases} \mathcal{D}(\pi^*) = \bigcap_{x\in\mathcal{A}} \mathcal{D}(\pi(x)^*) \\ \pi^*(x) = \pi(x^*)^*\lceil\mathcal{D}(\pi^*), \quad x \in \mathcal{A} \end{cases}$$

$$\begin{cases} \mathcal{D}(\pi^{**}) = \bigcap_{x\in\mathcal{A}} \mathcal{D}(\pi^*(x)^*) \\ \pi^{**}(x) = \pi^*(x^*)^*\lceil\mathcal{D}(\pi^{**}), \quad x \in \mathcal{A}. \end{cases}$$

Then we have the following

Proposition 1.8.5. π^* is a closed representation of $\mathcal{A}$ and π^{**} is a closed $*$-representation of $\mathcal{A}$ satisfying $\pi \subset \tilde{\pi} \subset \pi^{**} \subset \pi^*$. These representations $\pi, \tilde{\pi}, \pi^{**}$ and π^* don't coincide in general.

Proof. See Lemma 4.1 in Powers [1] and Proposition 8.1.12 in Schmüdgen [21].

Definition 1.8.6. A $*$-representation π of $\mathcal{A}$ is said to be *self-adjoint* (resp. *essentially self-adjoint* , *algebraically self-adjoint*) if $\pi = \pi^*$ (resp. $\tilde{\pi} = \pi^*, \pi^* = \pi^{**}$).

We remark that a $*$-representation π of $\mathcal{A}$ is closed (resp. self-adjoint, essentially self-adjoint, algebraically self-adjoint) iff the O*-algebra $\pi(\mathcal{A})$ is closed (resp. self-adjoint, essentially self-adjoint, algebraically self-adjoint).

Let π_1 and π_2 be representations of $\mathcal{A}$ on Hilbert spaces $\mathcal{H}_1$ and $\mathcal{H}_2$, respectively. We define an *intertwing space* $\mathbb{I}(\pi_1, \pi_2)$ of π_1 and π_2 which is an important tool in representation theory by

$$\mathbb{I}(\pi_1, \pi_2) = \{K \in \mathcal{B}(\mathcal{H}_1, \mathcal{H}_2); K\pi_1(x) \subset \pi_2(x)K, {}^\forall x \in \mathcal{A}\}.$$

Then we have the following

Proposition 1.8.7. Let π, π_1 and π_2 be $*$-reprentations of $\mathcal{A}$. Then the following statements hold:

(1) $\mathbb{I}(\pi, \pi) = \pi(\mathcal{A})'_s$ and $\mathbb{I}(\pi, \pi^*) = \pi(\mathcal{A})'_w$.

(2) $\mathbb{I}(\pi_1, \pi_2) \subset \mathbb{I}(\widetilde{\pi_1}, \widetilde{\pi_2}) \subset \mathbb{I}(\pi_1^{**}, \pi_2^{**})$,
$\mathbb{I}(\pi_1, \pi_2^*) \subset \mathbb{I}(\widetilde{\pi_1}, \pi_2^*) \subset \mathbb{I}(\pi_1^{**}, \pi_2^*)$.

(3) $\mathbb{I}(\pi_1, \pi_2)^* \subset \mathbb{I}(\pi_2^*, \pi_1^*)$,
$\mathbb{I}(\pi_1, \pi_2^*)^* \subset \mathbb{I}(\pi_2, \pi_1^*)$.

Proof. See Proposition 8.2.2, 8.2.3 in Schmüdgen [21].

Definition 1.8.8. Let π_1 and π_2 be $*$-representations of $\mathcal{A}$ on Hilbert spaces $\mathcal{H}_1$ and $\mathcal{H}_2$, respectively. If there exists an isometry U of $\mathcal{H}_1$ onto $\mathcal{H}_2$ such that $U\mathcal{D}(\pi_1) = \mathcal{D}(\pi_2)$ and $\pi_1(x)\xi = U^*\pi_2(x)U\xi$ for all $x \in \mathcal{A}$ and $\xi \in \mathcal{D}(\pi_1)$, then π_1 and π_2 are said to be *unitary equivalent* and denoted by $\pi_1 \cong \pi_2$.

For the unitary equivalence of two $*$-representations we have the following

Theorem 1.8.9. Let π_1 and π_2 be closed $*$-representations of a $*$-algebra $\mathcal{A}$ with identity in $\mathcal{H}_1$ and $\mathcal{H}_2$, respectively such that $\pi_i(\mathcal{A})'_{\mathrm{w}}\mathcal{D}(\pi_i) \subset \mathcal{D}(\pi_i)$ $(i = 1, 2)$. Suppose

(i) $\mathbb{I}(\pi_1, \pi_2^*) = \mathbb{I}(\pi_1, \pi_2)$ and $\mathbb{I}(\pi_2, \pi_1^*) = \mathbb{I}(\pi_2, \pi_1)$;
(ii) $\mathbb{I}(\pi_1, \pi_2)\mathcal{H}_1$ is total in $\mathcal{H}_2$ and $\mathbb{I}(\pi_2, \pi_1)\mathcal{H}_2$ is total in $\mathcal{H}_1$;
(iii) one of the following statements for von Neumann algebras $(\pi_1(\mathcal{A})'_{\mathrm{w}})'$ and $(\pi_2(\mathcal{A})'_{\mathrm{w}})'$ hold:

$(\mathrm{iii})_1$ $(\pi_1(\mathcal{A})'_{\mathrm{w}})'$ and $(\pi_2(\mathcal{A})'_{\mathrm{w}})'$ are standard von Neumann algebras, in particular, they are von Neumann algebras with cyclic and separating vectors.

$(\mathrm{iii})_2$ $\pi_1(\mathcal{A})'_{\mathrm{w}}$ and $\pi_2(\mathcal{A})'_{\mathrm{w}}$ are properly infinite and of coutable type.

$(\mathrm{iii})_3$ $\mathcal{H}_1$ and $\mathcal{H}_2$ are separable, and $(\pi_1(\mathcal{A})'_{\mathrm{w}})'$ and $(\pi_2(\mathcal{A})'_{\mathrm{w}})'$ are von Neumann algebras of type III.

Then π_1 and π_2 are unitarily equivalent.

Proof. See Theorem 3.2 and Corollary 3.6 in Ikeda-Inoue-Takakura[1].

1.9 Trace functionals on O*-algebras

Let $\mathcal{A}$ be a $*$-algebra. A linear functional f on $\mathcal{A}$ is said to be *positive* if $f(x^*x) \geq 0$ for all $x \in \mathcal{A}$. For a positive linear functional f on $\mathcal{A}$ we can construct the Gelfand-Naimark-Segal representation as follows:

Theorem 1.9.1. Let f be a positive linear functional on a $*$-algebra $\mathcal{A}$. Then there exists a closed $*$-representation π_f of $\mathcal{A}$ on a Hilbert space $\mathcal{H}_f$

and a linear map λ_f of $\mathcal{A}$ into the domain $\mathcal{D}(\pi_f)$ of π_f such that $\lambda_f(\mathcal{A})$ is dense in $\mathcal{D}(\pi_f)$ with respect to t_{π_f}, and $(\lambda_f(x)|\lambda_f(y)) = f(y^*x)$ and $\lambda_f(xy) = \pi_f(x)\lambda_f(y)$ for all $x, y \in \mathcal{A}$. The pair (π_f, λ_f) is uniquely determined by f up to unitary equivalence. We call the triple $(\pi_f, \lambda_f, \mathcal{H}_f)$ the GNS-construction for f.

Proof. We give simply the proof. From the same proof of the Schwartz inequality we have

$$|f(y^*x)|^2 \leq f(y^*y)f(x^*x), \quad x, y \in \mathcal{A}.$$

And so, $\mathcal{N}_f \equiv \{x \in \mathcal{A}; f(x^*x) = 0\}$ is a left ideal of $\mathcal{A}$ and the quotient space $\lambda_f(\mathcal{A}) = \{\lambda_f(x) \equiv x + \mathcal{N}_f; x \in \mathcal{A}\}$ is a pre-Hilbert space with inner product

$$(\lambda_f(x)|\lambda_f(y)) = f(y^*x), \quad x, y \in \mathcal{A}.$$

We denote by $\mathcal{H}_f$ the Hilbert space obtained by the completion of the pre-Hilbert space $\lambda_f(\mathcal{A})$. We can define a $*$-representation π_f° of $\mathcal{A}$ by

$$\pi_f^\circ(x)\lambda_f(y) = \lambda_f(xy), \quad x, y \in \mathcal{A}$$

and denote by π_f the closure of π_f°. Then λ_f is a linear map of $\mathcal{A}$ into $\mathcal{D}(\pi_f)$ satisfying $\lambda_f(\mathcal{A})$ is dense in $\mathcal{D}(\pi_f)[t_{\pi_f}]$ and $\lambda_f(xy) = \pi_f(x)\lambda_f(y)$ for all $x, y \in \mathcal{A}$. Let (π'_f, λ'_f) be a pair of a closed $*$-representation π'_f of $\mathcal{A}$ on a Hilbert space $\mathcal{H}'_f$ and a linear map λ'_f of $\mathcal{A}$ into $\mathcal{D}(\pi'_f)$ satisfying $\lambda'_f(\mathcal{A})$ is dense in $\mathcal{D}(\pi'_f)[t_{\pi'_f}]$, and $(\lambda'_f(x)|\lambda'_f(y)) = f(y^*x)$ and $\lambda'_f(xy) = \pi'_f(x)\lambda'_f(y)$ for all $x, y \in \mathcal{A}$. Here we put

$$U\lambda_f(x) = \lambda'_f(x), \quad x \in \mathcal{A}.$$

Then U is an isometry of $\lambda_f(\mathcal{A})$ onto $\lambda'_f(\mathcal{A})$, and so it can be extended to an isometry of $\mathcal{H}_f$ onto $\mathcal{H}'_f$. The extension is also denoted by U. Then it is easily shown that $U\mathcal{D}(\pi_f) = \mathcal{D}(\pi'_f)$ and $\pi_f(x) = U^*\pi'_f(x)U$ for all $x \in \mathcal{A}$. This completes the proof.

We consider positive linear functionals on O*-algebras. Let $\mathcal{M}$ be a closed O*-algebra on $\mathcal{D}$ in $\mathcal{H}$ with the identity operator I. We define two positive cones $\mathcal{P}(\mathcal{M})$ and $\mathcal{M}_+$ as follows:

$$\mathcal{P}(\mathcal{M}) = \{\sum_k X_k^\dagger X_k; X_k \in \mathcal{M}\},$$

$$\mathcal{M}_+ = \{X \in \mathcal{M}; X \geq 0\}.$$

A linear functional f on $\mathcal{M}$ is said to be *positive* (resp. *strongly positive*) if $f(X) \geq 0$ for each $X \in \mathcal{P}(\mathcal{M})$ (resp. $\mathcal{M}_+$). It is clear that every strongly positive linear functional on $\mathcal{M}$ is positive, but the converse does not hold in general as seen next:

Example 1.9.2. Let $\mathcal{M}$ be the O*-algebra generated by the position and momentum operators Q and P on the Schwartz space $\mathcal{S}(\mathbb{R})$ and put $A = \frac{1}{\sqrt{2}}(Q+iP)$. It can be checked that $(A^\dagger A - I)(A^\dagger A - 2I) \notin \mathcal{P}(\mathcal{M})$. Hence there exists a positive linear functional f on $\mathcal{M}$ such that

$$f((A^\dagger A - I)(A^\dagger A - 2I)) < 0. \tag{1.9.1}$$

Let $\{\xi_n\}$ be the orthonormal basis in $L^2(\mathbb{R})$ consisting of the eigenvectors of the number operator $N = AA^\dagger$. Since

$$((A^\dagger A - I)(A^\dagger A - 2I)\xi|\xi) = \sum_{n=0}^{\infty}(n-1)(n-2)|(\xi|\xi_n)|^2 \geq 0$$

for each $\xi \in \mathcal{S}(\mathbb{R})$, we have $(A^\dagger A - I)(A^\dagger A - 2I) \in \mathcal{M}_+$, which implies by (1.9.1) that f is not strongly positive.

Many important examples of states in quantum physics are trace functionals, that is, they are of the form $f(X) = \operatorname{tr} \overline{TX}$ with a certain positive trace class operator T. In this section we define and study trace functionals in detail for the preparation of considering quantum moment problem in Chapter IV : Given an O*-algebra $\mathcal{M}$, under what conditions is every strongly positive linear functional on $\mathcal{M}$ a trace functional?

Let $\mathfrak{S}_1(\mathcal{H})$ be the set of all trace class operators on $\mathcal{H}$. Every operator T in $\mathfrak{S}_1(\mathcal{H})$ can be represented as $T = \sum_{n=1}^{\infty} t_n \xi_n \otimes \overline{\eta_n}$, where $\{t_n\} \subset \mathbb{C}$, $\sum_{n=1}^{\infty} |t_n| < \infty$ and $\{\xi_n\}_{n\in\mathbb{N}'}$ and $\{\eta_n\}_{n\in\mathbb{N}'}$ are orthonormal sets in $\mathcal{H}$ with $\mathbb{N}' = \{n \in \mathbb{N}; t_n \neq 0\}$. Further, the trace norm $\nu(T) \equiv \operatorname{tr}|T|$ equals $\sum_n |t_n|$. In case $T^* = T$ we can have in addition that $t_n \in \mathbb{R}$ and $\xi_n = \eta_n$ for all $n \in \mathbb{N}'$ (see Köthe [1], §46.6). Further, we put $\xi_n = \eta_n = 0$ for all $n \in \mathbb{N}\setminus\mathbb{N}'$. If the preceding conditions are fulfilled, then we call the sum $\sum_{n=1}^{\infty} t_n\xi_n \otimes \overline{\eta_n}$ a *canonical representation* of T. We define some subsets of $\mathfrak{S}_1(\mathcal{H})$ as follows:

$$\mathfrak{S}_1(\mathcal{M}) = \{T \in \mathcal{B}(\mathcal{H}); T\mathcal{H} \subset \mathcal{D}, T^*\mathcal{H} \subset \mathcal{D} \text{ and } XT, XT^* \in \mathfrak{S}_1(\mathcal{H}) \text{ for all } X \in \mathcal{M}\},$$

$$\mathfrak{S}_1(\mathcal{M})_+ = \{T \in \mathfrak{S}_1(\mathcal{M}); T \geq 0\},$$

$${}_1\mathfrak{S}(\mathcal{M}) = \{T \in \mathcal{B}(\mathcal{H}); TX \text{ and } T^*X \text{ are closable and } \overline{TX}, \overline{T^*X} \in \mathfrak{S}_1(\mathcal{H}) \text{ for all } X \in \mathcal{M}\},$$

$${}_1\mathfrak{S}(\mathcal{M})_+ = \{T \in {}_1\mathfrak{S}(\mathcal{M}); T \geq 0\}.$$

We have the following

Lemma 1.9.3. The following statements hold:
(1) $\mathfrak{S}_1(\mathcal{M}) \subset {}_1\mathfrak{S}(\mathcal{M})$.
(2) $\mathfrak{S}_1(\mathcal{M}) = \{T \in \mathcal{B}(\mathcal{H}); T\mathcal{H} \subset \mathcal{D}, T^*\mathcal{H} \subset \mathcal{D}$ and $\overline{XTY} \in \mathfrak{S}_1(\mathcal{H})$ for all $X, Y \in \mathcal{M}\}$
and it is a $*$-subalgebra of $\mathcal{B}(\mathcal{H})$ satisfying $\mathcal{M}\mathfrak{S}_1(\mathcal{M}) = \mathfrak{S}_1(\mathcal{M})$.
(3) ${}_1\mathfrak{S}(\mathcal{M}) = \{T \in \mathcal{B}(\mathcal{H}); T\mathcal{H} \subset \mathcal{D}^*(\mathcal{M}), T^*\mathcal{H} \subset \mathcal{D}^*(\mathcal{M})$ and $X^*T, X^*T^* \in \mathfrak{S}_1(\mathcal{H})$ for all $X \in \mathcal{M}\}$,
and it is a $*$-subalgebra of $\mathcal{B}(\mathcal{H})$ satisfying ${}_1\mathfrak{S}(\mathcal{M})\mathcal{M} = {}_1\mathfrak{S}(\mathcal{M})$.

Let $\{\xi_n\}, \{\eta_n\} \in \mathcal{D}$. If $\sum_{n=1}^{\infty} \|X\xi_n\|\|Y\eta_n\| < \infty$ for all $X, Y \in \mathcal{M}$, then we say that *the series* $\sum_{n=1}^{\infty} \xi_n \otimes \overline{\eta_n}$ *converges absolutely with respect to* $\mathcal{M}$.

Lemma 1.9.4. (1) Let $\{\xi_n\}$ and $\{\eta_n\}$ in $\mathcal{D}$. Suppose the series $\sum_{n=1}^{\infty} \xi_n \otimes \overline{\eta_n}$ converges absolutely w.r.t. $\mathcal{M}$. Then $T \equiv \sum_{n=1}^{\infty} \xi_n \otimes \overline{\eta_n} \in \mathfrak{S}_1(\mathcal{M})$ and tr $\overline{XTY^\dagger} = \sum_n (X\xi_n|Y\eta_n)$ for all $X, Y \in \mathcal{M}$.

(2) Let $T \in \mathfrak{S}_1(\mathcal{M})$ and $T = \sum_{n=1}^{\infty} t_n \xi_n \otimes \overline{\eta_n}$ a canonical representation of T. Then $\sum_{n=1}^{\infty} (t_n\xi_n) \otimes \overline{\eta_n}$ converges absolutely w.r.t. $\mathcal{M}$.

Proof. (1) It follows from Köthe [1], §42, 5 that

$$T \equiv \sum_{n=1}^{\infty} \xi_n \otimes \overline{\eta_n} \in \mathfrak{S}_1(\mathcal{H}) \text{ and } \operatorname{tr} T = \sum_{n=1}^{\infty} (\xi_n|\eta_n). \tag{1.9.2}$$

Take arbitrary $X \in \mathcal{M}$ and $x \in \mathcal{H}$. Since $\sum_{n=1}^{\infty} \|X\xi_n\|\|Y\eta_n\| < \infty$ for all $X, Y \in \mathcal{M}$, it follows that

$$\{(\sum_{k=1}^{n} \xi_k \otimes \overline{\eta_k})x\} \subset \mathcal{D},$$

$$\lim_{n\to\infty} (\sum_{k=1}^{n} \xi_k \otimes \overline{\eta_k})x = Tx,$$

$$\lim_{n\to\infty} X(\sum_{k=1}^{n} \xi_k \otimes \overline{\eta_k})x = \sum_{k=1}^{\infty} (x|\eta_k)X\xi_k,$$

which implies by the closedness of $\mathcal{M}$ that $T\mathcal{H} \subset \mathcal{D}$ and $XT = \sum_{n=1}^{\infty} X\xi_n \otimes \overline{\eta_n}$. Similarly, we have $T^*\mathcal{H} \subset \mathcal{D}$. Hence it follows from (1.9.2) that $XT \in \mathfrak{S}_1(\mathcal{H})$ and $\operatorname{tr} XT = \sum_{n=1}^{\infty}(X\xi_n|\eta_n)$, which means that $T \in \mathfrak{S}_1(\mathcal{M})$.

(2) Since $X^\dagger XT$, $Y^\dagger YT^* \in \mathfrak{S}_1(\mathcal{H})$ for all $X, Y \in \mathcal{M}$, it follows that

$$\begin{aligned}\sum_{n\in\mathbb{N}} |t_n|\|X\xi_n\|^2 &= \sum_{n\in\mathbb{N}} |(X^\dagger X(t_n\xi_n)|\xi_n)| \\ &= \sum_{n\in\mathbb{N}} |(X^\dagger XT\xi_n|\xi_n)| \\ &< \infty\end{aligned}$$

and

$$\begin{aligned}\sum_{n\in\mathbb{N}} |t_n|\|Y\eta_n\|^2 &= \sum_{n\in\mathbb{N}} |(Y^\dagger Y(\overline{t_n}\eta_n)|\eta_n)| \\ &= \sum_{n\in\mathbb{N}} |(Y^\dagger YT^*\eta_n|\eta_n)| \\ &< \infty,\end{aligned}$$

which implies

$$\sum_n |t_n|\|X\xi_n\|\|Y\eta_n\| \leq (\sum_n |t_n|\|X\xi_n\|^2)^{1/2}(\sum_n |t_n|\|Y\eta_n\|^2)^{1/2}.$$

This completes the proof.

Lemma 1.9.5. (1) Every element T of $\mathfrak{S}_1(\mathcal{M})$ can be written as $T = (T_1 - T_2) + i(T_3 - T_4)$ with $T_i \in \mathfrak{S}_1(\mathcal{M})_+ (i = 1, \cdots, 4)$.

(2) Suppose $\mathcal{D}[t_{\mathcal{M}}]$ is sequentially complete. Then $\mathfrak{S}_1(\mathcal{M}) = \mathfrak{S}_1(\mathcal{L}^\dagger(\mathcal{D}))$, and so $\mathfrak{S}_1(\mathcal{M})\lceil\mathcal{D}$ is a two-sided ideal of $\mathcal{L}^\dagger(\mathcal{D})$.

Proof. (1) By the $*$-invariance of $\mathfrak{S}_1(\mathcal{M})$ we may assume that T is hermitian. Let $T = \sum_n t_n\xi_n \otimes \overline{\xi_n}$ be a canonical representation of T. We put

$$\begin{aligned}\mathbb{N}_+ &= \{n \in \mathbb{N}; t_n > 0\}, \\ T_+ &= \sum_{n\in\mathbb{N}_+} t_n\xi_n \otimes \overline{\xi_n}.\end{aligned}$$

Since $\sum_n (t_n\xi_n)\otimes\overline{\xi_n}$ converges absolutely w.r.t. $\mathcal{M}$, it follows that $\sum_{n\in\mathbb{N}_+} (t_n\xi_n)\otimes \overline{\xi_n}$ also converges absolutely w.r.t. $\mathcal{M}$, which implies by Lemma 1.9.4 that

$T_+ \in \mathfrak{S}_1(\mathcal{M})_+$. Since $T \equiv T_+ - T \geq 0$ and $T, T_+ \in \mathfrak{S}_1(\mathcal{M})$, we have $T_- \in \mathfrak{S}_1(\mathcal{M})_+$.

(2) Take arbitrary $T \in \mathfrak{S}_1(\mathcal{M})$ and $A, B \in \mathcal{L}^\dagger(\mathcal{D})$. Let $T = \sum_n t_n \xi_n \otimes \overline{\eta_n}$ be a canonical representation of T. By the closed graph theorem we have $t_{\mathcal{M}} = t_{\mathcal{L}^\dagger(\mathcal{D})}$. Hence we have

$$\|A\xi\| \leq \|X\xi\| \text{ and } \|B\xi\| \leq \|Y\xi\|, \quad \xi \in \mathcal{D}$$

for some $X, Y \in \mathcal{M}$, which implies that $\sum_n (t_n \xi_n) \otimes \overline{\eta_n}$ converges absolutely w.r.t. $\mathcal{L}^\dagger(\mathcal{D})$. By Lemma 1.9.4 we have $T \in \mathfrak{S}_1(\mathcal{L}^\dagger(\mathcal{D}))$. This completes the proof.

Lemma 1.9.6. The following statements are equivalent:
(i) $\mathcal{M}$ is self-adjoint.
(ii) $\mathfrak{S}_1(\mathcal{M}) = {}_1\mathfrak{S}(\mathcal{M})$.

Proof. (i) $\Rightarrow$ (ii) This follows from Lemma 1.9.3, (2).

(ii) $\Rightarrow$ (i) Take an arbitrary $\xi \in \mathcal{D}^*(\mathcal{M})$. Then we have $\xi \otimes \bar{\xi} \in {}_1\mathfrak{S}(\mathcal{M}) = \mathfrak{S}_1(\mathcal{M})$, which implies $\xi \in \mathcal{D}$. Hence $\mathcal{M}$ is self-adjoint.

For any $T \in {}_1\mathfrak{S}(\mathcal{M})$ we define two linear functionals on $\mathcal{M}$ by

$$\begin{aligned} {}_T f(X) &= \text{ tr } \overline{TX}, \\ f_T(X) &= \text{ tr } X^{\dagger *} T, \qquad X \in \mathcal{M}. \end{aligned}$$

Lemma 1.9.7. (1) $f_T = {}_T f$ for each $T \in \mathfrak{S}_1(\mathcal{M})$.
(2) f_T is a strongly positive linear functional on $\mathcal{M}$ for each $T \in \mathfrak{S}_1(\mathcal{M})_+$.
(3) Any $f_T, T \in \mathfrak{S}_1(\mathcal{M})$, is written as $f_T = f_{T_1} - f_{T_2} + i(f_{T_3} - f_{T_4})$ whereby $T_j \in \mathfrak{S}_1(\mathcal{M})_+, j = 1, \cdots, 4$.

Proof. (1) Let $T = \sum_n t_n \xi_n \otimes \overline{\eta_n}$ be a canonical representation of T. Since $T \in \mathfrak{S}_1(\mathcal{M})$, we have $\{\xi_n\}, \{\eta_n\} \subset \mathcal{D}$, so that $XT = \sum_n t_n X\xi_n \otimes \overline{\eta_n}$ and $\overline{TX} = \sum_n t_n \xi_n \otimes \overline{X^\dagger \eta_n}$ for each $X \in \mathcal{M}$. Hence we have

$$\begin{aligned} {}_T f(X) = \text{ tr } \overline{TX} &= \sum_n t_n (\xi_n | X^\dagger \eta_n) \\ &= \sum_n t_n (X\xi_n | \eta_n) \\ &= \text{ tr } XT \\ &= f_T(X) \end{aligned}$$

for each $X \in \mathcal{M}$.

(2) Let $T \in \mathfrak{S}_1(\mathcal{M})_+$ and $T = \sum_n t_n \xi_n \otimes \overline{\xi_n}$ a canonical representation of T. Then it follows that $t_n \geq 0$ and $\xi_n \in \mathcal{D}$ for each $n \in \mathbb{N}$ and $f_T(X) = \sum_n t_n (X\xi_n | \xi_n)$ for each $X \in \mathcal{M}$, which implies that f_T is strongly positive.

(3) This follows from (2) and Lemma 1.9.5, (1).

For any $T \in {}_1\mathfrak{S}(\mathcal{M})$ the linear functionals ${}_Tf$ and f_T are well-defined, but the assertions of Lemma 1.9.7 are not true in general as seen next.

Lemma 1.9.8. The following statements are equivalent:
(i) $\mathcal{M}$ is algebraically self-adjoint.
(ii) ${}_Tf = f_T$ for each $T \in {}_1\mathfrak{S}(\mathcal{M})$.

Proof. (i) $\Rightarrow$ (ii) Take an arbitrary $T \in {}_1\mathfrak{S}(\mathcal{M})$. Let $T = \sum_n t_n \xi_n \otimes \overline{\eta_n}$ be a canonical representation of T. Then $\{\xi_n\}, \{\eta_n\} \in \mathcal{D}^*(\mathcal{M})$ and

$$f_T(X) = \operatorname{tr} X^{\dagger *} T = \sum_n t_n (X^{\dagger *}\xi_n | \eta_n),$$

$${}_Tf(X) = \operatorname{tr} \overline{TX} = \sum_n t_n (\xi_n | X^* \eta_n)$$

for each $X \in \mathcal{M}$, which implies that $f_T = {}_Tf$.

(ii) $\Rightarrow$ (i) Take arbitrary $\xi, \eta \in \mathcal{D}^*(\mathcal{M})$. Then we have $T \equiv \xi \otimes \overline{\eta} \in {}_1\mathfrak{S}(\mathcal{M})$ and

$$(X^{\dagger *}\xi | \eta) = f_T(X) = {}_Tf(X) = (\xi | X^* \eta)$$

for each $X \in \mathcal{M}$, which implies $\xi \in \mathcal{D}^{**}(\mathcal{M})$. Hence we have $\mathcal{D}^*(\mathcal{M}) = \mathcal{D}^{**}(\mathcal{M})$. This completes the proof.

Remark 1.9.9. Suppose $\mathcal{M}$ is algebraically self-adjoint and $T \in {}_1\mathfrak{S}(\mathcal{M})_+$. Then f_T is a positive linear functional on $\mathcal{M}$. But, we don't know whether f_T is strongly positive.

We investigate the continuity of trace functionals f_T, $T \in \mathfrak{S}_1(\mathcal{M})$, with respect to some topologies:

Proposition 1.9.10. Every trace functional f_T, $T \in \mathfrak{S}_1(\mathcal{M})$, on $\mathcal{M}$ is continuous with respect to both topologies ρ and $t_{\sigma w}$.

Proof. By Lemma 1.9.7 it is sufficient to show the continuity of f_T, $T \in \mathfrak{S}_1(\mathcal{M})_+$. Let $T \in \mathfrak{S}_1(\mathcal{M})_+$ and $T = \sum_n t_n \xi_n \otimes \overline{\xi_n}$ a canonical representation of T. Since

$$f_T(X) = \sum_n (X(t_n\xi_n)|\xi_n), \quad X \in \mathcal{M},$$

it follows that f_T is continuous w.r.t. $t_{\sigma w}$. Take an arbitrary $A \in \mathcal{M}_+$. We put $\eta_n = \sqrt{t_n}\xi_n$, $n \in \mathbb{N}$. Then since $\sum_n \eta_n \otimes \overline{\eta_n}$ converges absolutely w.r.t. $\mathcal{M}$ and

$$\begin{aligned}|f_T(X)| \leq \sum_n |(X\eta_n|\eta_n)| &= \sum_n \frac{|(X\eta_n|\eta_n)|}{(A\eta_n|\eta_n)}(A\eta_n|\eta_n) \\ &\leq (\sum_n (A\eta_n|\eta_n))\rho_A(X)\end{aligned}$$

for all $X \in \mathfrak{A}_A \equiv \{X \in \mathcal{M}; \rho_A(X) < \infty\}$, it follows that f_T is continuous w.r.t. the topology ρ. This completes the proof.

We next investigate the continuity of trace functionals with respect to the uniform topology τ_u. We define the locally convex topology τ_c called *precompact topology* determined by the family of seminorms:

$$p_{\mathfrak{M},\mathfrak{N}}(X) = \sup_{\xi \in \mathfrak{M}, \eta \in \mathfrak{N}} |(X\xi|\eta)|, \quad X \in \mathcal{M},$$

where $\mathfrak{M}$ and $\mathfrak{N}$ range over the precompact subsets of $\mathcal{D}[t_\mathcal{M}]$.

Proposition 1.9.11. Suppose $\mathcal{D}[t_\mathcal{M}]$ is a Fréchet space. Then every trace functional f_T, $T \in \mathfrak{S}_1(\mathcal{M})$, is continuous with respect to the topology τ_c. Hence it is continuous with respect to τ_u.

Proof. The projective tensor product topology on $\mathcal{D}[t_\mathcal{M}] \otimes_\pi \mathcal{D}[t_\mathcal{M}]$ is defined by the family of seminorms $\{\| \ \|_X \otimes_\pi \| \ \|_Y; X, Y \in \mathcal{M}\}$. Here the seminorm $\| \ \|_X \otimes_\pi \| \ \|_Y$ is defined by

$$\| \ \|_X \otimes_\pi \| \ \|_Y(T) = \inf\{\sum_{k=1}^n \|X\xi_k\|\|Y\eta_k\|\},$$

where the infimum is taken over all representations of $T \in \mathcal{D}[t_\mathcal{M}] \otimes_\pi \mathcal{D}[t_\mathcal{M}] = \mathcal{F}(\mathcal{D})$ as a finite sum $T = \sum_{k=1}^n \xi_k \otimes \overline{\eta_k}$ with $\{\xi_1, \cdots, \xi_n\}$ and $\{\eta_1, \cdots, \eta_n\}$ in $\mathcal{D}$. Hence every element T of $\mathfrak{S}_1(\mathcal{M})$ belongs to the completion of the projective tensor product $\mathcal{D}[t_\mathcal{M}] \otimes_\pi \mathcal{D}[t_\mathcal{M}]$. Therefore, by the Grothendieck result (Köthe [1], §41, 4) T is represented as $T = \sum_n t_n\xi_n \otimes \overline{\eta_n}$, where $\sum_n |t_n| < \infty$, and $\{\xi_n\}$ and $\{\eta_n\}$ in $\mathcal{D}[t_\mathcal{M}]$ converge to 0. Since $\sum_n (t_n\xi_n) \otimes \overline{\eta_n}$ converges absolutely w.r.t. $\mathcal{M}$, we have

$$|f_T(X)| = |\sum_n (X(t_n\xi_n)|\eta_n)|$$
$$\leq (\sum_n |t_n|) P_{\mathfrak{M},\mathfrak{N}}(X), \quad X \in \mathcal{M},$$

where $\mathfrak{M} = \{\xi_n\}$ and $\mathfrak{N} = \{\eta_n\}$. Clearly, $\mathfrak{M}$ and $\mathfrak{N}$ are pre-compact in $\mathcal{D}[t_{\mathcal{M}}]$. Therefore f_T is continuous with respect to τ_c. This completes the proof.

Proposition 1.9.12. Suppose that $\mathcal{M}$ contains the restriction N of the inverse of a trace-class positive operator on $\mathcal{H}$. Then every trace functional f_T, $T \in \mathfrak{S}_1(\mathcal{M})$, is continuous with respect to τ_u.

For the proof of Proposition 1.9.12 we prepare the following lemmas:

Lemma 1.9.13. Suppose T is a positive bounded operator on $\mathcal{H}$ with $T\mathcal{H} \subset \mathcal{D}$. Then $T^\nu\mathcal{H} \subset \mathcal{D}$ for all $\nu > 0$.

Proof. Let $0 < \nu \leq 1$. Since $\mathcal{D} \subset \mathcal{D}(I + X^*\overline{X}) \subset \mathcal{D}(\overline{X})$ for each $X \in \mathcal{M}$ and $\mathcal{M}$ is closed, we have

$$\mathcal{D} = \bigcap_{X \in \mathcal{M}} \mathcal{D}(I + X^*\overline{X}). \tag{1.9.3}$$

Take an arbitrary $X \in \mathcal{M}$. Then we have

$$T\mathcal{H} \subset \mathcal{D} \subset \mathcal{D}((I + X^*\overline{X})^{\frac{1}{\nu}}) = (I + X^*\overline{X})^{-\frac{1}{\nu}}\mathcal{H},$$

which implies that $(I + X^*X)^{\frac{1}{\nu}}T$ is bounded. Hence there exists a constant $\gamma > 0$ such that

$$\|Tx\| \leq \gamma\|(I + X^*\overline{X})^{-\frac{1}{\nu}}x\|$$

for all $x \in \mathcal{H}$. Since $\mathcal{D}(I+X^*\overline{X}) = \mathcal{D}(\gamma(I+X^*\overline{X}))$, we can take $\gamma = 1$ without any loss of generality. Then it follows from the Kato-Heinz inequality (Kato [1]) that

$$\|T^\nu x\| \leq \|(I + X^*\overline{X})^{-1}x\|$$

for all $x \in \mathcal{H}$, which implies

$$T^\nu\mathcal{H} \subset (I + X^*\overline{X})^{-1}\mathcal{H} = \mathcal{D}(I + X^*\overline{X}).$$

Hence it follows from (1.9.3) that $T^\nu\mathcal{H} \subset \mathcal{D}$. Let $\nu > 1$. Choose a positive number n with $n > \nu$. Then $T^n\mathcal{H} \subset \mathcal{D}$, and so $T^\nu\mathcal{H} = (T^n)^{\nu/n}\mathcal{H} \subset \mathcal{D}$ by the above consequence. This completes the proof.

Lemma 1.9.14. Suppose $\mathcal{M}$ contains the restriction N of the inverse of a positive trace-class operator on $\mathcal{H}$ and T is a positive operator on $\mathcal{H}$ with $T\mathcal{H} \subset \mathcal{D}$. Then T^ν is a trace-class operator for each $0 < \nu \leq 1$.

Proof. Let $N = \sum_l n_l \xi_l \otimes \overline{\xi_l}$ be a canonical representation of N. Since N^{-1} is a positive trace-class operator, it follows that $\{\xi_k\}$ is an orthonormal basis in $\mathcal{H}$ contained in $\mathcal{D}$ and $\sum_l \frac{1}{n_l} < \infty$. Let $T = \sum_n t_n \eta_n \otimes \overline{\eta_n}$ be a canonical representation of T and U an isometry on $\mathcal{H}$ defined by $U\xi_n = \eta_n$ for each $n \in \mathbb{N}$. Take an arbitrary $k \in \mathbb{N}$. Since $T\mathcal{H} \subset \mathcal{D}$, it follows that $U^*TU\mathcal{H} \subset \mathcal{D}$, which implies $N^k U^*TU$ is bounded and we put $\gamma_k = \|N^k U^*TU\|$. Then we have

$$\begin{aligned} \|U^*TUx\| &= \|U^*TUN^kN^{-k}x\| \\ &\leq \gamma_k \|N^{-k}x\| \end{aligned}$$

for all $x \in \mathcal{H}$, which implies

$$t_l \leq \frac{\gamma_k}{n_l^k}, \qquad l \in \mathbb{N}.$$

Hence we have

$$t_l^\nu \leq \frac{\gamma_k^\nu}{n_l^{k\nu}} \leq \frac{\gamma_k^\nu}{n_l}$$

for suitable k, and so

$$\sum_l t_l^\nu \leq \gamma_k^\nu \sum_l \frac{1}{n_l} < \infty.$$

This means that T^ν is a trace-class operator on $\mathcal{H}$. This completes the proof.

The proof of Proposition 1.9.12: Take an arbitrary $T \in \mathfrak{S}_1(\mathcal{M})_+$. Let $T = \sum_n t_n \xi_n \otimes \overline{\xi_n}$ be a canonical representation of T. By Lemma 1.9.13 and 1.9.14 we have

$$\sum_n t_n^\nu < \infty \text{ for } 0 < \nu \leq 1 \quad \text{and} \quad T^\nu \mathcal{H} \subset \mathcal{D}. \tag{1.9.4}$$

Hence XT^ν is bounded for each $X \in \mathcal{M}$ and $0 < \nu \leq 1$, and so $T^\nu \mathcal{K}$ is a bounded subset in $\mathcal{D}[t_\mathcal{M}]$, where $\mathcal{K}$ is the unit ball in $\mathcal{H}$. We now have

$$\begin{aligned} |f_T(X)| &= \sum_n (XT\xi_n \mid \xi_n)| \\ &\leq \sum_n t_n^{\frac{1}{3}} |(XT^{\frac{1}{3}}\xi_n \mid T^{\frac{1}{3}}\xi_n)| \\ &\leq (\sum_n t_n^{\frac{1}{3}}) \|X\|_{T^{\frac{1}{3}}\mathcal{K}} \end{aligned}$$

for all $X \in \mathcal{M}$, which implies that f_T is continuous with respect to τ_u. By Lemma 1.9.7, (2), a general f_T, $T \in \mathfrak{S}_1(\mathcal{M})$, is continuous with respect to τ_u. This completes the proof.

Notes

1.1. A $*$-algebra of closable linear operators defined on a common dense domain in a Hilbert space was introduced by Lassner [1] and he called it an O_p^*-algebra, and we here call it an O*-algebra according to the Schmüdgen book [21]. The notion of closed O*-algebras was introduced independently by Lassner [1] and Powers [1]. The notion of self-adjoint O*-algebras was introduced by Powers [1] in analogy with a self-adjoint operator.

1.2. The weak commutant first appeared in papers on quantum field theory: cf. Ruelle [1] and Streater-Wightman [1]. In the context of O*-algebras or $*$-representations weak commutants were first studied by Vasiliev [1], Powers [1] and Uhlmann [1].

1.3. For O*-algebras $\mathcal{M}$ on $\mathcal{D}$ Powers [1] pointed out that there are some pathologies between $\mathcal{M}$-invariant subspaces $\mathfrak{M}$ of $\mathcal{D}$ and the projections $P_{\overline{\mathfrak{M}}}$, and he showed that if $\mathcal{M}$ is self-adjoint, then $\mathfrak{M}$ is essentially self-adjoint if and only if $P_{\overline{\mathfrak{M}}} \in \mathcal{M}'_{\mathrm{w}}$ and $P_{\overline{\mathfrak{M}}}\mathcal{D} = \overline{\mathfrak{M}}^{t_{\mathcal{M}}}$. After that, Ikeda-Inoue [1] extended this result in case of non-selfadjoint O*-algebras. Powers [1] defined two different notions of cyclic vectors and strongly cyclic vectors for O*-algebras. Schmüdgen [21] gave an example of a cyclic but not strongly cyclic vector, and in [21] he has called a cyclic (resp. strongly cyclic) vector a weakly cyclic (resp. cyclic) vector. Gudder-Scruggs [1] investigated the relations between them and the commutant.

1.4. Induced extensions of O*-algebras were studied by Borchers-Yngvason [1] in their approach to the decomposition theory, and introduced independently by Inoue-Ueda-Yamauchi [1] for the studies of the integrability of commutative O*-algebras and unbounded bicommutants of O*-algebras. After that, Schmüdgen [21] and Inoue-Kurose-Ôta [1] studied induced extensions in detail.

1.5. It was first shown by Powers [1] that a self-adjoint commutative O*-algebra $\mathcal{M}$ is integrable if and only if the von Neumann algebra $(\mathcal{M}'_{\mathrm{w}})'$ is commutative, and the further studies were done by Schmüdgen [21]. Let A and B be commuting symmetric operators in $\mathcal{L}^\dagger(\mathcal{D})$. The integrability of the polynomial algebra $\mathcal{P}(A, B)$ generated by A and B was first investigated by Inoue-Takesue [1], and Schmüdgen [21] studied the polynomial algebra $\mathcal{P}(A, B)$ in detail.

1.6. The uniform and quasi-uniform topology were introduced and studied by Lassner [1], and the ρ-topology and λ-topology were defined and studied by Jurzak [1] and Arnal-Jurzak [1]. The other topologies τ_{w}, τ_s, τ_s^*, $\tau_{\sigma s}$ and $\tau_{\sigma s}^*$ were introduced by Inoue [1] and Arnal-Jurzak [1]. More detailed studies about topologies of O*-algebras were done in Schmüdgen [21].

1.7. The notion of extended W*-algebras was first introduced by Dixon [1] and further studies was done by Inoue [1,2,3] for the study of unbounded Hilbert algebras. The notion of generalized von Neumann algebras was defined by Inoue [9].

1.8. The studies of $*$-representations of $*$-algebras were first done by Vasiliv [1] and Powers [1]. The study of unitary equivalence of two $*$-representations was done by Takesue [1], Inoue-Tkesue [1] and Ikeda-Inoue-Takakura [1].

1.9. The Gelfand-Naimark-Segal construction (Theorem 1.9.1) for positive linear functionals on general $*$-algebras was first introduced by Powers[1]. Example 1.9.2 is due to Counter example I in Woronowicz [1]. The spaces $\mathfrak{S}_1(\mathcal{M}), \mathfrak{S}_1(\mathcal{M})_+, {}_1\mathfrak{S}(\mathcal{M})$ and ${}_1\mathfrak{S}(\mathcal{M})_+$ and trace functionals f_T and ${}_Tf$, $T \in {}_1\mathfrak{S}(\mathcal{M})_+$ were introduced by Lassner and Timmermann [1]. Further studies were developed by Schmüdgen [2], [21]. Proposition 1.9.11 is due to Schmüdgen [2]. Proposition 1.9.12 is due to Lassner and Timmermann [1].

Let $\mathcal{D}$ be a dense subspace in a Hilbert space. We denote by $\mathcal{L}^\dagger(\mathcal{D},\mathcal{H})$ the set of linear operators X such that $\mathcal{D}(X) = \mathcal{D}$, $\mathcal{D}(X^*) \supset \mathcal{D}$. The set $\mathcal{L}^\dagger(\mathcal{D},\mathcal{H})$ is a partial $*$-algebra with respect to the following operators: the usual sum $X_1 + X_2$, the scalar multiplication λX, the involution $X \to X^\dagger = X^* \lceil \mathcal{D}$ and the partial multiplication $X_1 \Box X_2 = X_1^{\dagger *} X_2$, defined whenever $X_2\mathcal{D} \subset \mathcal{D}(X_1^{\dagger *})$ and $X_1^*\mathcal{D} \subset \mathcal{D}(X_2^*)$. A partial O*-algebra on $\mathcal{D}$ is a $*$-subalgebra $\mathcal{M}$ of $\mathcal{L}^\dagger(\mathcal{D},\mathcal{H})$, that is, $\mathcal{M}$ is a subspace of $\mathcal{L}^\dagger(\mathcal{D},\mathcal{H})$, containing the identity operator and such that $X^\dagger \in \mathcal{M}$ whenever $X \in \mathcal{M}$ and $X_1 \Box X_2 \in \mathcal{M}$ for any $X_1, X_2 \in \mathcal{M}$ such that $X_1 \Box X_2$ is well-defined. Partial O*-algebras have been studied in Antoine-Inoue-Trapani [1,2,3], Antoine-Karwowski [1] and Kürsten [1].

2. Standard systems and modular systems

In this chapter we introduce and study the notion of standard systems which makes it possible to develop the Tomita-Takesaki theory in O*-algebras. In Section 2.1 we define and study the notion of cyclic generalized vectors for an O*-algebra which is a generalization of that of cyclic vectors. Three commutants λ^C, λ' and λ^σ of a cyclic generalized vector λ are defined and investigated. In Section 2.2 we introduce the notions of three standard systems $(\mathcal{M}, \lambda, \lambda^C)$, $(\mathcal{M}, \lambda, \lambda')$ and $(\mathcal{M}, \lambda, \lambda^\sigma)$ consisting of an O*-algebra $\mathcal{M}$, a cyclic generalized vector λ and the commutants λ^C, λ' and λ^σ, and in these cases the modular automorphism group $\{\sigma_t^\lambda\}_{t\in\mathbb{R}}$ of $\mathcal{M}$ is defined and the quasi-weight φ_λ defined by λ satisfies the KMS-condition with respect tp $\{\sigma_t^\lambda\}_{t\in\mathbb{R}}$. The most important object in these is the standard system $(\mathcal{M}, \lambda, \lambda^C)$, and then λ is said to be a standard generalized vector for $\mathcal{M}$. In Section 2.3 we define and study the notions of quasi-standard generalized vectors and modular generalized vectors which are able to apply the Tomita-Takesaki theory to more examples. Such generalized vectors are extended to standard generalized vectors. In Section 2.4 we investigate the standardness and the modularity of generalized vectors in special cases: A. Standard systems with vectors. B. Standard tracial generalized vectors. C. Standard systems for semifinite O*-algebras. D. Standard generalized vectors in the Hilbert space of Hilbert-Schmidt operators. E. Standard systems constructed by von Neumann algebras with standard generalized vectors. In Section 2.5 we introduce the notion of semifiniteness of generalized vectors and generalize the Connes cocycle theorem to standard, semifinite generalized vectors for generalized von Neumann algebras. In Section 2.6 we generalize the Pedersen-Takesaki Radon-Nikodym theorem to standard, semifinite generalized vectors for generalized von Neumann algebras. We construct the standard, semifinite generalized vector λ_A associated with a given standard, semifinite generalized vector λ and a given positive self-adjoint operator A affiliated with the centralizer of λ, and consider when a standard, semifinite generalized vector μ is represented as such a λ_A. This is closely related to the invariance of μ for the modular automorphism group $\{\sigma_t^\lambda\}_{t\in\mathbb{R}}$ and the properties of the Connes cocycle $\{[D\mu : D\lambda]_t\}_{t\in\mathbb{R}}$.

2.1 Cyclic generalized vectors

In this section we define and study the notion of cyclic generalized vectors for an O*-algebra which is a generalization of that of cyclic vectors. Throughout this section let $\mathcal{M}$ be an O*-algebra on a dense subspace $\mathcal{D}$ in a Hilbert space $\mathcal{H}$.

Definition 2.1.1. A map λ of $\mathcal{M}$ into $\mathcal{D}$ is said to be a *generalized vector* for $\mathcal{M}$ if the following conditions hold:
(i) The domain $\mathcal{D}(\lambda)$ of λ is a left ideal of $\mathcal{M}$.
(ii) λ is a linear map of $\mathcal{D}(\lambda)$ into $\mathcal{D}$.
(iii) $\lambda(AX) = A\lambda(X)$ for all $A \in \mathcal{M}$ and $X \in \mathcal{D}(\lambda)$.
A generalized vector λ for $\mathcal{M}$ is said to be *cyclic* (resp. *strongly cyclic*) if $\lambda(\mathcal{D}(\lambda))$ is dense in $\mathcal{H}$ (resp. $\mathcal{D}[t_{\mathcal{M}}]$).

Let λ be a generalized vector for $\mathcal{M}$. Then the closure of the O*-algebra $\mathcal{M}\lceil_{\lambda(\mathcal{D}(\lambda))}$ on $\lambda(\mathcal{D}(\lambda))$ in $\mathcal{H}(\lambda) \equiv \overline{\lambda(\mathcal{D}(\lambda))}$ is called the *O*-algebra defined by* λ and denoted by $\mathcal{M}(\lambda)$. If λ is strongly cyclic, then $\mathcal{M}(\lambda) = \mathcal{M}$, but even if λ is cyclic, then $\mathcal{M}(\lambda) \neq \mathcal{M}$ in general. Let λ_1 and λ_2 be generalized vectors for $\mathcal{M}$. If $\mathcal{D}(\lambda_1) \subset \mathcal{D}(\lambda_2)$ and $\lambda_1(X) = \lambda_2(X)$ for each $X \in \mathcal{D}(\lambda_1)$, then λ_2 is said to be an *extension* of λ_1 and denoted by $\lambda_2 \supset \lambda_1$.

We give an important example of generalized vectors for O*-algebras.

Example 2.1.2. Let $\mathcal{M}$ be an O*-algebra on $\mathcal{D}$ in $\mathcal{H}$ and $\xi \in \mathcal{H}$. We put

$$\begin{cases} \mathcal{D}(\lambda_\xi) = \{X \in \mathcal{M} \ ; \ \xi \in \mathcal{D}(X^{\dagger *}) \text{ and } X^{\dagger *}\xi \in \mathcal{D}\}, \\ \lambda_\xi(X) = X^{\dagger *}\xi, \quad X \in \mathcal{D}(\lambda_\xi). \end{cases}$$

Then λ_ξ is a generalized vector for $\mathcal{M}$. It is clear that λ_ξ is cyclic (resp. strongly cyclic) if and only if $\{X^{\dagger *}\xi \ ; \ X \in \mathcal{D}(\lambda_\xi)\}$ is dense in $\mathcal{H}$ (resp. $\mathcal{D}[t_{\mathcal{M}}]$). We remark that putting

$$\begin{cases} \mathcal{D}(\lambda) = \{X \in \mathcal{M} \ ; \ \xi \in \mathcal{D}(\overline{X}) \text{ and } \overline{X}\xi \in \mathcal{D}\}, \\ \lambda(X) = \overline{X}\xi, \quad X \in \mathcal{D}(\lambda), \end{cases}$$

$\mathcal{D}(\lambda)$ is not necessarily a left ideal of $\mathcal{M}$ because $\overline{X+Y} \not\supset \overline{X} + \overline{Y}$ in general, and so λ is not a generalized vector for $\mathcal{M}$. Suppose $\xi \in \mathcal{D}$. Then it follows that $\mathcal{D}(\lambda_\xi) = \mathcal{M}$ and $\lambda_\xi(X) = X\xi$ for each $X \in \mathcal{M}$. Hence λ_ξ can be identical with the vector ξ; for example, λ_ξ is cyclic (resp. strongly cyclic) if and only if ξ is cyclic (resp. strongly cyclic).

Let us give an O*-algebra which has no strongly cyclic vector, but has a strongly cyclic generalized vector of this above type. Let

$$\mathcal{M} = \{\sum_{k=1}^{n} f_k \left(\frac{d}{dt}\right)^k \lceil C_0^\infty(\mathbb{R}) \ ; \ f_k \in C^\infty(\mathbb{R}), \ 0 \leq k \leq n, \ n \in \mathbb{N}\}.$$

Then $\mathcal{M}$ is a self-adjoint O*-algebra on $C_0^\infty(\mathbb{R})$ which has no strongly cyclic vector by the property of suport. Let $\xi_0(t) = e^{-t^2}, t \in \mathbb{R}$. Since $\mathcal{N} \equiv \{fd^n/dt^n \lceil C_0^\infty(\mathbb{R})\ ;\ f \in C_0^\infty(\mathbb{R}),\ n \in \mathbb{N}\} \subset \mathcal{D}(\lambda_{\xi_0})$ and $\mathcal{N}\xi_0 = C_0^\infty(\mathbb{R})$, it follows that λ_{ξ_0} is a strongly cyclic generalized vector for $\mathcal{M}$.

We next define commutants and bicommutants of cyclic generalized vectors. Let λ be a generalized vector for $\mathcal{M}$. Suppose

$(\mathrm{C})_1$ $\mathcal{M}'_{\mathrm{w}}\mathcal{D} \subset \mathcal{D}$,

$(\mathrm{C})_2$ $\lambda(\mathcal{D}(\lambda)^\dagger \mathcal{D}(\lambda))$ is total in $\mathcal{H}$.

We define three commutants of λ as follows:

$$\begin{cases} \mathcal{D}(\lambda') = \{K \in \mathcal{M}'_{\mathrm{w}}\ ;\ {}^\exists\xi_K \in \bigcap_{X \in \mathcal{D}(\lambda)} \mathcal{D}(\overline{X}) \text{ s.t. } K\lambda(X) = \overline{X}\xi_K \\ \qquad\qquad\qquad\qquad\qquad\qquad\qquad\qquad \text{for all } X \in \mathcal{D}(\lambda)\} \\ \lambda'(K) = \xi_K, \quad K \in \mathcal{D}(\lambda'). \end{cases}$$

$$\begin{cases} \mathcal{D}(\lambda^\sigma) = \{K \in \mathcal{M}'_{\mathrm{w}}\ ;\ {}^\exists\xi_K \in \bigcap_{X \in \mathcal{D}(\lambda)} \mathcal{D}(X^{\dagger *}) \text{ s.t. } K\lambda(X) = X^{\dagger *}\xi_K \\ \qquad\qquad\qquad\qquad\qquad\qquad\qquad\qquad \text{for all } X \in \mathcal{D}(\lambda)\} \\ \lambda^\sigma(K) = \xi_K, \quad K \in \mathcal{D}(\lambda^\sigma). \end{cases}$$

$$\begin{cases} \mathcal{D}(\lambda^C) = \{K \in \mathcal{M}'_{\mathrm{w}}\ ;\ {}^\exists\xi_K \in \mathcal{D} \text{ s.t. } K\lambda(X) = X\xi_K \text{ for all } X \in \mathcal{D}(\lambda)\} \\ \lambda^C(K) = \xi_K, \quad K \in \mathcal{D}(\lambda^C). \end{cases}$$

Then we have the following

Proposition 2.1.3. λ', λ^σ and λ^C are generalized vectors for the von Neumann algebra $\mathcal{M}'_{\mathrm{w}}$ and $\lambda^C \subset \lambda' \subset \lambda^\sigma$.

Proof. Since $\lambda(\mathcal{D}(\lambda)^\dagger \mathcal{D}(\lambda))$ is total in $\mathcal{H}$, it follows that ξ_K is uniquely determined for $K \in \mathcal{D}(\lambda')$, $\mathcal{D}(\lambda')$ is a subspace of $\mathcal{M}'_{\mathrm{w}}$ and λ' is a linear map of $\mathcal{D}(\lambda')$ into $\bigcap_{X \in \mathcal{D}(\lambda)} \mathcal{D}(\overline{X})$. Take arbitrary $C \in \mathcal{M}'_{\mathrm{w}}$ and $K \in \mathcal{D}(\lambda')$. Since $\overline{X}$ is affiliated with $(\mathcal{M}'_{\mathrm{w}})'$ for each $X \in \mathcal{D}(\lambda)$ (by $(\mathrm{C})_1$), it follows that $C\lambda'(K) \in \bigcap_{X \in \mathcal{D}(\lambda)} \mathcal{D}(\overline{X})$ and $CK\lambda(X) = C\overline{X}\lambda'(K) = \overline{X}C\lambda'(K)$ for all $X \in \mathcal{D}(\lambda)$, which implies $CK \in \mathcal{D}(\lambda')$ and $\lambda'(CK) = C\lambda'(K)$. Therefore, $\mathcal{D}(\lambda')$ is a left ideal of $\mathcal{M}'_{\mathrm{w}}$ and λ' is a generalized vector for $\mathcal{M}'_{\mathrm{w}}$. We can similarly show that λ^σ and λ^C are generalized vectors for $\mathcal{M}'_{\mathrm{w}}$. It is clear that $\lambda^C \subset \lambda' \subset \lambda^\sigma$.

We define the notion of regular systems, and characterize it.

Definition 2.1.4. A system $(\mathcal{M}, \lambda, \lambda')$ (resp. $(\mathcal{M}, \lambda, \lambda^\sigma)$, $(\mathcal{M}, \lambda, \lambda^C)$) is said to be *regular* if it satisfies the conditions $(\mathrm{C})_1$, $(\mathrm{C})_2$ and moreover

(R) $\mathcal{D}(\lambda')^* \cap \mathcal{D}(\lambda')$ (resp. $\mathcal{D}(\lambda^\sigma)^* \cap \mathcal{D}(\lambda^\sigma)$, $\mathcal{D}(\lambda^C)^* \cap \mathcal{D}(\lambda^C)$) is a nondegenerate $*$-subalgebra of $\mathcal{M}'_w$.

A system $(\mathcal{M}, \lambda, \lambda')$ (resp. $(\mathcal{M}, \lambda, \lambda^\sigma)$, $(\mathcal{M}, \lambda, \lambda^C)$) is said to be *strongly regular* if the conditions $(C)_1$, $(C)_2$ and the following condition (SR) hold:

(SR) There exists a net $\{K_\alpha\}$ in $\mathcal{D}(\lambda')^* \cap \mathcal{D}(\lambda')$ (resp. $\mathcal{D}(\lambda^\sigma)^* \cap \mathcal{D}(\lambda^\sigma)$, $\mathcal{D}(\lambda^C)^* \cap \mathcal{D}(\lambda^C)$) for which $K_\alpha \uparrow I$ and $K_\alpha K_\beta = K_\beta K_\alpha$ for each α, β.

When $(\mathcal{M}, \lambda, \lambda^C)$ is regular (resp. strongly regular), λ is simply said to be *regular* (resp. *strongly regular*) .

Since $\mathcal{D}(\lambda^C) \subset \mathcal{D}(\lambda') \subset \mathcal{D}(\lambda^\sigma)$, it follows that if $(\mathcal{M}, \lambda, \lambda^C)$ is regular (resp. strongly regular) then $(\mathcal{M}, \lambda, \lambda')$ and $(\mathcal{M}, \lambda, \lambda^\sigma)$ are regular (resp. strongly regular).

Proposition 2.1.5. Let $\mathcal{M}$ be a closed O*-algebra on $\mathcal{D}$ in $\mathcal{H}$ such that $\mathcal{M}'_w \mathcal{D} \subset \mathcal{D}$ and λ a strongly cyclic generalized vector for $\mathcal{M}$ such that $\lambda(\mathcal{D}(\lambda)^\dagger \mathcal{D}(\lambda))$ is total in $\mathcal{H}$. Then $(\mathcal{M}, \lambda, \lambda^C)$(resp. $(\mathcal{M}, \lambda, \lambda')$, $(\mathcal{M}, \lambda, \lambda^\sigma)$) is regular if and only if there exists a net $\{\xi_\alpha\}$ in $\mathcal{D}$ (resp. $\bigcap_{X \in \mathcal{D}(\lambda)} \mathcal{D}(\overline{X})$, $\bigcap_{X \in \mathcal{D}(\lambda)} \mathcal{D}(X^{\dagger *})$) such that $\|\lambda(X)\| = \sup_\alpha \|X^{\dagger *} \xi_\alpha\|$ for each $X \in \mathcal{D}(\lambda)$.

Proof. Suppose $(\mathcal{M}, \lambda, \lambda^C)$ is regular. Then there exists a net $\{K_\alpha\}$ in $\mathcal{D}(\lambda^C)^* \cap \mathcal{D}(\lambda^C)$ such that $0 \leq K_\alpha \leq I$, $^\forall \alpha$ and $\{K_\alpha\}$ converges strongly to I. We put $\xi_\alpha = \lambda^C(K_\alpha)$, $^\forall \alpha$. Then we have

$$\sup_\alpha \|X\xi_\alpha\| = \sup_\alpha \|K_\alpha \lambda(X)\| = \|\lambda(X)\|, \quad ^\forall X \in \mathcal{D}(\lambda).$$

Conversely suppose that there exists a net $\{\xi_\alpha\}$ in $\mathcal{D}$ such that $\|\lambda(X)\| = \sup_\alpha \|X\xi_\alpha\|$, $^\forall X \in \mathcal{D}(\lambda)$. Then since λ is strongly cyclic, there exists a net $\{C_\alpha\}$ in $\mathcal{M}'_w$ such that $C_\alpha \lambda(X) = X\xi_\alpha$ for each $X \in \mathcal{D}(\lambda)$ and α. We now put $K_\alpha = |C_\alpha|$, $^\forall \alpha$. Then it follows that

$$K_\alpha \in \mathcal{D}(\lambda^C)^* \cap \mathcal{D}(\lambda^C), \quad \|K_\alpha\| \leq 1 \text{ and}$$
$$\sup_\alpha \|K_\alpha \lambda(X)\| = \sup_\alpha \|X\xi_\alpha\| = \|\lambda(X)\|, \quad ^\forall X \in \mathcal{D}(\lambda),$$

which implies that $\{K_\alpha\}$ converges strongly to I, and so $\mathcal{D}(\lambda^C)^* \cap \mathcal{D}(\lambda^C)$ is nondegenerate. Similarly we obtain the similar results for $(\mathcal{M}, \lambda, \lambda')$ and $(\mathcal{M}, \lambda, \lambda^\sigma)$.

We next define the notion of cyclic and separating systems which is a generalization of that of cyclic and separating vectors.

Definition 2.1.6. $(\mathcal{M}, \lambda, \lambda')$ (resp. $(\mathcal{M}, \lambda, \lambda^\sigma)$, $(\mathcal{M}, \lambda, \lambda^C)$) is said to be a *cyclic and separating system* if it satisfies the conditions $(C)_1$, $(C)_2$ and moreover

(C)$_3$ $\lambda'(\mathcal{D}(\lambda')^*\mathcal{D}(\lambda'))$ (resp. $\lambda^\sigma(\mathcal{D}(\lambda^\sigma)^*\mathcal{D}(\lambda^\sigma)), \lambda^C(\mathcal{D}(\lambda^C)^*\mathcal{D}(\lambda^C))$) is total in $\mathcal{H}$.

A cyclic and separating system $(\mathcal{M},\lambda,\lambda')$, $((\mathcal{M},\lambda,\lambda^\sigma),(\mathcal{M},\lambda,\lambda^C))$ is said to be *strongly cyclic* if λ is strongly cyclic.

In case of $\mathcal{M}=\mathcal{M}_b$ we have $\lambda'=\lambda^\sigma=\lambda^C$, and so we simply call $(\mathcal{M},\lambda)$ a *cyclic and separating system* provided $(\mathcal{M},\lambda,\lambda')$ is cyclic and separating.

We remark that if $(\mathcal{M},\lambda,\lambda'),(\mathcal{M},\lambda,\lambda^\sigma)$ and $(\mathcal{M},\lambda,\lambda^C)$ are cyclic and separating systems then they are regular, respectively. In fact, it is easily shown that $|C|\in\mathcal{D}(\lambda')$ whenever $C\in\mathcal{D}(\lambda')$, so that $(\mathcal{D}(\lambda')^*\cap\mathcal{D}(\lambda'))\mathcal{H}\supset\{|C|\xi\ ;\ C\in\mathcal{D}(\lambda'),\xi\in\mathcal{H}\}\supset\lambda'(\mathcal{D}(\lambda')^*\mathcal{D}(\lambda'))$. Hence, $(\mathcal{M},\lambda,\lambda')$ is regular. Cases of $(\mathcal{M},\lambda,\lambda^\sigma)$ and $(\mathcal{M},\lambda,\lambda^C)$ are similarly proved.

Let $(\mathcal{M},\lambda,\lambda')$ be a cyclic and separating system. Then λ' is a cyclic generalized vector for the von Neumann algebra $\mathcal{M}'_{\rm w}$ such that $\lambda'(\mathcal{D}(\lambda')^*\mathcal{D}(\lambda'))$ is total in $\mathcal{H}$, and so by Proposition 2.1.3 three commutants $(\lambda')'$, $(\lambda')^\sigma$ and $(\lambda')^C$ of λ' are well-defined and they coincide. To emphasize the commutant of λ' we use the notation $(\lambda')'$ (simply λ'') as the commutant of λ', and then λ'' is defined as follows:

$$\begin{cases}\mathcal{D}(\lambda'')=\{A\in(\mathcal{M}'_{\rm w})'\ ;\ ^\exists\xi_A\in\mathcal{H}\text{ s.t. }A\lambda'(K)=K\xi_A,\ ^\forall K\in\mathcal{D}(\lambda')\},\\ \lambda''(A)=\xi_A,\quad A\in\mathcal{D}(\lambda'').\end{cases}$$

When $(\mathcal{M},\lambda,\lambda^\sigma)$ (resp. $(\mathcal{M},\lambda,\lambda^C)$) is a cyclic and separating system, we can similarly define the commutant $\lambda^{\sigma\sigma}$ (resp. λ^{CC}) of λ^σ (resp. λ^C) as follows:

$$\begin{cases}\mathcal{D}(\lambda^{\sigma\sigma})=\{A\in(\mathcal{M}'_{\rm w})'\ ;\ ^\exists\xi_A\in\mathcal{H}\text{ s.t. }A\lambda^\sigma(K)=K\xi_A,\ ^\forall K\in\mathcal{D}(\lambda^\sigma)\},\\ \lambda^{\sigma\sigma}(A)=\xi_A,\quad A\in\mathcal{D}(\lambda^{\sigma\sigma}).\end{cases}$$

$$\begin{cases}\mathcal{D}(\lambda^{CC})=\{A\in(\mathcal{M}'_{\rm w})'\ ;\ ^\exists\xi_A\in\mathcal{H}\text{ s.t. }A\lambda^C(K)=K\xi_A,\ ^\forall K\in\mathcal{D}(\lambda^C)\},\\ \lambda^{CC}(A)=\xi_A,\quad A\in\mathcal{D}(\lambda^{CC}).\end{cases}$$

Then we have the following

Proposition 2.1.7. (1) Suppose $(\mathcal{M},\lambda,\lambda')$ is a cyclic and separating system. Then $((\mathcal{M}'_{\rm w})',\lambda'')$ is a cyclic and separating system satisfying $\lambda'''\equiv(\lambda'')'=\lambda'$.

(2) Suppose $(\mathcal{M},\lambda,\lambda^\sigma)$ is a cyclic and separating system. Then $((\mathcal{M}'_{\rm w})',\lambda^{\sigma\sigma})$ is a cyclic and separating system satisfying $\lambda^{\sigma\sigma\sigma}\equiv(\lambda^{\sigma\sigma})^\sigma=\lambda^\sigma$.

(3) Suppose $(\mathcal{M},\lambda,\lambda^C)$ is a cyclic and separating system. Then $((\mathcal{M}'_{\rm w})',\lambda^{CC})$, $((\mathcal{M}'_{\rm w})',\lambda'')$ and $((\mathcal{M}'_{\rm w})',\lambda^{\sigma\sigma})$ are cyclic and separating systems satisfying $\lambda^{\sigma\sigma}\subset\lambda''\subset\lambda^{CC}$ and $\lambda^C\subset\lambda^{CCC}\equiv(\lambda^{CC})^C\subset\lambda'\subset\lambda^\sigma$.

Proof. (1) It is easily shown that λ'' is a generalized vector for $(\mathcal{M}'_{\rm w})'$.

We show that $\lambda''(\mathcal{D}(\lambda'')^*\mathcal{D}(\lambda''))$ is dense in $\mathcal{H}$. Take an arbitrary $X \in \mathcal{D}(\lambda)$. Let $\overline{X} = U|\overline{X}|$ be the polar decomposition of $\overline{X}$ and $|\overline{X}| = \int_0^\infty t dE(t)$ the spectral resolution of $|\overline{X}|$. We put $E_n = \int_0^n dE(t)$ and $X_n = \overline{X}E_n$ for $n \in \mathbb{N}$. Since $\overline{X}$ is affiliated with $(\mathcal{M}'_{\mathrm{w}})'$, it follows that $U, X_n \in (\mathcal{M}'_{\mathrm{w}})'$ and

$$\begin{aligned} X_n\lambda'(K) = U|\overline{X}|E_n\lambda'(K) &= UE_nU^*\overline{X}\lambda'(K) \\ &= UE_nU^*K\lambda(X) \\ &= KUE_nU^*\lambda(X) \end{aligned}$$

for all $K \in \mathcal{D}(\lambda')$ and $n \in \mathbb{N}$. Hence we have

$$X_n \in \mathcal{D}(\lambda'') \text{ and } \lambda''(X_n) = UE_nU^*\lambda(X), \quad n \in \mathbb{N}. \tag{2.1.1}$$

Further, since

$$\begin{aligned} (UU^*\lambda(X)|\lambda'(K_1^*K_2)) &= (UU^*K_1\lambda(X)|\lambda'(K_2)) \\ &= (UU^*\overline{X}\lambda'(K_1)|\lambda'(K_2)) \\ &= (\overline{X}\lambda'(K_1)|\lambda'(K_2)) \\ &= (\lambda(X)|\lambda'(K_1^*K_2)) \end{aligned}$$

for all $K_1, K_2 \in \mathcal{D}(\lambda')$ and $\lambda'(\mathcal{D}(\lambda')^*\mathcal{D}(\lambda'))$ is total in $\mathcal{H}$, it follows that $UU^*\lambda(X) = \lambda(X)$, which implies by (2.1.1) that

$$\lim_{n\to\infty} \lambda''(X_n) = \lambda(X). \tag{2.1.2}$$

Further, by the definition of X_n we have

$$\lim_{n\to\infty} X_n\xi = \overline{X}\xi, \quad {}^\forall\xi \in \mathcal{D}(\overline{X}) \text{ and } \lim_{n\to\infty} X_n^*\eta = X^*\eta, \quad {}^\forall\eta \in \mathcal{D}(X^*),$$

and so $\{A^*\lambda(Y)\ ;\ A \in \mathcal{D}(\lambda''),\ Y \in \mathcal{D}(\lambda)\}$ is dense for $\lambda(\mathcal{D}(\lambda)^\dagger\mathcal{D}(\lambda))$ and further, $\lambda''(\mathcal{D}(\lambda'')^*\mathcal{D}(\lambda''))$ is dense for $\{A^*\lambda(Y)\ ;\ A \in \mathcal{D}(\lambda''),\ Y \in \mathcal{D}(\lambda)\}$ by (2.1.1) and (2.1.2). Hence it follows that $\lambda''(\mathcal{D}(\lambda'')^*\mathcal{D}(\lambda''))$ is total in $\mathcal{H}$, so that $\lambda''' \equiv (\lambda'')'$ is well-defined by

$$\begin{cases} \mathcal{D}(\lambda''') = \{K \in \mathcal{M}'_{\mathrm{w}}\ ;\ {}^\exists\xi_K \in \mathcal{H} \text{ s.t. } K\lambda''(A) = A\xi_K, \quad {}^\forall A \in \mathcal{D}(\lambda'')\}, \\ \lambda'''(K) = \xi_K, \quad K \in \mathcal{D}(\lambda'''). \end{cases}$$

It is clear that $\lambda' \subset \lambda'''$. Hence, $((\mathcal{M}'_{\mathrm{w}})', \lambda'')$ is a cyclic and separating system. We finally show $\lambda''' = \lambda'$. Take arbitrary $K \in \mathcal{D}(\lambda''')$ and $X \in \mathcal{D}(\lambda)$. Then it follows from (2.1.1) and (2.1.2) that

$$\lim_{n\to\infty} E_n\lambda'''(K) = \lambda'''(K) \text{ and } \lim_{n\to\infty} \overline{X}E_n\lambda'''(K) = \lim_{n\to\infty} K\lambda'''(X_n) = K\lambda(X).$$

Therefore, $\lambda'''(K) \in \bigcap_{X \in \mathcal{D}(\lambda)} \mathcal{D}(\overline{X})$ and $\overline{X}\lambda'''(K) = K\lambda(X)$ for all $X \in \mathcal{D}(\lambda)$, and so $K \in \mathcal{D}(\lambda')$.

(2) This is proved similarly to the proof of (1) considering the polar decomposition of $X^{\dagger *}$ and the spectral decomposition of $|X^{\dagger *}|, X \in \mathcal{D}(\lambda)$.

(3) It is clear that λ^{cc} is a generalized vector for $(\mathcal{M}'_{\mathrm{w}})'$. It follows from Proposition 2.1.3 that $(\mathcal{M}, \lambda, \lambda')$ and $(\mathcal{M}, \lambda, \lambda^{\sigma})$ are cyclic and separating systems, which implies by (1) and (2) that $((\mathcal{M}'_{\mathrm{w}})', \lambda'')$ and $((\mathcal{M}'_{\mathrm{w}})', \lambda^{\sigma\sigma})$ are cyclic and separating systems. Further, since $\lambda^{\sigma\sigma} \subset \lambda'' \subset \lambda^{cc}$, it follows that $((\mathcal{M}'_{\mathrm{w}})', \lambda^{cc})$ is a cyclic and separating system. By (1) and (2) we have $\lambda^{c} \subset \lambda^{ccc} \subset \lambda' \subset \lambda^{\sigma}$. This completes the proof.

We remark that $\lambda''' = \lambda'$ and $\lambda^{\sigma\sigma\sigma} = \lambda^{\sigma}$, but $\lambda^{ccc} \neq \lambda^{c}$ in general as will be seen in Proposition 2.1.11, (6) later.

We difine and study the fullness of a cyclic and separating system.

Proposition 2.1.8. (1) Suppose $(\mathcal{M}, \lambda, \lambda^{c})$ is a cyclic and separating system and put

$$\begin{cases} \mathcal{D}(\lambda_e) = \{X \in \mathcal{M}\,;\ {}^{\exists}\xi_X \in \mathcal{D} \text{ s.t. } X\lambda^{c}(K) = K\xi_X,\ {}^{\forall}K \in \mathcal{D}(\lambda^{c})\}, \\ \lambda_e(X) = \xi_X, \quad X \in \mathcal{D}(\lambda_e). \end{cases}$$

Then $(\mathcal{M}, \lambda_e, \lambda_e^{c})$ is a cyclic and separating system such that $\lambda \subset \lambda_e$, $\lambda^{c} = \lambda_e^{c}$ and

$$\begin{cases} \mathcal{D}(\lambda_e) = \{X \in \mathcal{M}\,;\ {}^{\exists}\{A_\alpha\} \subset \mathcal{D}(\lambda^{cc}) \text{ and } {}^{\exists}\xi_X \in \mathcal{D} \text{ s.t. } A_\alpha\xi \to X\xi,\ {}^{\forall}\xi \in \mathcal{D} \\ \qquad\qquad\qquad \text{and } \lambda^{cc}(A_\alpha) \to \xi_X\}; \\ \lambda_e(X) = \xi_X, \quad X \in \mathcal{D}(\lambda_e). \end{cases}$$

(2) Suppose $(\mathcal{M}, \lambda, \lambda^{\sigma})$ is a cyclic and separating system and put

$$\begin{cases} \mathcal{D}(\lambda_e) = \{X \in \mathcal{M}\,;\ \lambda^{\sigma}(\mathcal{D}(\lambda^{\sigma})) \subset \mathcal{D}(X^{\dagger *}) \\ \qquad\qquad \text{and } {}^{\exists}\xi_X \in \mathcal{D} \text{ s.t. } X^{\dagger *}\lambda^{\sigma}(K) = K\xi_X,\ {}^{\forall}K \in \mathcal{D}(\lambda^{\sigma})\}, \\ \lambda_e(X) = \xi_X, \quad X \in \mathcal{D}(\lambda_e). \end{cases}$$

Then $(\mathcal{M}, \lambda_e, \lambda_e^{\sigma})$ is a cyclic and separating system such that $\lambda \subset \lambda_e$, $\lambda^{\sigma} = \lambda_e^{\sigma}$ and

$$\begin{cases} \mathcal{D}(\lambda_e) = \{X \in \mathcal{M}\,;\ {}^{\exists}\{A_\alpha\} \subset \mathcal{D}(\lambda^{\sigma\sigma}) \text{ and } {}^{\exists}\xi_X \in \mathcal{D} \text{ s.t. } A_\alpha\xi \to X^{\dagger *}\xi, \\ \qquad\qquad {}^{\forall}\xi \in \mathcal{D}(X^{\dagger *}) \text{ and } \lambda^{\sigma\sigma}(A_\alpha) \to \xi_X \in \mathcal{D}\}, \\ \lambda_e(X) = \xi_X, \quad X \in \mathcal{D}(\lambda_e). \end{cases}$$

Proof. (2) Since $\lambda^{\sigma}(\mathcal{D}(\lambda^{\sigma})^{*}\mathcal{D}(\lambda^{\sigma}))$ is total in $\mathcal{H}$, it follows that λ_e is well-defined. Further, since $(A+B)^{\dagger *} \supset A^{\dagger *} + B^{\dagger *}$ and $(AB)^{\dagger *} \supset A^{\dagger *}B^{\dagger *}$ for each $A, B \in \mathcal{M}$, it follows that $\mathcal{D}(\lambda_e)$ is a subspace of $\mathcal{M}$ and λ_e is a linear map of $\mathcal{D}(\lambda_e)$ into $\mathcal{D}$. Take arbitrary $A \in \mathcal{M}$ and $X \in \mathcal{D}(\lambda_e)$. Then we have

$$(AX)^{\dagger *}\lambda^{\sigma}(K) = A^{\dagger *}X^{\dagger *}\lambda^{\sigma}(K) = AK\lambda_e(X) = KA\lambda_e(X)$$

for each $K \in \mathcal{D}(\lambda^{\sigma})$, and so $AX \in \mathcal{D}(\lambda_e)$ and $\lambda_e(AX) = A\lambda_e(X)$. Hence λ_e is a generalized vector for $\mathcal{M}$. It is clear that $\lambda \subset \lambda_e$ and $\lambda^{\sigma} = \lambda_e^{\sigma}$, so that $(\mathcal{M}, \lambda_e, \lambda_e^{\sigma})$ is a cyclic and separating system. We put

$$\begin{cases} \mathcal{D}(\mu) = \{X \in \mathcal{M}; {}^{\exists}\{A_\alpha\} \subset \mathcal{D}(\lambda^{\sigma\sigma}) \text{ and } {}^{\exists}\xi_X \in \mathcal{D} \text{ s.t.} \\ \qquad\qquad A_\alpha\xi \to X^{\dagger *}\xi, {}^{\forall}\xi \in \mathcal{D}(X^{\dagger *}) \text{ and } \lambda^{\sigma\sigma}(A_\alpha) \to \xi_X\}, \\ \mu(X) = \xi_X, \quad X \in \mathcal{D}(\mu). \end{cases}$$

We show $\lambda_e = \mu$. Take an arbitrary $X \in \mathcal{D}(\lambda_e)$. Since $(\mathcal{M}, \lambda_e, \lambda_e^{\sigma})$ is a cyclic and separating system such that $\lambda_e^{\sigma} = \lambda^{\sigma}$, we can show similarly to the proof of Proposition 2.1.5 that $X \in \mathcal{D}(\mu)$ and $\mu(X) = \lambda_e(X)$. The converse is easily shown. Similarly we can prove the statement (1). This completes the proof.

Remark 2.1.9. Let $(\mathcal{M}, \lambda, \lambda')$ be a cyclic and separating system. We put

$$\begin{cases} \mathcal{D}(\lambda_e) = \{X \in \mathcal{M} \,;\, {}^{\exists}\xi_X \in \mathcal{D} \text{ s.t. } \overline{X}\lambda'(K) = K\xi_X, {}^{\forall}K \in \mathcal{D}(\lambda')\}, \\ \lambda_e(X) = \xi_X, \quad X \in \mathcal{D}(\lambda_e). \end{cases}$$

Then λ_e is not necessarily a generalized vector for $\mathcal{M}$. In fact, we don't know even which $\mathcal{D}(\lambda_e)$ is a subspace of $\mathcal{M}$ because $\overline{X_1 + X_2} \not\supset \overline{X_1} + \overline{X_2}$ for $X_1, X_2 \in \mathcal{D}(\lambda_e)$. We can prove that if $\mathcal{M}$ is integrable, then $\lambda' = \lambda^{\sigma}$ and so $(\mathcal{M}, \lambda_e, \lambda_e')$ is a cyclic and separating system such that $\lambda \subset \lambda_e$ and $\lambda' = \lambda_e'$.

Definition 2.1.10. A cyclic and separating system $(\mathcal{M}, \lambda, \lambda^C)$ $((\mathcal{M}, \lambda, \lambda^{\sigma}),$ $(\mathcal{M}, \lambda, \lambda'))$ is said to be *full* if $\lambda = \lambda_e$. In particular, if $(\mathcal{M}, \lambda, \lambda^C)$ is a full cyclic and separating system, then λ is said to be *full* .

We remark that if $(\mathcal{M}, \lambda, \lambda^C)$ and $(\mathcal{M}, \lambda, \lambda^{\sigma})$ are cyclic and separating systems, then they can be extended to the full cyclic and separating systems $(\mathcal{M}, \lambda_e, \lambda_e^C)$ and $(\mathcal{M}, \lambda_e, \lambda_e^{\sigma})$, respectively, but this doesn't hold in case of $(\mathcal{M}, \lambda, \lambda')$.

For the commutants $\lambda_\xi^{\sigma}, \lambda_\xi'$ and λ_ξ^C of the generalized vector λ_ξ for $\mathcal{M}$ associated with $\xi \in \mathcal{H}$ we have the following

Proposition 2.1.11. Let $\xi \in \mathcal{H} \setminus \mathcal{D}$. Suppose
(i) $\mathcal{M}'_{\mathrm{w}}\mathcal{D} \subset \mathcal{D}$,
(ii) ξ is a cyclic and separating vector for $(\mathcal{M}'_{\mathrm{w}})'$,
(iii) $\{X^{\dagger}Y^{\dagger *}\xi \,;\, X, Y \in \mathcal{D}(\lambda_\xi)\}$ is total in $\mathcal{H}$.
Then the following statements hold:

(1) $\mathcal{D}(\lambda_\xi^\sigma) = \mathcal{M}'_{\mathrm{w}}$ and $\lambda_\xi^\sigma(C) = C\xi$ for each $C \in \mathcal{M}'_{\mathrm{w}}$;
$\mathcal{D}(\lambda_\xi^{\sigma\sigma}) = (\mathcal{M}'_{\mathrm{w}})'$ and $\lambda_\xi^{\sigma\sigma}(A) = A\xi$ for each $A \in (\mathcal{M}'_{\mathrm{w}})'$.

Hence $(\mathcal{M}, \lambda_\xi, \lambda_\xi^\sigma)$ is a full cyclic and separating system.

(2) $\mathcal{D}(\lambda'_\xi) = \{K \in \mathcal{M}'_{\mathrm{w}}\ ;\ K\xi \in \bigcap_{X \in \mathcal{D}(\lambda_\xi)} \mathcal{D}(\overline{X})\}$,
$\mathcal{D}(\lambda_\xi^c) = \{K \in \mathcal{M}'_{\mathrm{w}}\ ;\ K\xi \in \mathcal{D}\}$.

(3) $\lambda_\xi^c \subsetneqq \lambda'_\xi \subsetneqq \lambda_\xi^\sigma$ in general.

(4) The following statements are equivalent:
(a) $(\mathcal{M}, \lambda_\xi, \lambda'_\xi)$ is a cyclic and separating system.
(b) $\xi \in \bigcap_{X \in \mathcal{D}(\lambda_\xi)} \mathcal{D}(\overline{X})$.
(c) $\mathcal{D}(\lambda'_\xi) = \mathcal{M}'_{\mathrm{w}}$.
(d) $\lambda'_\xi = \lambda_\xi^\sigma$.

If this is true, then $(\mathcal{M}, \lambda_\xi, \lambda'_\xi)$ is full.

(5) Suppose $(\mathcal{M}, \lambda_\xi, \lambda_\xi^c)$ is a cyclic and separating system. Then λ_ξ is a full generalized vector satisfying $\lambda_\xi^{cc} = \lambda''_\xi = \lambda_\xi^{\sigma\sigma}$ and $\lambda_\xi^c \subsetneqq \lambda_\xi^{ccc} = \lambda'_\xi = \lambda_\xi^\sigma$.

Proof. The statements (1) and (2) are easily shown.

(3) Suppose $\xi \in \bigcap_{X \in \mathcal{D}(\lambda_\xi)} \mathcal{D}(\overline{X})$. Then $I \in \mathcal{D}(\lambda'_\xi)$, but $I \notin \mathcal{D}(\lambda_\xi^c)$. Hence we have $\lambda_\xi^c \subsetneqq \lambda'_\xi$. Suppose $\xi \notin \bigcap_{X \in \mathcal{D}(\lambda_\xi)} \mathcal{D}(\overline{X})$. Then $I \notin \mathcal{D}(\lambda'_\xi)$, and so $\lambda'_\xi \subsetneqq \lambda_\xi^\sigma$.

(4) (a) $\Rightarrow$ (d) Since $\lambda''_\xi \supset \lambda_\xi^{\sigma\sigma}$ by (3), it follows from (1) that $\lambda''_\xi = \lambda_\xi^{\sigma\sigma}$, which implies by Proposition 2.1.7 that $\lambda'_\xi = \lambda_\xi^\sigma$.

(d) $\Rightarrow$ (c) This follows from (1).

(c) $\Rightarrow$ (b) This is trivial.

(b) $\Rightarrow$ (a) Suppose $\xi \in \bigcap_{X \in \mathcal{D}(\lambda_\xi)} \mathcal{D}(\overline{X})$. Then $I \in \mathcal{D}(\lambda'_\xi)$, and so $\lambda'_\xi = \lambda_\xi^\sigma$. Hence it follows from (1) that $(\mathcal{M}, \lambda_\xi, \lambda'_\xi)$ is a cyclic and separating system.

(5) By (3) we have $\lambda_\xi^{\sigma\sigma} \subset \lambda''_\xi \subset \lambda_\xi^{cc}$, and so we show $\lambda_\xi^{cc} \subset \lambda_\xi^{\sigma\sigma}$. Take an arbitrary $A \in \mathcal{D}(\lambda_\xi^{cc})$. Since $\mathcal{D}(\lambda_\xi^c)^* \cap \mathcal{D}(\lambda_\xi^c)$ is a nondegenerate $*$-subalgebra of $\mathcal{M}'_{\mathrm{w}}$, there exists a net $\{K_\alpha\}$ in $\mathcal{D}(\lambda_\xi^c)^* \cap \mathcal{D}(\lambda_\xi^c)$ such that $0 \le K_\alpha \le I$ for each α and $\{K_\alpha\}$ converges strongly to I. Then it follows that

$$\begin{aligned}
&\{K_\beta C K_\alpha\} \subset \mathcal{D}(\lambda_\xi^c)^* \cap \mathcal{D}(\lambda_\xi^c),\\
&\lim_{\alpha,\beta} \lambda_\xi^c(K_\beta C K_\alpha) = \lim_{\alpha,\beta} K_\beta C K_\alpha \xi = C\xi = \lambda_\xi^\sigma(C),\\
&\lim_{\alpha,\beta} \lambda_\xi^c((K_\beta C K_\alpha)^*) = \lim_{\alpha,\beta} K_\alpha C^* K_\beta \xi = C^*\xi = \lambda_\xi^\sigma(C^*)
\end{aligned}$$

for each $C \in \mathcal{M}'_{\mathrm{w}}$, which implies

$$A\lambda_\xi^\sigma(C) = \lim_{\alpha\beta} A\lambda_\xi^c(K_\beta C K_\alpha) = \lim_{\alpha\beta} K_\beta C K_\alpha \lambda^{cc}(A) = C\lambda^{cc}(A)$$

for each $C \in \mathcal{D}(\lambda_\xi^\sigma) = \mathcal{M}'_{\mathrm{w}}$. Hence we have $A \in \mathcal{D}(\lambda_\xi^{\sigma\sigma})$ and $\lambda_\xi^{\sigma\sigma}(A) = \lambda_\xi^{CC}(\mathrm{A})$, and so $\lambda_\xi^{CC} \subset \lambda_\xi^{\sigma\sigma}$. Thus we have $\lambda_\xi^{CC} = \lambda_\xi'' = \lambda_\xi^{\sigma\sigma}$, and further by Proposition 2.1.7 $\lambda_\xi^C \subsetneqq \lambda_\xi^{CCC} = \lambda_\xi' = \lambda_\xi^\sigma$. We finally show $(\mathcal{M}, \lambda_\xi, \lambda_\xi^C)$ is full. Take an arbitrary $X \in \mathcal{D}((\lambda_\xi)_e)$, that is, $^\exists \xi_X \in \mathcal{D}$ s.t.$XK\xi = K\xi_X, {}^\forall K \in \mathcal{D}(\lambda_\xi^C)$. Then we have

$$(X^\dagger \eta|\xi) = \lim_\alpha (X^\dagger \eta|K_\alpha \xi) = \lim_\alpha (\eta|XK_\alpha \xi) = \lim_\alpha (\eta|K_\alpha \xi_X) = (\eta|\xi_X)$$

for each $\eta \in \mathcal{D}$, and so $\xi \in \mathcal{D}(X^{\dagger *})$ and $X^{\dagger *}\xi = \xi_X \in \mathcal{D}$. Hence we have $X \in \mathcal{D}(\lambda_\xi)$, which implies λ_ξ is full. This completes the proof.

2.2 Standard systems and standard generalized vectors

In this section we study standard systems which are able to develop the Tomita-Takesaki theory in O*-algebras. Throughout this section let $(\mathcal{M}, \lambda)$ be a pair of a closed O*-algebra $\mathcal{M}$ on a dense subspace $\mathcal{D}$ in a Hilbert space $\mathcal{H}$ and a generalized vector λ for $\mathcal{M}$ satisfying

$(\mathrm{S})_1$ $\mathcal{M}'_{\mathrm{w}}\mathcal{D} \subset \mathcal{D}$,

$(\mathrm{S})_2$ $\lambda((\mathcal{D}(\lambda)^\dagger \cap \mathcal{D}(\lambda))^2)$ is total in $\mathcal{H}$.

Then it follows from $(\mathrm{S})_1$, $(\mathrm{S})_2$ and Proposition 2.1.3 that λ' and λ^C are generalized vectors for the von Neumann algebra $\mathcal{M}'_{\mathrm{w}}$ and

$$\begin{aligned} \mathcal{D}(\lambda') = \{K \in \mathcal{M}'_{\mathrm{w}};\, {}^\exists \xi_K \in \bigcap_{X \in \mathcal{D}(\lambda)} \mathcal{D}(\overline{X}) \text{ s.t.} \\ K\lambda(X) = \overline{X}\xi_K, {}^\forall X \in \mathcal{D}(\lambda)^\dagger \cap \mathcal{D}(\lambda)\}, \quad (2.2.1) \\ \mathcal{D}(\lambda^C) = \{K \in \mathcal{M}'_{\mathrm{w}};\, {}^\exists \xi_K \in \mathcal{D} \text{ s.t.} \\ K\lambda(X) = X\xi_K, {}^\forall X \in \mathcal{D}(\lambda)^\dagger \cap \mathcal{D}(\lambda)\}. \quad (2.2.2) \end{aligned}$$

In fact, the inclusion $\subset$ is trivial. Conversely suppose $K \in \mathcal{M}'_{\mathrm{w}}$ such that $K\lambda(X) = \overline{X}\xi_K$, ${}^\forall X \in \mathcal{D}(\lambda)^\dagger \cap \mathcal{D}(\lambda)$ for some $\xi_K \in \bigcap_{X \in \mathcal{D}(\lambda)} \mathcal{D}(\overline{X})$. Take an arbitrary $X \in \mathcal{D}(\lambda)$. Since $Y^\dagger X \in \mathcal{D}(\lambda)^\dagger \cap \mathcal{D}(\lambda)$ for each $Y \in \mathcal{D}(\lambda)^\dagger \cap \mathcal{D}(\lambda)$, it follows that

$$(K\lambda(X)|\lambda(YZ)) = (K\lambda(Y^\dagger X)|\lambda(Z)) = (\overline{Y^\dagger X}\xi_K|\lambda(Z)) = (\overline{X}\xi_K|\lambda(YZ))$$

for each $Y, Z \in \mathcal{D}(\lambda)^\dagger \cap \mathcal{D}(\lambda)$, which implies by $(\mathrm{S})_2$ that $K \in \mathcal{D}(\lambda')$ and $\lambda'(K) = \xi_K$. Therefore the statement (2.2.1) holds. Similarly the statement (2.2.2) is shown.

By Proposition 2.1.7 we have the following

Lemma 2.2.1. Suppose

(S)$'_3$ $\lambda'((\mathcal{D}(\lambda')^* \cap \mathcal{D}(\lambda'))^2)$ is total in $\mathcal{H}$.
Then the following statements hold:

(1) $\lambda'(\mathcal{D}(\lambda')^* \cap \mathcal{D}(\lambda'))$ is an achieved right Hilbert algebra in $\mathcal{H}$ equipped with the multiplication and the involution:

$$\lambda'(K_1)\lambda'(K_2) = \lambda'(K_2K_1), \quad \lambda'(K)^\flat = \lambda'(K^*), \qquad K, K_1, K_2 \in \mathcal{D}(\lambda')^* \cap \mathcal{D}(\lambda'),$$

and its right von Neumann algebra equals $\mathcal{M}'_{\text{w}}$.

(2) $\lambda''(\mathcal{D}(\lambda'')^* \cap \mathcal{D}(\lambda''))$ is an achieved left Hilbert algebra in $\mathcal{H}$ equipped with the multiplication and the involution:

$$\lambda''(A)\lambda''(B) = \lambda''(AB), \quad \lambda''(A)^\# = \lambda''(A^*), \qquad A, B \in \mathcal{D}(\lambda'')^* \cap \mathcal{D}(\lambda''),$$

and it equals the commutant of the right Hilbert algebra $\lambda'(\mathcal{D}(\lambda')^* \cap \mathcal{D}(\lambda'))$.

(3) $\mathcal{D}(\lambda'') = \{A \in (\mathcal{M}'_{\text{w}})'; {}^\exists\xi_A \in \mathcal{H} \text{ s.t. } A\lambda'(K) = K\xi_A, {}^\forall K \in \mathcal{D}(\lambda')^* \cap \mathcal{D}(\lambda')\}$ and $\lambda''(A) = \xi_A$ for each $A \in \mathcal{D}(\lambda'')$.

(4) $\mathcal{D}(\lambda') = \mathcal{D}(\lambda''') = \{K \in \mathcal{M}'_{\text{w}}; {}^\exists\xi_K \in \mathcal{H} \text{ s.t. } K\lambda''(A) = A\xi_K, {}^\forall A \in \mathcal{D}(\lambda'')^* \cap \mathcal{D}(\lambda'')\}$ and $\lambda'(K) = \xi_K$ for each $K \in \mathcal{D}(\lambda')$.

Suppose (S)$'_3$ holds. By Lemma 2.2.1 the map $\lambda''(A) \to \lambda''(A^*)$, $A \in \mathcal{D}(\lambda'')^* \cap \mathcal{D}(\lambda'')$ is a closable conjugate-linear operator in $\mathcal{H}$, and so its closure is denoted by $S_{\lambda''}$. Since

$$(\lambda(X)|\lambda'(K_1^*K_2)) = (\lambda'(K_2^*K_1)|\lambda(X^\dagger))$$

for all $X \in \mathcal{D}(\lambda)^\dagger \cap \mathcal{D}(\lambda)$ and $K_1, K_2 \in \mathcal{D}(\lambda')^* \cap \mathcal{D}(\lambda')$, it follows that the map $\lambda(X) \to \lambda(X^\dagger)$, $X \in \mathcal{D}(\lambda)^\dagger \cap \mathcal{D}(\lambda)$ is a closable conjugate-linear operator in $\mathcal{H}$, and so its closure is denoted by S_λ. Let $S_{\lambda''} = J_{\lambda''}\Delta_{\lambda''}^{1/2}$ and $S_\lambda = J_\lambda\Delta_\lambda^{1/2}$ be the polar decompositions of $S_{\lambda''}$ and S_λ, respectively.

Lemma 2.2.2. Suppose $\lambda'((\mathcal{D}(\lambda')^* \cap \mathcal{D}(\lambda'))^2)$ is total in $\mathcal{H}$. Then $S_\lambda \subset S_{\lambda''}$.

Proof. Take an arbitrary $X \in \mathcal{D}(\lambda)^\dagger \cap \mathcal{D}(\lambda)$. Let $\overline{X} = U|\overline{X}|$ be the polar decomposition of $\overline{X}$, $|\overline{X}| = \int_0^\infty t dE(t)$ the spectral resolution of $|\overline{X}|$ and $E_n = \int_0^n dE(t), n \in \mathbb{N}$. We put $X_n = \overline{X}E_n, n \in \mathbb{N}$. Then, by (2.1.1) and (2.1.2) we have

$$X_n \in \mathcal{D}(\lambda'') \text{ and } \lim_{n\to\infty} \lambda''(X_n) = \lambda(X). \tag{2.2.3}$$

Further, since

$$X_n^*\lambda'(K) = E_n\overline{X^\dagger}\lambda'(K) = KE_n\lambda(X^\dagger)$$

for all $K \in \mathcal{D}(\lambda')$, it follows that

$$X_n^* \in \mathcal{D}(\lambda'') \text{ and } \lambda''(X_n^*) = E_n\lambda(X^\dagger), \quad n \in \mathbb{N}. \tag{2.2.4}$$

By (2.2.3) and (2.2.4) we have

$$X_n \in \mathcal{D}(\lambda'')^* \cap \mathcal{D}(\lambda''), \ \lim_{n\to\infty} \lambda''(X_n) = \lambda(X) \text{ and } \lim_{n\to\infty} \lambda''(X_n^*) = \lambda(X^\dagger). \tag{2.2.5}$$

Therefore, $\lambda(X) \in \mathcal{D}(S_{\lambda''})$ and $S_{\lambda''}\lambda(X) = \lambda(X^\dagger) = S_\lambda\lambda(X)$. This completes the proof.

By the Tomita fundamental theorem (Takesaki [1], Tomita [1]) we have

$$J_{\lambda''}(\mathcal{M}'_{\mathrm{w}})' J_{\lambda''} = \mathcal{M}'_{\mathrm{w}}; \tag{2.2.6}$$

$$\sigma_t^{\lambda''}(A) \equiv \Delta_{\lambda''}^{it} A \Delta_{\lambda''}^{-it} \in (\mathcal{M}'_{\mathrm{w}})' \quad (A \in (\mathcal{M}'_{\mathrm{w}})', t \in \mathbb{R}); \tag{2.2.7}$$

$$\sigma_t^{\lambda''}(\mathcal{D}(\lambda'')^* \cap \mathcal{D}(\lambda'')) = \mathcal{D}(\lambda'')^* \cap \mathcal{D}(\lambda'') \text{ and}$$
$$\lambda''(\sigma_t^{\lambda''}(B)) = \Delta_{\lambda''}^{it}\lambda''(B) \quad (B \in \mathcal{D}(\lambda'')^* \cap \mathcal{D}(\lambda''), \ t \in \mathbb{R}); \tag{2.2.8}$$

$$\sigma_t^{\lambda''}(C) \equiv \Delta_{\lambda''}^{it} C \Delta_{\lambda''}^{-it} \in \mathcal{M}'_{\mathrm{w}} \quad (C \in \mathcal{M}'_{\mathrm{w}}, t \in \mathbb{R}); \tag{2.2.9}$$

$$\sigma_t^{\lambda''}(\mathcal{D}(\lambda')^* \cap \mathcal{D}(\lambda')) = \mathcal{D}(\lambda')^* \cap \mathcal{D}(\lambda') \text{ and}$$
$$\lambda'(\sigma_t^{\lambda''}(K)) = \Delta_{\lambda''}^{it}\lambda'(K) \quad (K \in \mathcal{D}(\lambda')^* \cap \mathcal{D}(\lambda'), t \in \mathbb{R}). \tag{2.2.10}$$

Further, we have

$$\sigma_t^{\lambda''}(\mathcal{D}(\lambda'')) = \mathcal{D}(\lambda'') \text{ and } \lambda''(\sigma_t^{\lambda''}(B)) = \Delta_{\lambda''}^{it}\lambda''(B)$$
$$(B \in \mathcal{D}(\lambda''), t \in \mathbb{R}); \tag{2.2.11}$$

$$\sigma_t^{\lambda''}(\mathcal{D}(\lambda')) = \mathcal{D}(\lambda') \text{ and } \lambda'(\sigma_t^{\lambda''}(K)) = \Delta_{\lambda''}^{it}\lambda'(K)$$
$$(K \in \mathcal{D}(\lambda'), t \in \mathbb{R}). \tag{2.2.12}$$

In fact, the statement (2.2.11) follows from Lemma 2.2.1, (3) and by (2.2.10)

$$\begin{aligned} \sigma_t^{\lambda''}(B)\lambda'(K) &= \Delta_{\lambda''}^{it} B \Delta_{\lambda''}^{-it}\lambda'(K) \\ &= \Delta_{\lambda''}^{it} B \lambda'(\sigma_{-t}^{\lambda''}(K)) \\ &= \Delta_{\lambda''}^{it} \sigma_{-t}^{\lambda''}(K)\lambda''(B) \\ &= K \Delta_{\lambda''}^{it}\lambda''(B) \end{aligned}$$

for all $B \in \mathcal{D}(\lambda''), K \in \mathcal{D}(\lambda')^* \cap \mathcal{D}(\lambda')$ and $t \in \mathbb{R}$. The statement (2.2.12) follows from $\lambda''' = \lambda'$ and (2.2.11). By (2.2.7) the unitary group $\{\Delta_{\lambda''}^{it}\}_{t\in\mathbb{R}}$ implements a one-parameter group of $*$-automorphisms of the von Neumann algebra $(\mathcal{M}'_{\mathrm{w}})'$. But, we don't know how it acts on the O*-algebra $\mathcal{M}$, and so we define a system which has the best condition:

Definition 2.2.3. A triple $(\mathcal{M}, \lambda, \lambda')$ is said to be a *standard system* if it satisfies the above conditions $(\mathrm{S})_1 \sim (\mathrm{S})_3$ and the following conditions $(\mathrm{S})'_4$, $(\mathrm{S})'_5$ and $(\mathrm{S})'_6$:

(S)$'_4$ $\Delta_{\lambda''}^{it}\mathcal{D} \subset \mathcal{D}$, $t \in \mathbb{R}$.
(S)$'_5$ $\Delta_{\lambda''}^{it}\mathcal{M}\Delta_{\lambda''}^{-it} = \mathcal{M}$, $t \in \mathbb{R}$.
(S)$'_6$ $\sigma_t^{\lambda''}(\mathcal{D}(\lambda)^\dagger \cap \mathcal{D}(\lambda)) = \mathcal{D}(\lambda)^\dagger \cap \mathcal{D}(\lambda)$, $t \in \mathbb{R}$.

Theorem 2.2.4. Suppose $(\mathcal{M}, \lambda, \lambda')$ is a standard system. Then the following statements hold:

(1) $S_\lambda = S_{\lambda''}$.

(2) $\sigma_t^\lambda(X) \equiv \Delta_\lambda^{it} X \Delta_\lambda^{-it} = \sigma_t^{\lambda''}(X)$ for each $X \in \mathcal{M}$ and $t \in \mathbb{R}$ and $\{\sigma_t^\lambda\}_{t\in\mathbb{R}}$ is a one-parameter group of $*$-automorphisms of $\mathcal{M}$.

(3) λ satisfies the KMS-condition with respect to $\{\sigma_t^\lambda\}_{t\in\mathbb{R}}$, that is, for each $X, Y \in \mathcal{D}(\lambda)^\dagger \cap \mathcal{D}(\lambda)$ there exists an element $f_{X,Y}$ of $A(0,1)$ such that

$$f_{X,Y}(t) = (\lambda(\sigma_t^\lambda(X))|\lambda(Y)) \text{ and } f_{X,Y}(t+i) = (\lambda(Y^\dagger)|\lambda(\sigma_t^\lambda(X^\dagger)))$$

for all $t \in \mathbb{R}$, where $A(0,1)$ is the set of all complex-valued functions, bounded and continous on $0 \leq \operatorname{Im} z \leq 1$ and analytic in the interior.

Proof. Take arbitrary $X, Y \in \mathcal{D}(\lambda)^\dagger \cap \mathcal{D}(\lambda)$. By (2.2.5) there exist sequences $\{X_n\}$ and $\{Y_n\}$ in $\mathcal{D}(\lambda'')^* \cap \mathcal{D}(\lambda'')$ such that

$$\begin{aligned} &\lim_{n\to\infty} \lambda''(X_n) = \lambda(X), \quad \lim_{n\to\infty} \lambda''(X_n^*) = \lambda(X^\dagger), \\ &\lim_{n\to\infty} \lambda''(Y_n) = \lambda(Y), \quad \lim_{n\to\infty} \lambda''(Y_n^*) = \lambda(Y^\dagger). \end{aligned} \tag{2.2.13}$$

By Theorem 10.17 in Stratila-Zsido [1] and (2.2.8), for any $n \in \mathbb{N}$ there exists an element f_n of $A(0,1)$ such that

$$\begin{aligned} f_n(t) &= (\lambda''(\sigma_t^{\lambda''}(X_n))|\lambda''(Y_n)) = (\Delta_{\lambda''}^{it}\lambda''(X_n)|\lambda''(Y_n)), \\ f_n(t+i) &= (\lambda''(Y_n^*)|\lambda''(\sigma_t^{\lambda''}(X_n^*))) = (\lambda''(Y_n^*)|\Delta_{\lambda''}^{it}\lambda''(X_n^*)) \end{aligned} \tag{2.2.14}$$

for all $t \in \mathbb{R}$. Since $\lambda(\sigma_t^{\lambda''}(X))$ is well-defined by (S)$'_4$ and (S)$'_5$ and by (2.2.10)

$$\begin{aligned} (\lambda(\sigma_t^{\lambda''}(X))|\lambda'(K_1^*K_2)) &= (\overline{\sigma_t^{\lambda''}(X)}\lambda'(K_1)|\lambda'(K_2)) \\ &= (\Delta_{\lambda''}^{it}\overline{X}\lambda'(\sigma_{-t}^{\lambda''}(K_1))|\lambda'(K_2)) \\ &= (K_1\Delta_{\lambda''}^{it}\lambda(X)|\lambda'(K_2)) \\ &= (\Delta_{\lambda''}^{it}\lambda(X)|\lambda'(K_1^*K_2)) \end{aligned}$$

for all $K_1, K_2 \in \mathcal{D}(\lambda')^* \cap \mathcal{D}(\lambda')$, it follows that $\lambda(\sigma_t^{\lambda''}(X)) = \Delta_{\lambda''}^{it}\lambda(X)$ for all $t \in \mathbb{R}$, which implies by (2.2.13) and (2.2.14) that

$$\sup_{t\in\mathbb{R}}|f_n(t)-(\lambda(\sigma_t^{\lambda''}(X))|\lambda(Y))|$$
$$\leq \|\lambda''(X_n)-\lambda(X)\|\|\lambda''(Y_n)\|+\|\lambda(X)\|\|\lambda''(Y_n)-\lambda(Y)\|$$
$$\to 0(n\to\infty),$$
$$\sup_{t\in\mathbb{R}}|f_n(t+i)-(\lambda(Y^\dagger)|\lambda(\sigma_t^{\lambda''}(X^\dagger)))|$$
$$\leq \|\lambda''(Y_n^*)-\lambda(Y^\dagger)\|\|\lambda''(X_n^*)\|+\|\lambda(Y^\dagger)\|\|\lambda''(X_n^*)-\lambda(X^\dagger)\|$$
$$\to 0(n\to\infty).$$

Hence there exists an element $f_{X,Y}$ of $A(0,1)$ such that

$$f_{X,Y}(t)=(\lambda(\sigma_t^{\lambda''}(X))|\lambda(Y)) \text{ and } f_{X,Y}(t+i)=(\lambda(Y^\dagger)|\lambda(\sigma_t^{\lambda''}(X^\dagger))), \quad t\in\mathbb{R}. \tag{2.2.15}$$

We next show $S_\lambda = S_{\lambda''}$. Let $\mathcal{K}$ be the closure of $\{\lambda(X); X^\dagger = X \in \mathcal{D}(\lambda)^\dagger \cap \mathcal{D}(\lambda)\}$ in $\mathcal{H}$. Then $\mathcal{K}$ is a closed real subspace of $\mathcal{H}$. Since $\lambda(\mathcal{D}(\lambda)^\dagger\cap\mathcal{D}(\lambda)) \subset \mathcal{K}+i\mathcal{K}$ and it is dense in $\mathcal{H}$, we have $(\mathcal{K}+i\mathcal{K})^\perp=\{0\}$. Further, we have $\mathcal{K}\cap i\mathcal{K}=\{0\}$. In fact, take an arbitrary $\xi\in\mathcal{K}\cap i\mathcal{K}$. Then there exist sequences $\{A_n\}$ and $\{B_n\}$ in $\mathcal{D}(\lambda)^\dagger\cap\mathcal{D}(\lambda)$ such that $A_n^\dagger=A_n$, $B_n^\dagger=B_n$, $\lim_{n\to\infty}\lambda(A_n)=\xi$ and $\lim_{n\to\infty}\lambda(B_n)=-i\xi$, and then we have

$$\begin{aligned}(\xi|\lambda'(K_1^*K_2)) = \lim_{n\to\infty}(\lambda(A_n)|\lambda'(K_1^*K_2)) &= \lim_{n\to\infty}(\overline{A_n}\lambda'(K_1)|\lambda'(K_2))\\ &= \lim_{n\to\infty}(\lambda'(K_1)|K_2\lambda(A_n))\\ &= (\lambda'(K_2^*K_1)|\xi)\end{aligned}$$

for all $K_1, K_2\in\mathcal{D}(\lambda')^*\cap\mathcal{D}(\lambda')$, which implies

$$\begin{aligned}-i(\xi|\lambda'(K_1^*K_2)) &= \lim_{n\to\infty}(\lambda(B_n)|\lambda'(K_1^*K_2))\\ &= (\lambda'(K_2^*K_1)|-i\xi)\\ &= i(\lambda'(K_2^*K_1)|\xi)\\ &= i(\xi|\lambda'(K_1^*K_2)),\end{aligned}$$

so that by the totality of $\lambda'((\mathcal{D}(\lambda')^*\cap\mathcal{D}(\lambda'))^2)$ we have $\xi=0$. Therefore it follows that S_λ equals the closed operator S defined by

$$S(\xi+i\eta)=\xi-i\eta, \quad \xi,\eta\in\mathcal{K}. \tag{2.2.16}$$

Further, it follows from (2.2.15) and $(\mathrm{S})_6'$ that the one-parameter group $\{\Delta_{\lambda''}^{it}\}_{t\in\mathbb{R}}$ of unitary operators satisfies the KMS-condition with respect to $\mathcal{K}$ in the sence of Definition 3.4 in Rieffel-Van Daele [1] and $\Delta_{\lambda''}^{it}\mathcal{K}\subset\mathcal{K}$ for all $t\in\mathbb{R}$, so that by Theorem 3.8 in Rieffel-Van Daele [1] and (2.2.16) $\Delta_{\lambda''}^{it}=\Delta_\lambda^{it}$ for all $t\in\mathbb{R}$. Therefore it follows that $S_{\lambda''}=S_\lambda$, which implies by $(S)'_6$ and (2.2.15) that λ satisfies the KMS-condition with respect to the one-parameter group $\{\sigma_t^{\lambda}\}_{t\in\mathbb{R}}$ of $*$-automorphisms of $\mathcal{M}$. This completes the proof.

We next proceed to the standardness of the system $(\mathcal{M}, \lambda, \lambda^{c})$. We almost have the same results as Lemma 2.2.1 and (2.2.6) $\sim$ (2.2.12) for λ^{cc}.

Lemma 2.2.5. Suppose
$(S)_3^c$ $\lambda^c((\mathcal{D}(\lambda^c)^* \cap \mathcal{D}(\lambda^c))^2)$ is total in $\mathcal{H}$.
Then the following statements hold:

(1) $\lambda^c(\mathcal{D}(\lambda^c)^* \cap \mathcal{D}(\lambda^c))$ is a right Hilbert subalgebra of the right Hilbert algebra $\lambda'(\mathcal{D}(\lambda')^* \cap \mathcal{D}(\lambda'))$.

(2) $\lambda^{cc}(\mathcal{D}(\lambda^{cc})^* \cap \mathcal{D}(\lambda^{cc}))$ is an achived left Hilbert algebra in $\mathcal{H}$ containing $\lambda''(\mathcal{D}(\lambda'')^* \cap \mathcal{D}(\lambda''))$.

(3) Let $S_{\lambda^{cc}}$ be the closure of the involution $\lambda^{cc}(A) \to \lambda^{cc}(A^*)$ $(A \in \mathcal{D}(\lambda^{cc})^* \cap \mathcal{D}(\lambda^{cc}))$ and let $S_{\lambda^{cc}} = J_{\lambda^{cc}} \Delta_{\lambda^{cc}}^{1/2}$ be the polar decomposition of $S_{\lambda^{cc}}$. Then $S_\lambda \subset S_{\lambda''} \subset S_{\lambda^{cc}}$.

(4) $J_{\lambda^{cc}}(\mathcal{M}'_{\mathrm{w}})' J_{\lambda^{cc}} = \mathcal{M}'_{\mathrm{w}}$.

(5) $\sigma_t^{\lambda^{cc}}(A) \equiv \Delta_{\lambda^{cc}}^{it} A \Delta_{\lambda^{cc}}^{-it} \in (\mathcal{M}'_{\mathrm{w}})'$ for each $A \in (\mathcal{M}'_{\mathrm{w}})'$ and $t \in \mathbb{R}$,

$\sigma_t^{\lambda^{cc}}(\mathcal{D}(\lambda^{cc})) = \mathcal{D}(\lambda^{cc})$ and $\lambda^{cc}(\sigma_t^{\lambda^{cc}}(B)) = \Delta_{\lambda^{cc}}^{it}\lambda^{cc}(B)$ for each $B \in \mathcal{D}(\lambda^{cc})$ and $t \in \mathbb{R}$,

$\sigma_t^{\lambda^{cc}}(\mathcal{D}(\lambda^{cc})^* \cap \mathcal{D}(\lambda^{cc})) = \mathcal{D}(\lambda^{cc})^* \cap \mathcal{D}(\lambda^{cc})$ and $\lambda^{cc}(\sigma_t^{\lambda^{cc}}(B)) = \Delta_{\lambda^{cc}}^{it}\lambda^{cc}(B)$ for each $B \in \mathcal{D}(\lambda^{cc})^* \cap \mathcal{D}(\lambda^{cc})$ and $t \in \mathbb{R}$.

(6) $\sigma_t^{\lambda^{cc}}(C) \equiv \Delta_{\lambda^{cc}}^{it} C \Delta_{\lambda^{cc}}^{-it} \in \mathcal{M}'_{\mathrm{w}}$ for each $C \in \mathcal{M}'_{\mathrm{w}}$ and $t \in \mathbb{R}$,

$\sigma_t^{\lambda^{cc}}(\mathcal{D}(\lambda^{ccc})) = \mathcal{D}(\lambda^{ccc})$ and $\lambda^{ccc}(\sigma_t^{\lambda^{ccc}}(K)) = \Delta_{\lambda^{ccc}}^{it}\lambda^{ccc}(K)$ for each $K \in \mathcal{D}(\lambda^{ccc})$ and $t \in \mathbb{R}$,

$\sigma_t^{\lambda^{ccc}}(\mathcal{D}(\lambda^{ccc})^* \cap \mathcal{D}(\lambda^{ccc})) = \mathcal{D}(\lambda^{ccc})^* \cap \mathcal{D}(\lambda^{ccc})$ and $\lambda^{ccc}(\sigma_t^{\lambda^{ccc}}(K)) = \Delta_{\lambda^{ccc}}^{it}\lambda^{ccc}(K)$ for each $K \in \mathcal{D}(\lambda^{ccc})^* \cap \mathcal{D}(\lambda^{ccc})$ and $t \in \mathbb{R}$.

Remark 2.2.6. Let $\xi \in \mathcal{H}$. Suppose the conditions $(S)_1$ and $(S)_2$ for the generalized vector λ_ξ and the condition $(S)_3^c$ for λ_ξ^c hold. Then $(\mathcal{M}, \lambda_\xi, \lambda_\xi^c)$ is a cyclic and separating system, and so by Proposition 2.1.11 $\lambda''_\xi = \lambda_\xi^{cc}$. Hence the right Hilbert algebras $\lambda^c(\mathcal{D}(\lambda^c)^* \cap \mathcal{D}(\lambda^c))$ and $\lambda'(\mathcal{D}(\lambda')^* \cap \mathcal{D}(\lambda'))$ are equivalent. But, for general generalized vector λ we don't know whether their right Hilbert algebras are equivalent, or not.

By Lemma 2.2.5, (5) the unitary group $\{\Delta_{\lambda^{cc}}^{it}\}_{t\in\mathbb{R}}$ implements a one-parameter group $\{\sigma_t^{\lambda^{cc}}\}$ of $*$-automorphisms of the von Neumann algebra $(\mathcal{M}'_{\mathrm{w}})'$, but we don't know how it acts on the O*-algebra $\mathcal{M}$, and so we define a system which has the best condition:

Definition 2.2.7. A triple $(\mathcal{M}, \lambda, \lambda^{\mathcal{C}})$ is said to be an *essentially standard system* if it satisfies the conditions $(\mathrm{S})_1$, $(\mathrm{S})_2$, $(\mathrm{S})_3^{\mathcal{C}}$ and the following conditions $(\mathrm{S})_4^{\mathcal{C}}$ and $(\mathrm{S})_5^{\mathcal{C}}$:

$(\mathrm{S})_4^{\mathcal{C}}$ $\Delta_{\lambda^{CC}}^{it}\mathcal{D} \subset \mathcal{D}, \quad {}^\forall t \in \mathbb{R}.$

$(\mathrm{S})_5^{\mathcal{C}}$ $\Delta_{\lambda^{CC}}^{it}\mathcal{M}\Delta_{\lambda^{CC}}^{-it} = \mathcal{M}, \quad {}^\forall t \in \mathbb{R}.$

Further, if

$(\mathrm{S})_6^{\mathcal{C}}$ $\sigma_t^{\lambda^{CC}}(\mathcal{D}(\lambda)^\dagger \cap \mathcal{D}(\lambda)) = \mathcal{D}(\lambda)^\dagger \cap \mathcal{D}(\lambda), \quad {}^\forall t \in \mathbb{R},$

then $(\mathcal{M}, \lambda, \lambda^{\mathcal{C}})$ is said to be a *standard system* .

Theorem 2.2.8. (1) Suppose $(\mathcal{M}, \lambda, \lambda^{\mathcal{C}})$ is a standard system. Then $(\mathcal{M}, \lambda, \lambda')$ is a standard system and $S_\lambda = S_{\lambda''} = S_{\lambda^{CC}}$.

(2) Suppose $(\mathcal{M}, \lambda, \lambda^{\mathcal{C}})$ is an essentially standard system. Then $(\mathcal{M}, \lambda_e, \lambda_e^{\mathcal{C}})$ is a standard system, where λ_e is the extension of λ in Proposition 2.1.8.

Proof. (1) We can prove $S_\lambda = S_{\lambda^{CC}}$ in the same way as in Theorem 2.2.4, and so by Lemma 2.2.5, (3) we have $S_\lambda = S_{\lambda''} = S_{\lambda^{CC}}$. Therefore $(\mathcal{M}, \lambda, \lambda')$ is a standard system.

(2) Since $\lambda \subset \lambda_e$ and $\lambda^{\mathcal{C}} = \lambda_e^{\mathcal{C}}$ by Proposition 2.1.8, it follows that $(\mathcal{M}, \lambda_e, \lambda_e^{\mathcal{C}})$ is an essentially standard system. We show $\sigma_t^{\lambda^{CC}}(\mathcal{D}(\lambda_e)^\dagger \cap \mathcal{D}(\lambda_e)) \subset \mathcal{D}(\lambda_e)^\dagger \cap \mathcal{D}(\lambda_e)$ for each $t \in \mathbb{R}$. Take an arbitrary $X \in \mathcal{D}(\lambda_e)^\dagger \cap \mathcal{D}(\lambda_e)$. Since $\lambda_e'' \subset \lambda_e^{CC} = \lambda^{CC}$, it follows from (2.2.5) that there exists a sequence $\{X_n\}$ in $\mathcal{D}(\lambda^{CC})^* \cap \mathcal{D}(\lambda^{CC})$ such that $X_n \underset{\tau_s^*}{\to} X$, $\lim_{n\to\infty} \lambda^{CC}(X_n) = \lambda_e(X)$ and $\lim_{n\to\infty} \lambda^{CC}(X_n^*) = \lambda_e(X^\dagger)$. By Lemma 2.2.5, (5) we have that for each $t \in \mathbb{R}$

$$
\begin{aligned}
&\{\sigma_t^{\lambda^{CC}}(X_n)\} \subset \mathcal{D}(\lambda^{CC})^* \cap \mathcal{D}(\lambda^{CC}), \quad \Delta_{\lambda^{CC}}^{it}\mathcal{D} \subset \mathcal{D},\\
&\sigma_t^{\lambda^{CC}}(X_n) \underset{\tau_s^*}{\to} \sigma_t^{\lambda^{CC}}(X),\\
&\lim_{n\to\infty} \lambda^{CC}(\sigma_t^{\lambda^{CC}}(X_n)) = \lim_{n\to\infty} \Delta_{\lambda^{CC}}^{it}\lambda^{CC}(X_n) = \Delta_{\lambda^{CC}}^{it}\lambda_e(X),\\
&\lim_{n\to\infty} \lambda^{CC}(\sigma_t^{\lambda^{CC}}(X_n)^*) = \Delta_{\lambda^{CC}}^{it}\lambda_e(X^\dagger),
\end{aligned}
$$

which implies by Proposition 2.1.8 that $\sigma_t^{\lambda^{CC}}(X) \in \mathcal{D}(\lambda_e)^\dagger \cap \mathcal{D}(\lambda_e)$ and $\lambda_e(\sigma_t^{\lambda^{CC}}(X)) = \Delta_{\lambda^{CC}}^{it}\lambda_e(X)$. Therefore it follows that $\sigma_t^{\lambda^{CC}}(\mathcal{D}(\lambda_e)^\dagger \cap \mathcal{D}(\lambda_e)) \subset \mathcal{D}(\lambda_e)^\dagger \cap \mathcal{D}(\lambda_e)$ for each $t \in \mathbb{R}$, which implies $(\mathcal{M}, \lambda_e, \lambda_e^{\mathcal{C}})$ is a standard system. This completes the proof.

Remark 2.2.9. (1) It seems meaningless to define the notion of essentially standardness of systems $(\mathcal{M}, \lambda, \lambda')$ (iff it satisfies the conditions $(\mathrm{S})_1$, $(\mathrm{S})_2$ and $(\mathrm{S})_3' \sim (\mathrm{S})_5'$ without the condition $(\mathrm{S})_6'$). Because we can't construct a standard extension of $(\mathcal{M}, \lambda, \lambda')$ as the standard extension $(\mathcal{M}, \lambda_e, \lambda_e^{\mathcal{C}})$ in general (Remark 2.1.9).

(2) It seems also meaningless to consider the standardness of $(\mathcal{M}, \lambda, \lambda^\sigma)$ by the following reason: Suppose $\lambda^\sigma((\mathcal{D}(\lambda^\sigma)^* \cap \mathcal{D}(\lambda^\sigma))^2)$ is total in $\mathcal{H}$. As seen in Proposition 2.1.8 the extension λ_e for $(\mathcal{M}, \lambda, \lambda^\sigma)$ is possible, but the closed operator $S_{\lambda^{\sigma\sigma}}$ defined as the closure of the involution $\lambda^{\sigma\sigma}(A) \to \lambda^{\sigma\sigma}(A^*)$ and S_λ don't have any relations in general.

As seen in Remark 2.2.9 it is more natural and useful to treat with the standard system $(\mathcal{M}, \lambda, \lambda^C)$ than to do the systems $(\mathcal{M}, \lambda, \lambda')$ and $(\mathcal{M}, \lambda, \lambda^\sigma)$, and so we mainly study the standard system $(\mathcal{M}, \lambda, \lambda^C)$, and introduce the following notions:

Definition 2.2.10. When $(\mathcal{M}, \lambda, \lambda^C)$ is a standard (resp. essentially standard) system, we call simply λ the *standard* (resp. *essentially standard*) generalized vector.

2.3 Modular generalized vectors

Weakening the conditions $(S)_5^C$ and $(S)_6^C$ in Definition 2.2.7, we define and study the notions of quasi-standard generalized vectors and modular generalized vectors which are able to apply the Tomita-Takesaki theory to more examples. Throughout this section let $(\mathcal{M}, \lambda)$ be a pair of an O*-algebra $\mathcal{M}$ on $\mathcal{D}$ in $\mathcal{H}$ and a generalized vector for $\mathcal{M}$.

Definition 2.3.1. A system $(\mathcal{M}, \lambda, \lambda^C)$ is said to be *quasi-standard* if
$(S)_1$ $\mathcal{M}'_{\mathrm{w}}\mathcal{D} \subset \mathcal{D}$,
$(S)_2$ $\lambda((\mathcal{D}(\lambda)^\dagger \cap \mathcal{D}(\lambda))^2)$ is total in $\mathcal{H}$,
$(S)_3^C$ $\lambda^C((\mathcal{D}(\lambda^C)^* \cap \mathcal{D}(\lambda^C))^2)$ is total in $\mathcal{H}$,
$(S)_4^C$ $\Delta_{\lambda^{CC}}^{it}\mathcal{D} \subset \mathcal{D}$ for each $t \in \mathbb{R}$.
And then λ is said to be a *quasi-standard generalized vector* for $\mathcal{M}$.

Theorem 2.3.2. Suppose $(\mathcal{M}, \lambda, \lambda^C)$ is a quasi-standard system and then put

$$\begin{cases} \mathcal{D}(\overline{\lambda}) = \{X \in \mathcal{M}''_{\mathrm{wc}};\ {}^\exists\xi_X \in \mathcal{D} \text{ s.t. } X\lambda^C(K) = K\xi_X,\ {}^\forall K \in \mathcal{D}(\lambda^C)\} \\ \overline{\lambda}(X) = \xi_X, \quad X \in \mathcal{D}(\overline{\lambda}). \end{cases}$$

Then $\overline{\lambda}$ is a standard generalized vector for the generalized von Neumann algebra $\mathcal{M}''_{\mathrm{wc}}$ such that $\lambda \subset \overline{\lambda}$, $\lambda^C = \overline{\lambda}^C$ and

$$\begin{cases} \mathcal{D}(\overline{\lambda}) = \{X \in \mathcal{M}''_{\mathrm{wc}};\ {}^\exists\{A_\alpha\} \subset \mathcal{D}(\lambda^{CC}) \text{ and } {}^\exists\xi_X \in \mathcal{D} \\ \qquad\qquad \text{s.t. } A_\alpha\xi \to X\xi,\ {}^\forall\xi \in \mathcal{D} \text{ and } \lambda^{CC}(A_\alpha) \to \xi_X\} \\ \overline{\lambda}(X) = \xi_X, \quad X \in \mathcal{D}(\overline{\lambda}). \end{cases}$$

Proof. It is shown similarly to the proof of Proposition 2.1.8 that $\overline{\lambda}$ is a generalized vector for $\mathcal{M}''_{\mathrm{wc}}$ such that $\lambda \subset \overline{\lambda}$, $\lambda^C = \overline{\lambda}^C$ and

$$\begin{cases} \mathcal{D}(\overline{\lambda}) = \{X \in \mathcal{M}''_{\mathrm{wc}}; {}^\exists\{A_\alpha\} \subset \mathcal{D}(\lambda^{CC}) \text{ and } {}^\exists \xi_X \in \mathcal{D} \\ \qquad\qquad \text{s.t. } A_\alpha \xi \to X\xi, {}^\forall \xi \in \mathcal{D} \text{ and } \lambda^{CC}(A_\alpha) \to \xi_X\} \\ \overline{\lambda}(X) = \xi_X, \quad X \in \mathcal{D}(\overline{\lambda}). \end{cases}$$

Hence $(\mathcal{M}, \overline{\lambda}, \overline{\lambda}^C)$ satisfies the condition $(S)_1$, $(S)_2$ and $(S)_3^C$. Further, since $\Delta^{it}_{\lambda^{CC}}\mathcal{D} \subset \mathcal{D}$ and $\sigma_t^{\lambda^{CC}}(\mathcal{M}'_{\mathrm{w}}) \subset \mathcal{M}'_{\mathrm{w}}$ for each $t \in \mathbb{R}$, it follows that

$$\Delta^{it}_{\lambda^{CC}} X \Delta^{-it}_{\lambda^{CC}} C\xi = \Delta^{it}_{\lambda^{CC}} X \sigma_{-t}^{\lambda^{CC}}(C) \Delta^{-it}_{\lambda^{CC}}\xi = C\Delta^{it}_{\lambda^{CC}} X \Delta^{-it}_{\lambda^{CC}}\xi$$

for each $X \in \mathcal{M}''_{\mathrm{wc}}$, $C \in \mathcal{M}'_{\mathrm{w}}$, $\xi \in \mathcal{D}$ and $t \in \mathbb{R}$, which implies $\Delta^{it}_{\lambda^{CC}} X \Delta^{-it}_{\lambda^{CC}} \in \mathcal{M}''_{\mathrm{wc}}$ for each $X \in \mathcal{M}''_{\mathrm{wc}}$ and $t \in \mathbb{R}$. Hence we have $\sigma_t^{\lambda^{CC}}(\mathcal{M}''_{\mathrm{wc}}) = \mathcal{M}''_{\mathrm{wc}}$ for each $t \in \mathbb{R}$. It follows from the definition of $\overline{\lambda}$ that $\overline{\lambda}$ is full, and hence $\sigma_t^{\lambda^{CC}}(\mathcal{D}(\overline{\lambda})^\dagger \cap \mathcal{D}(\overline{\lambda})) = \mathcal{D}(\overline{\lambda})^\dagger \cap \mathcal{D}(\overline{\lambda})$ for all $t \in \mathbb{R}$. Thus $\overline{\lambda}$ is a standard generalized vector for $\mathcal{M}''_{\mathrm{wc}}$. This completes the proof.

We next define another generalization of standard generalized vectors as follows:

Definition 2.3.3. A system $(\mathcal{M}, \lambda, \lambda^C)$ is said to be *modular* if the conditions $(S)_1$, $(S)_2$ and $(S)_3^C$ in Definition 2.3.1 and the following condition (M) hold:

(M) There exists a dense subspace $\mathcal{E}$ of $\mathcal{D}[t_\mathcal{M}]$ such that

$(M)_1$ $\lambda(\mathcal{D}(\lambda)^\dagger \cap \mathcal{D}(\lambda)) \subset \mathcal{E}$,

$(M)_2$ $\{\lambda^C(K_1K_2); K_i \in \mathcal{D}(\lambda^C)^* \cap \mathcal{D}(\lambda^C)$ s.t. $\lambda^C(K_i), \lambda^C(K_i^*) \in \mathcal{E}, i = 1, 2\}$ is total in the Hilbert space $\mathcal{D}(S^*_{\lambda^C})$,

$(M)_3$ $\mathcal{M}\mathcal{E} \subset \mathcal{E}$,

$(M)_4$ $\Delta^{it}_{\lambda^{CC}}\mathcal{E} \subset \mathcal{E}$ for all $t \in \mathbb{R}$.

And then λ is said to be a *modular generalized vector* for $\mathcal{M}$.

Lemma 2.3.4. Suppose $(\mathcal{M}, \lambda, \lambda^C)$ is a modular system and then denote by $\mathcal{D}^M_\lambda$ the subspace of $\mathcal{D}$ generated by $\bigcup_{\mathcal{E} \in \mathcal{F}} \mathcal{E}$, where $\mathcal{F}$ is the set of all subspaces of $\mathcal{D}$ satisfying the conditions $(M)_1 \sim (M)_4$ of Definition 2.3.3. Then the following statements hold:

(1) $\mathcal{M}\lceil_{\mathcal{D}^M_\lambda}$ is an O*-algebra on $\mathcal{D}^M_\lambda$ such that $(\mathcal{M}\lceil_{\mathcal{D}^M_\lambda})'_{\mathrm{w}} = \mathcal{M}'_{\mathrm{w}}$ and $\mathcal{M}'_{\mathrm{w}}\mathcal{D}^M_\lambda \subset \mathcal{D}^M_\lambda$.

(2) $(\mathcal{M}\lceil_{\mathcal{D}^M_\lambda})''_{\mathrm{wc}}$ is a generalized von Neumann algebra on $\mathcal{D}^M_\lambda$ over $(\mathcal{M}'_{\mathrm{w}})'$.

(3) $\{\sigma_t^{\lambda^{CC}}\}_{t\in\mathbb{R}}$ is a one-parameter group of $*$-automorphisms of $(\mathcal{M}\lceil_{\mathcal{D}^M_\lambda})''_{\mathrm{wc}}$.

Proof. (1) It is clear that $\mathcal{D}^M_\lambda$ is a subspace of $\mathcal{D}$ which is the largest element of $\mathcal{F}$, so that $\mathcal{M}\lceil_{\mathcal{D}^M_\lambda}$ is an O*-algebra on $\mathcal{D}^M_\lambda$. Since $\mathcal{D}^M_\lambda$ is dense in

$\mathcal{D}[t_{\mathcal{M}}]$, we have $(\mathcal{M}\lceil_{\mathcal{D}_\lambda^M})'_{\rm w} = \mathcal{M}'_{\rm w}$. Since $\mathcal{M}'_{\rm w}\mathcal{D} \subset \mathcal{D}$ and $\sigma_t^{\lambda^{cc}}(\mathcal{M}'_{\rm w}) = \mathcal{M}'_{\rm w}$ for each $t \in \mathbb{R}$, it follows that the subspace of $\mathcal{D}$ generated by $\mathcal{M}'_{\rm w}\mathcal{D}_\lambda^M$ satisfies the conditions $(\mathrm{M})_1 \sim (\mathrm{M})_4$, which implies by the maximum of $\mathcal{D}_\lambda^M$ that $\mathcal{M}'_{\rm w}\mathcal{D}_\lambda^M = \mathcal{D}_\lambda^M$.

(2) It follows from (1) and Lemma 2.4 in Inoue [9] that $(\mathcal{M}\lceil_{\mathcal{D}_\lambda^M})''_{\rm wc}$ is an O*-algebra on $\mathcal{D}_\lambda^M$ such that

$$(\mathcal{M}\lceil_{\mathcal{D}_\lambda^M})''_{\rm wc} = \{X \in \mathcal{L}^\dagger(\mathcal{D}_\lambda^M)\ ;\ \overline{X} \text{ is affiliated with } (\mathcal{M}'_{\rm w})'\}$$

and

$$\Delta_{\lambda^{CC}}^{it}(\mathcal{M}\lceil_{\mathcal{D}_\lambda^M})''_{\rm wc}\Delta_{\lambda^{CC}}^{-it} = (\mathcal{M}\lceil_{\mathcal{D}_\lambda^M})''_{\rm wc} \text{ for all } t \in \mathbb{R}. \tag{2.3.1}$$

We show that the O*-algebra $(\mathcal{M}\lceil_{\mathcal{D}_\lambda^M})''_{\rm wc}$ on $\mathcal{D}_\lambda^M$ is closed. In fact, it is easily shown that the completion $\widetilde{\mathcal{D}_\lambda^M}$ of the locally convex space $\mathcal{D}[t_{(\mathcal{M}\lceil_{\mathcal{D}_\lambda^M})''_{\rm wc}}]$ satisfies the conditions $(\mathrm{M})_1 \sim (\mathrm{M})_4$, so that we have $\widetilde{\mathcal{D}_\lambda^M} = \mathcal{D}_\lambda^M$ by the maximum of $\mathcal{D}_\lambda^M$. Thus $(\mathcal{M}\lceil_{\mathcal{D}_\lambda^M})''_{\rm wc}$ is a generalized von Neumann algebra on $\mathcal{D}_\lambda^M$ over $(\mathcal{M}'_{\rm w})'$.

(3) This follows from (2.3.1).

Lemma 2.3.5. Suppose $(\mathcal{M}, \lambda, \lambda^C)$ is a modular system and then put

$$\begin{cases} \mathcal{D}(\lambda_{\mathcal{D}_\lambda^M}) = \{X\lceil_{\mathcal{D}_\lambda^M}; X \in \mathcal{D}(\lambda) \text{ s.t. } \lambda(X) \in \mathcal{D}_\lambda^M\} \\ \lambda_{\mathcal{D}_\lambda^M}(X\lceil_{\mathcal{D}_\lambda^M}) = \lambda(X), \quad X\lceil_{\mathcal{D}_\lambda^M} \in \mathcal{D}(\lambda_{\mathcal{D}_\lambda^M}). \end{cases}$$

Then $\lambda_{\mathcal{D}_\lambda^M}$ is a quasi-standard generalized vector for the O*-algebra $\mathcal{M}\lceil_{\mathcal{D}_\lambda^M}$ satisfying

(i) $\lambda_{\mathcal{D}_\lambda^M}((\mathcal{D}(\lambda_{\mathcal{D}_\lambda^M})^\dagger \cap \mathcal{D}(\lambda_{\mathcal{D}_\lambda^M}))^2)$ is total in $\mathcal{H}$;

(ii) $\mathcal{D}(\lambda^C_{\mathcal{D}_\lambda^M}) = \{K \in \mathcal{D}(\lambda^C); \lambda^C(K) \in \mathcal{D}_\lambda^M\}$ and $\lambda^C_{\mathcal{D}_\lambda^M}(K) = \lambda^C(K)$ for each $K \in \mathcal{D}(\lambda^C_{\mathcal{D}_\lambda^M})$;

(iii) $\lambda^{CC}_{\mathcal{D}_\lambda^M}$ is well-defined and $\mathcal{D}(\lambda^{CC}_{\mathcal{D}_\lambda^M})^* \cap \mathcal{D}(\lambda^{CC}_{\mathcal{D}_\lambda^M}) = \mathcal{D}(\lambda^{CC})^* \cap \mathcal{D}(\lambda^{CC})$.

Proof. It is clear that $\lambda_{\mathcal{D}_\lambda^M}$ is a generalized vector for $\mathcal{M}\lceil_{\mathcal{D}_\lambda^M}$. By $(\mathrm{M})_1$ we have

$$\{X\lceil_{\mathcal{D}_\lambda^M}; X \in \mathcal{D}(\lambda)^\dagger \cap \mathcal{D}(\lambda)\} \subset \mathcal{D}(\lambda_{\mathcal{D}_\lambda^M})^\dagger \cap \mathcal{D}(\lambda_{\mathcal{D}_\lambda^M}), \tag{2.3.2}$$

and so $\lambda_{\mathcal{D}_\lambda^M}((\mathcal{D}(\lambda_{\mathcal{D}_\lambda^M})^\dagger \cap \mathcal{D}(\lambda_{\mathcal{D}_\lambda^M}))^2)$ is total in $\mathcal{H}$. The statement (ii) follows from (2.2.4) and (2.3.2), and the statement (iii) follows from (ii) and $(\mathrm{M})_2$. Further, it follows from the above (i) $\sim$ (iii) that $\lambda_{\mathcal{D}_\lambda^M}$ is quasi-standard. This completes the proof.

By Theorem 2.3.2 and Lemma 2.3.4, 2.3.5 we have the following

Theorem 2.3.6. Suppose $(\mathcal{M}, \lambda, \lambda^{C})$ is a modular system and put

$$\begin{cases} \mathcal{D}(\lambda_S) \equiv \mathcal{D}(\overline{\lambda_{\mathcal{D}_\lambda^M}}) \\ \qquad = \{X \in (\mathcal{M}\lceil_{\mathcal{D}_\lambda^M})''_{\mathrm{wc}};\ ^\exists \xi_X \in \mathcal{D}_\lambda^M \text{ s.t. } X\lambda^{C}_{\mathcal{D}_\lambda^M}(K) = K\xi_X, \\ \qquad\qquad\qquad\qquad\qquad\qquad\qquad\qquad {}^\forall K \in \mathcal{D}(\lambda^{C}_{\mathcal{D}_\lambda^M})\} \\ \lambda_S(X) \equiv \overline{\lambda_{\mathcal{D}_\lambda^M}}(X) = \xi_X, \quad X \in \mathcal{D}(\lambda_S). \end{cases}$$

Then λ_S is a standard generalized vector for the generalized von Neumann algebra $(\mathcal{M}\lceil_{\mathcal{D}_\lambda^M})''_{\mathrm{wc}}$ on $\mathcal{D}_\lambda^M$ over $(\mathcal{M}'_{\mathrm{w}})'$ such that $\lambda_{\mathcal{D}_\lambda^M} \subset \lambda_S$, $\lambda^{C}_{\mathcal{D}_\lambda^M} = \lambda^{C}_S$ and $\mathcal{D}(\lambda_S^{CC})^* \cap \mathcal{D}(\lambda_S^{CC}) = \mathcal{D}(\lambda^{CC})^* \cap \mathcal{D}(\lambda^{CC})$.

We remark that it is meaningless to consider the notion of modularity of systems $(\mathcal{M}, \lambda, \lambda')$ because the extension theory for $(\mathcal{M}, \lambda, \lambda')$ does not succeed as seen in Remark 2.1.9.

2.4 Special cases

A. Standard systems associated with vectors

Let $\mathcal{M}$ be a closed O*-algebra on $\mathcal{D}$ in $\mathcal{H}$ and $\xi \in \mathcal{H}$. We consider when $(\mathcal{M}, \lambda_\xi, \lambda_\xi^{C})$ is a standard (or modular) system.

Proposition 2.4.1. Suppose

$(\mathrm{S})_1$ $\mathcal{M}'_{\mathrm{w}}\mathcal{D} \subset \mathcal{D}$,

$(\mathrm{S})_2$ $\{X_1^\dagger X_2^{\dagger *}\xi; \xi \in \mathcal{D}(X_i^{\dagger *}) \cap \mathcal{D}(X_i^*)$ and $X_i^{\dagger *}\xi, X_i^*\xi \in \mathcal{D}, i = 1, 2\}$ is total in $\mathcal{H}$,

$(\mathrm{S})_3^{C}$ $\{K_1K_2\xi; K_i \in \mathcal{M}'_{\mathrm{w}}$ s.t. $K_i\xi, K_i^*\xi \in \mathcal{D}, i = 1, 2\}$ is total in $\mathcal{H}$,

$(\mathrm{S})_4^{C}$ ${\Delta''_\xi}^{it}\mathcal{D} \subset \mathcal{D}$, $\ {}^\forall t \in \mathbb{R}$, which Δ''_ξ is the modular operator of the achieved left Hilbert algebra $(\mathcal{M}'_{\mathrm{w}})'\xi$.

Then λ_ξ is a quasi-standard generalized vector for $\mathcal{M}$. Further, suppose

$(\mathrm{S})_5^{C}$ ${\Delta''_\xi}^{it}\mathcal{M}{\Delta''_\xi}^{-it} = \mathcal{M}$, $\ {}^\forall t \in \mathbb{R}$.

Then λ_ξ is a standard generalized vector for $\mathcal{M}$.

Proof. It follows from $(\mathrm{S})_1$, $(\mathrm{S})_2$, $(\mathrm{S})_3^{C}$ and Proposition 2.1.11 that $(\mathcal{M}, \lambda_\xi, \lambda_\xi^{C})$ is a cyclic and separating system such that $\mathcal{D}(\lambda_\xi^{CC}) = (\mathcal{M}'_{\mathrm{w}})'$ and $\lambda_\xi^{CC}(A) = A\xi$ for each $A \in (\mathcal{M}'_{\mathrm{w}})'$, and so $(\mathcal{M}'_{\mathrm{w}})'\xi$ is an achieved left Hilbert algebra in $\mathcal{H}$ and $\Delta''_\xi = \Delta_{\lambda_\xi^{CC}}$. Hence the condition $(\mathrm{S})_4^{C}$ implies that λ_ξ is a quasi-standard generalized vector for $\mathcal{M}$. Further, suppose the condition $(\mathrm{S})_5^{C}$ holds. Since λ_ξ is full, it follows that λ_ξ is standard.

Proposition 2.4.2. Suppose the conditions $(\mathrm{S})_1$, $(\mathrm{S})_2$ and the following condition (M) hold, then λ_ξ is a modular generalized vector for $\mathcal{M}$:

(M) There exists a dense subspace $\mathcal{E}$ of $\mathcal{D}[t_{\mathcal{M}}]$ such that
$(\mathrm{M})_1$ $\{X^{\dagger *}\xi; \xi \in \mathcal{D}(X^{\dagger *}) \cap \mathcal{D}(X^*)$ and $X^{\dagger *}\xi, X^*\xi \in \mathcal{D}\} \subset \mathcal{E}$;
$(\mathrm{M})_2$ $\{K_1K_2\xi; K_i \in \mathcal{M}'_{\mathrm{w}}$ s.t. $K_i\xi, K_i^*\xi \in \mathcal{E}, i = 1,2\}$ is total in $\mathcal{H}$;
$(\mathrm{M})_3$ $\Delta_\xi^{''it}\mathcal{E} \subset \mathcal{E}, \quad {}^\forall t \in \mathbb{R}$;
$(\mathrm{M})_4$ $\mathcal{M}\mathcal{E} \subset \mathcal{E}$.

Proof. We put

$$\mathfrak{A} = \{K\xi; K \in \mathcal{M}'_{\mathrm{w}} \text{ s.t. } K\xi, K^*\xi \in \mathcal{D}_{\mathcal{D}^{\mathcal{M}}_{\lambda_\xi}}\}.$$

Then it follows from $(\mathrm{M})_2$ that $\mathfrak{A}$ is a right Hilbert algebra in $\mathcal{H}$ whose commutant $\mathfrak{A}'$ equals the achieved left Hilbert algebra $(\mathcal{M}'_{\mathrm{w}})'\xi$, which implies $\mathfrak{A}^2$ is total in the Hilbert space $\mathcal{D}(S^*_{\lambda_\xi^{cc}})(= \mathcal{D}(S^*_{(\mathcal{M}'_{\mathrm{w}})'\xi}))$. Therefore λ_ξ is a modular generalized vector for $\mathcal{M}$.

Definition 2.4.3. Let $\xi \in \mathcal{D}$. If λ_ξ is standard (resp. quasi-standard, modular), then ξ is said to be a *standard* (resp. *quasi-standard, modular*) *vector* for $\mathcal{M}$.

For the standard (quasi-standard, modular) vectors we have the following

Corollary 2.4.4. Let $\xi \in \mathcal{D}$. If the below conditions $(\mathrm{S})_1$, $(\mathrm{S})_2$, $(\mathrm{S})_3$ and $(\mathrm{S})_4$ hold, then ξ is a quasi-standard vector for $\mathcal{M}$; and if the further condition $(\mathrm{S})_5$ holds, then ξ is a standard vector. If the below conditions $(\mathrm{S})_1$, $(\mathrm{S})_2$, $(\mathrm{S})_3$ and (M) hold, then ξ is a modular vector for $\mathcal{M}$:
$(\mathrm{S})_1$ $\mathcal{M}'_{\mathrm{w}}\mathcal{D} \subset \mathcal{D}$.
$(\mathrm{S})_2$ $\mathcal{M}\xi$ is dense in $\mathcal{H}$.
$(\mathrm{S})_3$ $\mathcal{M}'_{\mathrm{w}}\xi$ is dense in $\mathcal{H}$.
$(\mathrm{S})_4$ $\Delta_\xi^{''it}\mathcal{D} \subset \mathcal{D}, \quad {}^\forall t \in \mathbb{R}$.
$(\mathrm{S})_5$ $\Delta_\xi^{''it}\mathcal{M}\Delta_\xi^{''-it} = \mathcal{M}, \quad {}^\forall t \in \mathbb{R}$.
(M) There exists a dense subspace $\mathcal{E}$ of $\mathcal{D}[t_{\mathcal{M}}]$ such that
$(\mathrm{M})_3$ $\Delta_\xi^{''it}\mathcal{E} \subset \mathcal{E} \quad {}^\forall t \in \mathbb{R}$;
$(\mathrm{M})_4$ $\mathcal{M}\mathcal{E} \subset \mathcal{E}$.

Proof. Suppose the conditions $(\mathrm{S})_1$, $(\mathrm{S})_2$, $(\mathrm{S})_3$ and (M) hold. Then it is easily shown that the linear span of $\mathcal{M}'_{\mathrm{w}}\mathcal{E}$ satisfies all of the conditions $(\mathrm{M})_1$ $\sim$ $(\mathrm{M})_4$ in Proposition 2.4.2, so that λ_ξ is modular. Hence ξ is a modular vector for $\mathcal{M}$. The other assertions follow from Proposition 2.4.1.

B. Standard tracial generalized vectors

Let $\mathcal{M}$ be a closed O*-algebra on $\mathcal{D}$ in $\mathcal{H}$ such that $\mathcal{M}'_{\mathrm{w}}\mathcal{D} \subset \mathcal{D}$. A generalized vector μ for $\mathcal{M}$ is said to be *tracial* if $(\mu(X)|\mu(Y)) = (\mu(Y^\dagger)|\mu(X^\dagger))$ for each $X, Y \in \mathcal{D}(\mu)^\dagger \cap \mathcal{D}(\mu)$. Here we consider when a tracial generalized vector μ for $\mathcal{M}$ is standard. We first introduce standard tracial generalized vectors constructed by the Segal L^p-spaces.

Example 2.4.5. Let $\mathcal{M}_0$ be a von Neumann algebra on a Hilbert space $\mathcal{H}$ and μ_0 a standard tracial generalized vector for $\mathcal{M}_0$. Then $\mu_0(\mathcal{D}(\mu_0)^* \cap \mathcal{D}(\mu_0))$ is an achieved Hilbert algebra in $\mathcal{H}$, and so the natural trace φ_{μ_0} on $(\mathcal{M}_0)_+$ can be defined by

$$\varphi_{\mu_0}(A) = \begin{cases} \|\mu_0(B)\|^2 & \text{if } A = B^*B \text{ for some } B \in \mathcal{D}(\mu_0)^* \cap \mathcal{D}(\mu_0), \\ \infty & \text{if otherwise.} \end{cases}$$

We denote by $L^p(\varphi_{\mu_0})$ $(1 \leq p \leq \infty)$ the Segal L^p-space with respect to φ_{μ_0} (Segal [1]). For each $\xi \in \mathcal{H}$ we put

$$\pi_0(\xi)\mu_0(A) \equiv \pi_0'(\mu_0(A))\xi = J_{\mu_0}A^*J_{\mu_0}\xi, \quad A \in \mathcal{D}(\mu_0)^* \cap \mathcal{D}(\mu_0),$$

where π_0' is the right regular representation of the Hilbert algebra $\mu_0(\mathcal{D}(\mu_0)^* \cap \mathcal{D}(\mu_0))$ and J_{μ_0} is the unitary involution on $\mathcal{H}$ defined by $J_{\mu_0}\mu_0(A) = \mu_0(A^*)$, $A \in \mathcal{D}(\mu_0)^* \cap \mathcal{D}(\mu_0)$. It is well known in Inoue [2] and Segal [1] that

$$\overline{\pi_0(\xi)}^* = \overline{\pi_0(J_{\mu_0}\xi)}, \quad \xi \in \mathcal{H}, \tag{2.4.1}$$

which implies

$$\begin{aligned} \overline{\pi_0(\xi)} \dotplus \overline{\pi_0(\eta)} &\equiv \overline{\overline{\pi_0(\xi)} + \overline{\pi_0(\eta)}} = \overline{\pi_0(\xi+\eta)}, \\ \lambda \cdot \overline{\pi_0(\xi)} &\equiv \begin{cases} \lambda\overline{\pi_0(\xi)} \left(= \overline{\pi(\lambda\xi)}\right), & \lambda \neq 0 \\ 0, & \lambda = 0 \end{cases} \end{aligned} \tag{2.4.2}$$

for each $\xi, \eta \in \mathcal{H}$ and $\lambda \in \mathbb{C}$. We now put

$$\begin{aligned} \mathcal{H}^p_{\mu_0} &= \{\xi \in \mathcal{H} \; ; \; \overline{\pi_0(\xi)} \in L^p(\varphi_{\mu_0})\}, \; 1 \leq p \leq \infty \; ; \\ \mathcal{H}^\omega_{\mu_0} &= \bigcap_{2 \leq p < \infty} \mathcal{H}^p_{\mu_0}. \end{aligned}$$

Take arbitrary $\xi, \eta \in \mathcal{H}^\omega (\equiv \mathcal{H}^\omega_{\mu_0})$. By (2.4.1) we have $\overline{\pi_0(\xi)} \cdot \overline{\pi_0(\eta)} \equiv \overline{\overline{\pi_0(\xi)}\,\overline{\pi_0(\eta)}} \in L^\omega_2(\varphi_{\mu_0}) \equiv \bigcap_{2 \leq p < \infty} L^p(\varphi_{\mu_0})$, and so $\overline{\pi_0(\xi)} \cdot \overline{\pi_0(\eta)} = \overline{\pi_0(\zeta)}$ for some $\zeta \in \mathcal{H}^\omega$, which implies $\eta \in \mathcal{D}(\overline{\pi_0(\xi)})$ and $\zeta = \overline{\pi_0(\xi)}\eta$. Hence it follows from (2.4.2) that $\mathcal{H}^\omega$ is a subspace of $\mathcal{H}$ and it is a $*$-algebra equipped with the multiplication and the involution :

$$\xi\eta \equiv \overline{\pi_0(\xi)}\eta \quad \text{and} \quad \xi^* = J_{\mu_0}\xi$$

for $\xi, \eta \in \mathcal{H}^\omega$, so that we put

$$\pi^\omega(\xi)\eta = \xi\eta, \quad \xi, \eta \in \mathcal{H}^\omega.$$

By (2.4.1) the closure $\widetilde{\pi^\omega}$ of π^ω is an integrable representation of $\mathcal{H}^\omega$ and $\mathcal{M}^\omega \equiv \widetilde{\pi^\omega}(\mathcal{H}^\omega)$ is an integrable O*-algebra on $\mathcal{D}(\widetilde{\pi^\omega})$, and further the O*-algebra $\mathcal{M}^\omega_\infty$ generated by $\mathcal{M}_0\lceil_{\mathcal{D}(\widetilde{\pi^\omega})}$ and $\mathcal{M}^\omega$ is an EW*-algebra on $\mathcal{D}(\widetilde{\pi^\omega})$ over $\mathcal{M}_0$. We define a generalized vector μ^ω_∞ for $\mathcal{M}^\omega_\infty$ by

$$\begin{cases} \mathcal{D}(\mu^\omega_\infty) = \mathcal{M}^\omega \\ \mu^\omega_\infty(\widetilde{\pi^\omega}(\xi)) = \xi, \quad \xi \in \mathcal{H}^\omega \end{cases}$$

and define a generalized vector μ^ω for $\mathcal{M}^\omega$ by the restriction $\mu^\omega \equiv \mu^\omega_\infty\lceil_{\mathcal{M}^\omega}$ of μ^ω_∞ to $\mathcal{M}^\omega$. Then it follows that μ^ω and μ^ω_∞ are standard tracial generalized vectors for $\mathcal{M}^\omega$ and $\mathcal{M}^\omega_\infty$, respectively, such that $(\mu^\omega)^{cc} = (\mu^\omega_\infty)^{cc} = \mu_0$.

Proposition 2.4.6. Let $\mathcal{M}$ be a closed O*-algebra on $\mathcal{D}$ in $\mathcal{H}$ such that $\mathcal{M}'_{\mathrm{w}}\mathcal{D} \subset \mathcal{D}$. Suppose μ is a tracial generalized vector for $\mathcal{M}$ such that $\mu((\mathcal{D}(\mu)^\dagger \cap \mathcal{D}(\mu))^2)$ is total in $\mathcal{H}$. Then the following statements are equivalent:

(i) μ is standard, that is, $(\mathcal{M}, \mu, \mu^c)$ is a standard system.
(ii) $\mu^c((\mathcal{D}(\mu^c)^* \cap \mathcal{D}(\mu^c))^2)$ is total in $\mathcal{H}$.
(iii) $(\mathcal{M}, \mu, \mu')$ is a standard system.
(iv) $\mu'((\mathcal{D}(\mu')^* \cap \mathcal{D}(\mu'))^2)$ is total in $\mathcal{H}$.
(v) $J_\mu(\mathcal{M}'_{\mathrm{w}})'J_\mu = \mathcal{M}'_{\mathrm{w}}$, where J_μ is the unitary involution on $\mathcal{H}$ defined by $J_\mu\mu(Y) = \mu(Y^\dagger)$ for each $Y \in \mathcal{D}(\mu)^\dagger \cap \mathcal{D}(\mu)$.

If this is true, then $J_\mu = J_{\mu^{cc}} = J_{\mu''}$ and $\Delta_\mu = \Delta_{\mu^{cc}} = \Delta_{\mu''} = I$, and further $Y^{\dagger *} = \overline{Y}$ for each $Y \in \mathcal{D}(\mu)$.

Proof. (ii) $\Rightarrow$ (i) Since $\mu^{cc}(\mathcal{D}(\mu^{cc})^* \cap \mathcal{D}(\mu^{cc}))$ is an achieved left Hilbert algebra in $\mathcal{H}$ and μ is tracial, it follows that $S_{\mu^{cc}} = J_{\mu^{cc}}\Delta^{\frac{1}{2}}_{\mu^{cc}} \supset S_\mu = J_\mu$, which implies that $\Delta_{\mu^{cc}} = I$ and $J_{\mu^{cc}} = J_\mu$. Hence μ is standard.

(i) $\Rightarrow$ (iii) This follows from Theorem 2.2.8.

(iii) $\Rightarrow$ (iv) This is trivial.

(iv) $\Rightarrow$ (v) We can prove in the same way as in the proof of (ii) $\Rightarrow$ (i) that $\Delta_{\mu''} = I$ and $J_{\mu''} = J_\mu$, which implies that $J_\mu(\mathcal{M}'_{\mathrm{w}})'J_\mu = \mathcal{M}'_{\mathrm{w}}$.

(v) $\Rightarrow$ (ii) Let $X \in \mathcal{M}$, and let $\overline{X} = U_X|\overline{X}|$ the polar decomposition of $\overline{X}$ and $|\overline{X}| = \int_0^\infty t\, dE_X(t)$ the spectral resolution of $|\overline{X}|$. We put

$$E_X(n) = \int_0^n dE_X(t) \text{ and } X_n = \overline{X}E_X(n), \; n \in \mathbb{N}.$$

Then we have $U_X, E_X(n), X_n \in (\mathcal{M}'_{\mathrm{w}})'$ for each $n \in \mathbb{N}$. Take arbitrary $Y \in \mathcal{D}(\mu)$ and $n \in \mathbb{N}$. Then, by the assumption (iv) we have

$$J_\mu Y_n^* J_\mu \mu(Z) = J_\mu E_Y(n)\mu(Y^\dagger Z^\dagger) = ZJ_\mu E_Y(n)J_\mu\mu(Y)$$

for each $Z \in \mathcal{D}(\mu)^\dagger \cap \mathcal{D}(\mu)$, which implies

$$J_\mu Y_n^* J_\mu \in \mathcal{D}(\mu^C) \text{ and } \mu^C(J_\mu Y_n^* J_\mu) = J_\mu E_Y(n) J_\mu \mu(Y), \tag{2.4.3}$$

and hence

$$\lim_{n\to\infty} \mu^C(J_\mu Y_n^* J_\mu) = \mu(Y). \tag{2.4.4}$$

For any $Y \in \mathcal{D}(\mu)^\dagger \cap \mathcal{D}(\mu)$ we have

$$\begin{aligned} J_\mu Y_n J_\mu \mu(Z) &= J_\mu U_Y E_Y(n) U_Y^* Y J_\mu \mu(Z) \\ &= Z J_\mu U_Y E_Y(n) U_Y^* J_\mu \mu(Y^\dagger) \end{aligned}$$

for each $Z \in \mathcal{D}(\mu)^\dagger \cap \mathcal{D}(\mu)$, and hence by (2.4.3)

$$\begin{aligned} &J_\mu Y_n J_\mu \in \mathcal{D}(\mu^C)^* \cap \mathcal{D}(\mu^C), \\ &\mu^C(J_\mu Y_n J_\mu) = J_\mu U_Y E_Y(n) U_Y^* J_\mu \mu(Y^\dagger), \\ &\mu^C(J_\mu Y_n^* J_n) = J_\mu E_Y(n) J_\mu \mu(Y). \end{aligned} \tag{2.4.5}$$

Further, by (2.4.5) we have

$$\begin{aligned} (J_\mu U_Y U_Y^* \mu(Y) | \mu(Z_1^\dagger Z_2)) &= \lim_{n\to\infty} (\mu^C(J_\mu Y_n J_\mu) | \mu(Z_1^\dagger Z_2)) \\ &= \lim_{n\to\infty} (J_\mu Y_n J_\mu \mu(Z_1) | \mu(Z_2)) \\ &= (J_\mu \mu(Y Z_1^\dagger) | \mu(Z_2)) \\ &= (\mu(Y^\dagger) | \mu(Z_1^\dagger Z_2)) \end{aligned}$$

for each $Z_1, Z_2 \in \mathcal{D}(\mu)^\dagger \cap \mathcal{D}(\mu)$. Since $\mu((\mathcal{D}(\mu)^\dagger \cap \mathcal{D}(\mu))^2)$ is total in $\mathcal{H}$, it follows that $U_Y U_Y^* \mu(Y) = \mu(Y)$, which implies by (2.4.5) that

$$\lim_{n\to\infty} \mu^C(J_\mu Y_n J_\mu) = \mu(Y^\dagger). \tag{2.4.6}$$

Since $\mu((\mathcal{D}(\mu)^\dagger \cap \mathcal{D}(\mu))^2)$ is total in $\mathcal{H}$, it follows from (2.4.3) $\sim$ (2.4.6) that $\mu^C((\mathcal{D}(\mu^C)^* \cap \mathcal{D}(\mu^C))^2)$ is total in $\mathcal{H}$. Thus the statements (i) $\sim$ (v) are equivalent and $\Delta_{\mu''} = \Delta_{\mu^{CC}} = I$ and $J_{\mu''} = J_{\mu^{CC}} = J_\mu$.

We finally show that $Y^{\dagger *} = \overline{Y}$ for each $Y \in \mathcal{D}(\mu)$. We first show

$$\begin{aligned} &\mathcal{D}(\mu^C)^* \cap \mathcal{D}(\mu^C) = \{J_\mu A^* J_\mu \ ; \ A \in \mathfrak{A}\}, \\ &\mu^C(J_\mu A^* J_\mu) = \mu^{CC}(A), \quad \mu^C(J_\mu A J_\mu) = \mu^{CC}(A^*), \ A \in \mathfrak{A}, \end{aligned} \tag{2.4.7}$$

where

$$\mathfrak{A} \equiv \{A \in \mathcal{D}(\mu^{CC})^* \cap \mathcal{D}(\mu^{CC}) \ ; \ \mu^{CC}(A), \mu^{CC}(A^*) \in \mathcal{D}\}. \tag{2.4.8}$$

In fact, take an arbitrary $A \in \mathfrak{A}$. Then we have $J_\mu A J_\mu, J_\mu A^* J_\mu \in \mathcal{M}'_{\mathrm{w}}$ and by (2.4.5) and (2.4.6)

$$
\begin{aligned}
(J_\mu A J_\mu)\mu(Y) &= \lim_{n\to\infty} J_\mu A \mu^C(J_\mu Y_n J_\mu) \\
&= \lim_{n\to\infty} J_\mu J_\mu Y_n J_\mu \mu^{CC}(A) \\
&= \lim_{n\to\infty} Y_n \mu^{CC}(A^*) \\
&= Y\mu^{CC}(A^*),
\end{aligned}
$$

$$
\begin{aligned}
(J_\mu A^* J_\mu)\mu(Y) &= \lim_{n\to\infty} J_\mu A^* \mu^C(J_\mu Y_n J_\mu) \\
&= Y\mu^{CC}(A)
\end{aligned}
$$

for each $Y \in \mathcal{D}(\mu)^\dagger \cap \mathcal{D}(\mu)$, which implies that $J_\mu A J_\mu \in \mathcal{D}(\mu^C)^* \cap \mathcal{D}(\mu^C)$, $\mu^C(J_\mu A J_\mu) = \mu^{CC}(A^*)$ and $\mu^C(J_\mu A^* J_\mu) = \mu^{CC}(A)$. Conversely, take an arbitrary $K \in \mathcal{D}(\mu^C)^* \cap \mathcal{D}(\mu^C)$ and put $A = J_\mu K^* J_\mu$. Then we have $A, A^* \in (\mathcal{M}'_{\mathrm{w}})'$,

$$
A\mu^C(K_1) = J_\mu K^* J_\mu \mu^C(K_1) = K_1 \mu^C(K) \text{ and } A^*\mu^C(K_1) = K_1\mu^C(K^*)
$$

for each $K_1 \in \mathcal{D}(\mu^C)^* \cap \mathcal{D}(\mu^C)$, which implies that $A \in \mathfrak{A}$, $\mu^{CC}(A) = \mu^C(K)$ and $\mu^{CC}(A^*) = \mu^C(K^*)$. Thus the statement (2.4.7) holds. Take an arbitrary $Y \in \mathcal{D}(\mu)$. By (2.4.7) we have

$$
\pi_0(\mu(Y))\mu^{CC}(A) = \pi'_0(\mu^{CC}(A))\mu(Y) = J_\mu A^* J_\mu \mu(Y) = Y\mu^{CC}(A),
$$

and further by (2.4.4)

$$
\begin{aligned}
\pi_0(J_\mu \mu(Y))\mu^{CC}(A) = J_\mu A^* \mu(Y) &= \lim_{n\to\infty} J_\mu A^* \mu^C(J_\mu Y_n^* J_\mu) \\
&= \lim_{n\to\infty} Y_n^* \mu^{CC}(A) \\
&= Y^\dagger \mu^{CC}(A)
\end{aligned}
$$

for each $A \in \mathfrak{A}$, and further since $\mathfrak{A}$ is a Hilbert algebra in $\mathcal{H}$ by (2.4.7), it follows that $\overline{\pi_0(\mu(Y))} \subset \overline{Y}$ and $\overline{\pi_0(J_\mu \mu(Y))} \subset \overline{Y^\dagger}$, which implies by (2.4.1) that

$$
\overline{\pi_0(\mu(Y))} \subset \overline{Y} \subset Y^{\dagger *} \subset \pi_0(J_\mu \mu(Y))^* = \overline{\pi_0(\mu(Y))}.
$$

Hence we have $Y^{\dagger *} = \overline{Y} = \overline{\pi_0(\mu(Y))}$ for each $Y \in \mathcal{D}(\mu)$. This completes the proof.

By Proposition 2.4.6 we have the following

Corollary 2.4.7. Let $\mathcal{M}$ be a closed O*-algebra on $\mathcal{D}$ in $\mathcal{H}$ such that $\mathcal{M}'_{\mathrm{w}}\mathcal{D} \subset \mathcal{D}$ and $\xi_0 \in \mathcal{D}$. Suppose ξ_0 is a cyclic tracial vector for $\mathcal{M}$. Then the following statements are equivalent:

(i) ξ_0 is standard.

(ii) $\mathcal{M}'_{\mathrm{w}}\xi_0$ is dense in $\mathcal{H}$.

(iii) $J_{\xi_0}(\mathcal{M}'_{\mathrm{w}})'J_{\xi_0} = \mathcal{M}'_{\mathrm{w}}$.

If this is true, then $J_{\xi_0} = J''_{\xi_0}$, $\Delta_{\xi_0} = \Delta''_{\xi_0} = I$ and $\mathcal{M}$ is an integrable O*-algebra on $\mathcal{D}$. Further, $\overline{\mathcal{M}}$ is a $*$-subalgebra of the $*$-algebra $L^\omega(\omega_{\xi_0}) \equiv \bigcap_{1\leq p<\infty} L^p(\omega_{\xi_0})$ equipped with the strong sum, strong scalar multiplication, strong product and adjoint, where $L^p(\omega_{\xi_0})$ is the Segal L^p-space with respect to the vector trace ω_{ξ_0} on $(\mathcal{M}'_{\mathrm{w}})'$.

C. Standard systems for semifinite O*-algebras

We treat with a standard system $(\mathcal{M}, K'\mu, (K'\mu)')$ constracted by a standard tracial generalized vector μ and a non-singular positive self-adjoint operator K', and consider when a standard system $(\mathcal{M}, \lambda, \lambda')$ is unitarily equivalent to such a standard system $(\mathcal{N}, K'\mu, (K'\mu)')$. Let $\mathcal{M}$ be a closed O*-algebra on $\mathcal{D}$ in $\mathcal{H}$ such that $\mathcal{M}'_{\mathrm{w}}\mathcal{D} \subset \mathcal{D}$, μ a standard tracial generalized vector for $\mathcal{M}$ and K' a non-singular positive self-adjoint operator in $\mathcal{H}$ affiliated with $\mathcal{M}'_{\mathrm{w}}$ whose domain $\mathcal{D}(K')$ contains $\mu(\mathcal{D}(\mu))$. Let $K' = \int_0^\infty t dE'(t)$ and $K \equiv J_\mu K' J_\mu = \int_0^\infty t dE(t)$ be the spectral resolutions of K' and K, respectively and let $E'(n) = \int_0^n dE'(t)$ and $E(n) = \int_0^n dE(t)$ for $n \in \mathbb{N}$. We here put

$$\begin{cases} \mathcal{D}(K'\mu) = \mathcal{D}(\mu), \\ (K'\mu)(X) = K'\mu(X), \quad X \in \mathcal{D}(\mu). \end{cases}$$

Then $K'\mu$ is a generalized vector for $\mathcal{M}$. In fact, let $K'_n = K'E'(n)$ and $K_n = KE(n)$, $n \in \mathbb{N}$. Then, for each $A \in \mathcal{M}$ and $X \in \mathcal{D}(\mu)$ we have

$$\lim_{n\to\infty} K'_n\mu(X) = K'\mu(X) \text{ and } \lim_{n\to\infty} AK'_n\mu(X) = K'\mu(AX).$$

Hence we have

$$K'\mu(X) \in \mathcal{D} \text{ and } AK'\mu(X) = K'A\mu(X),$$

which implies that $K'\mu$ is a generalized vector for $\mathcal{M}$. For the standardness of the generalized vector $K'\mu$ we have the following

Proposition 2.4.8. Let $\mathcal{M}$ be a closed O*-algebra on $\mathcal{D}$ in $\mathcal{H}$ such that $\mathcal{M}'_{\mathrm{w}}\mathcal{D} \subset \mathcal{D}$ and μ a standard tracial generalized vector for $\mathcal{M}$. Suppose K' is a non-singular positive self-adjoint operator in $\mathcal{H}$ affiliated with $\mathcal{M}'_{\mathrm{w}}$ such that $\mu(\mathcal{D}(\mu)) \subset \mathcal{D}(K')$, $K'\mu((\mathcal{D}(\mu)^\dagger \cap \mathcal{D}(\mu))^2)$ is total in $\mathcal{H}$ and $K'\mu(\mathcal{D}(\mu)^\dagger \cap \mathcal{D}(\mu))$ is dense in the Hilbert space $\mathcal{D}(K \cdot K'^{-1})$. Then $K'\mu$ is a generalized vector for $\mathcal{M}$ satisfying the following conditions:

(i) $(K'\mu)'((\mathcal{D}((K'\mu)')^* \cap \mathcal{D}((K'\mu)'))^2)$ is total in $\mathcal{H}$.

(ii) $S_{K'\mu} = S_{(K'\mu)''} = J_\mu K \cdot K'^{-1}$.

Further, $(\mathcal{M}, K'\mu, (K'\mu)')$ is a standard system if and only if $K^{it}\mathcal{D} \subset \mathcal{D}$ and $K^{it}YK^{-it}\lceil\mathcal{D} \in \mathcal{D}(\mu)$ for all $Y \in \mathcal{D}(\mu)$ and $t \in \mathbb{R}$.

Proof. Since μ is standard, there exists a net $\{A_\alpha\}$ in $\mathcal{D}(\mu^{cc})^* \cap \mathcal{D}(\mu^{cc})$ which strongly* converges to I. Let $Y \in \mathcal{D}(\mu)^\dagger \cap \mathcal{D}(\mu)$ and let $Y_n = \overline{Y}E_Y(n), n \in \mathbb{N}$. Then it follows from Proposition 2.1.7 that $\{Y_n\} \subset \mathcal{D}(\mu^{cc})^* \cap \mathcal{D}(\mu^{cc}), Y_n \to Y$ strongly*, $\lim_{n\to\infty} \mu^{cc}(Y_n) = \mu(Y)$ and $\lim_{n\to\infty} \mu^{cc}(Y_n^*) = \mu(Y^\dagger)$. Take arbitrary $C \in \mathcal{D}(\mu^c)$, $Y \in \mathcal{D}(\mu)^\dagger \cap \mathcal{D}(\mu)$ and $m, n \in \mathbb{N}$. Then we have

$$\begin{aligned}
E'(n)CE'(m)(K'\mu)(Y) &= \lim_{k\to\infty} E'(n)CE'(m)K'\mu^{cc}(Y_k) \\
&= \lim_{k\to\infty} E'(n)CJ_\mu KE(m)\mu^{cc}(Y_k^*) \\
&= \lim_{k\to\infty} \lim_\alpha E'(n)CJ_\mu A_\alpha KE(m)\mu^{cc}(Y_k^*) \\
&= \lim_{k\to\infty} \lim_\alpha E'(n)CJ_\mu \mu^{cc}(A_\alpha KE(m)Y_k^*) \\
&= \lim_{k\to\infty} \lim_\alpha E'(n)C\mu^{cc}(Y_k KE(m)A_\alpha^*) \\
&= \lim_{k\to\infty} \lim_\alpha Y_k E'(n)KE(m)A_\alpha^* \mu^c(C) \\
&= \lim_{k\to\infty} Y_k E'(n)KE(m)\mu^c(C),
\end{aligned}$$

and so

$$\begin{aligned}
(Y^\dagger\eta|E'(n)KE(m)\mu^c(C)) &= \lim_{k\to\infty} (Y_k^*\eta|E'(n)KE(m)\mu^c(C)) \\
&= \lim_{k\to\infty} (\eta|Y_k E'(n)KE(m)\mu^c(C)) \\
&= (\eta|E'(n)CE'(m)(K'\mu)(Y)),
\end{aligned}$$

which implies $E'(n)KE(m)\mu^c(C) \in \mathcal{D}(Y^{\dagger *}) = \mathcal{D}(\overline{Y})$ and

$$\overline{Y}E'(n)KE(m)\mu^c(C) = E'(n)CE'(m)(K'\mu)(Y).$$

Hence we have

$$\begin{aligned}
&E'(n)CE'(m) \in \mathcal{D}((K'\mu)')^* \cap \mathcal{D}((K'\mu)'), \\
&(K'\mu)'(E'(n)CE'(m)) = E'(n)KE(m)\mu^c(C), \\
&\qquad\qquad {}^\forall C \in \mathcal{D}(\mu^c)^* \cap \mathcal{D}(\mu^c), {}^\forall m, n \in \mathbb{N}. \qquad (2.4.9)
\end{aligned}$$

Similarly, we have

$$\begin{aligned}
&E'(n)CE'(m)K'^{-1} \in \mathcal{D}((K'\mu)')^* \cap \mathcal{D}((K'\mu)'), \\
&(K'\mu)'(E'(n)CE'(m)K'^{-1}) = E'(n)E(m)\mu^c(C), \\
&(K'\mu)'(E'(m)K'^{-1}C^*E'(n)) = E'(m)K'^{-1}E(n)K\mu^c(C) \\
&\qquad\qquad {}^\forall C \in \mathcal{D}(\mu^c)^* \cap \mathcal{D}(\mu^c),\ m, n \in \mathbb{N}. \qquad (2.4.10)
\end{aligned}$$

By (2.4.9) and (2.4.10) we have

$$E'(n)C_1E'(l)C_2E'(m)K'^{-1} \in (\mathcal{D}((K'\mu)')^* \cap \mathcal{D}((K'\mu)'))^2$$

and

$$\begin{aligned}\lim_{m,n\to\infty} & (K'\mu)'(E'(n)C_1E'(l)C_2E'(m)K'^{-1}) \\ &= \lim_{m,n\to\infty} E'(n)C_1E'(l)(K'\mu)'(E'(l)C_2E'(m)K'^{-1}) \\ &= \lim_{m,n\to\infty} E'(n)C_1E'(l)E(m)\mu^c(C_2) \\ &= \mu^c(C_1C_2)\end{aligned}$$

for each $C_1, C_2 \in \mathcal{D}(\mu^c)^* \cap \mathcal{D}(\mu^c)$, which implies since $\mu^c((\mathcal{D}(\mu^c)^* \cap \mathcal{D}(\mu^c))^2)$ is total in $\mathcal{H}$ that $(K'\mu)'(\mathcal{D}((K'\mu)')^* \cap \mathcal{D}((K'\mu)'))^2)$ is total in $\mathcal{H}$. We show $J_\mu K \cdot K'^{-1} = S_{K'\mu} = S_{(K'\mu)''}$. Since

$K'\mu(\mathcal{D}(\mu)^\dagger \cap \mathcal{D}(\mu))$ is densely contained in the Hilbert space $\mathcal{D}(K \cdot K'^{-1})$

and further

$$J_\mu K \cdot K'^{-1}K'\mu(Y) = J_\mu K\mu(Y) = K'\mu(Y^\dagger) = S_{K'\mu}(K'\mu)(Y)$$

for each $Y \in \mathcal{D}(\mu)^\dagger \cap \mathcal{D}(\mu)$, we have $J_\mu K \cdot K'^{-1} \subset S_{K'\mu}$. By Lemma 2.2.2 we generally have $S_{K'\mu} \subset S_{(K'\mu)''}$, and hence $J_\mu K \cdot K'^{-1} \subset S_{K'\mu} \subset S_{(K'\mu)''}$. Conversely we show $S_{(K'\mu)''} \subset J_\mu K \cdot K'^{-1}$. It is shown similarly to (2.4.9) that

$$\begin{aligned}& E'(n)CE'(m) \in \mathcal{D}((K'\mu)')^* \cap \mathcal{D}((K'\mu)'), \\ & (K'\mu)'(E'(n)CE'(m)) = E'(n)KE'(m)\mu^{ccc}(C), \\ & \qquad\qquad {}^\forall C \in \mathcal{D}(\mu^{ccc})^* \cap \mathcal{D}(\mu^{ccc}), \quad {}^\forall m, n \in \mathbb{N}\end{aligned}$$

and clearly

$$\begin{aligned}J_\mu A^* J_\mu \in \mathcal{D}(\mu^{ccc}) \text{ and } \mu^{ccc}(J_\mu A^* J_\mu) = \mu^{cc}(A), \\ {}^\forall A \in \mathcal{D}(\mu^{cc})^* \cap \mathcal{D}(\mu^{cc}),\end{aligned}$$

so that

$$\begin{aligned}& E'(n)J_\mu A^* J_\mu E'(m) \in \mathcal{D}((K'\mu)')^* \cap \mathcal{D}((K'\mu)'), \\ & (K'\mu)'(E'(n)J_\mu A^* J_\mu E'(m)) = KE(m)E'(n)\mu^{cc}(A), \\ & \qquad\qquad {}^\forall A \in \mathcal{D}(\mu^{cc})^* \cap \mathcal{D}(\mu^{cc}).\end{aligned}$$

Hence we have

$$
\begin{aligned}
&(KK'^{-1}K'E'(n)E(m)\mu^{cc}(A)|(K'\mu)''(B)) \\
&= ((K'\mu)'(E'(n)J_\mu A^* J_\mu E'(m))|(K'\mu)''(B)) \\
&= (S_{(K'\mu)''}(K'\mu)''(B)|S^*_{(K'\mu)''}(K'\mu)'(E'(n)J_\mu A^* J_\mu E'(m))) \\
&= (S_{(K'\mu)''}(K'\mu)''(B)|(K'\mu)'(E'(m)J_\mu A J_\mu E'(n))) \\
&= (S_{(K'\mu)''}(K'\mu)''(B)|KE(n)E'(m)\mu^{cc}(A^*)) \\
&= (S_{(K'\mu)''}(K'\mu)''(B)|J_\mu K'E'(n)E(m)\mu^{cc}(A)) \\
&= (K'E'(n)E(m)\mu^{cc}(A)|J_\mu S_{(K'\mu)''}(K'\mu)''(B))
\end{aligned}
$$

for each $A \in \mathcal{D}(\mu^{cc})^* \cap \mathcal{D}(\mu^{cc})$ and $B \in \mathcal{D}((K'\mu)'')^* \cap \mathcal{D}((K'\mu)'')$, and further it follows from (2.4.11) that $\{K'E'(n)E(m)\mu^{cc}(A)\ ;\ A \in \mathcal{D}(\mu^{cc})^* \cap \mathcal{D}(\mu^{cc})$ and $m, n \in \mathbb{N}\}$ is total in the Hilbert space $\mathcal{D}(K \cdot K'^{-1})$, which implies $S_{(K'\mu)''} \subset J_\mu K \cdot K'^{-1}$. Thus we have $S_{K'\mu} = S_{(K'\mu)''} = J_\mu K \cdot K'^{-1}$, and hence

$$J_{K'\mu} = J_{(K'\mu)''} = J_\mu \text{ and } \Delta_{K'\mu} = \Delta_{(K'\mu)''} = K \cdot K'^{-1}. \qquad (2.4.11)$$

It follows from (2.4.12) that $(\mathcal{M}, K'\mu, (K'\mu)')$ is a standard system if and only if $K^{it}\mathcal{D} \subset \mathcal{D}$ and $K^{it}YK^{-it}\lceil\mathcal{D} \in \mathcal{D}(\mu)$ for all $Y \in \mathcal{D}(\mu)$ and $t \in \mathbb{R}$. This completes the proof.

We consider when the condition in Proposition 2.4.8 "$K^{it}\mathcal{D} \subset \mathcal{D}$ and $K^{it}YK^{-it}\lceil\mathcal{D} \in \mathcal{D}(\mu)$ for all $Y \in \mathcal{D}(\mu)$ and $t \in \mathbb{R}$" holds.

Corollary 2.4.9. Let $(\mathcal{M}, \mu, K')$ be given in Proposition 2.4.8. Suppose μ is full and $K^{it}\lceil\mathcal{D} \in \mathcal{M}$ for all $t \in \mathbb{R}$. Then $(\mathcal{M}, K'\mu, (K'\mu)')$ is a standard system.

Proof. Take arbitrary $Y \in \mathcal{D}(\mu)$ and $t \in \mathbb{R}$. Then we have

$$
\begin{aligned}
(K^{it}YK^{-it}\mu^c(C)|\xi) &= (J_\mu K'^{-it}\mu^c(C^*)|Y^\dagger K^{-it}\xi) \\
&= \lim_\alpha (J_\mu C_\alpha K'^{-it}\mu^c(C^*)|Y^\dagger K^{-it}\xi) \\
&= \lim_\alpha (Y\mu^c(CK'^{it}C^*_\alpha)|K^{-it}\xi) \\
&= \lim_\alpha (CK'^{-it}C^*_\alpha\mu(Y)|K^{-it}\xi) \\
&= (CK^{it}K'^{-it}\mu(Y)|\xi)
\end{aligned}
$$

for all $C \in \mathcal{D}(\mu^c)^* \cap \mathcal{D}(\mu^c)$ and $\xi \in \mathcal{D}$, where $\{C_\alpha\}$ is a net in $\mathcal{D}(\mu^c)^* \cap \mathcal{D}(\mu^c)$ which converges strongly* to I. Hence it follows form the fullness of μ that $K^{it}YK^{-it} \in \mathcal{D}(\mu_e) = \mathcal{D}(\mu)$. By Proposition 2.4.8 $(\mathcal{M}, K'\mu, (K'\mu)')$ is a standard system.

We next consider the converse of Proposition 2.4.8:

When is a standard system $(\mathcal{M}, \lambda, \lambda')$ unitarily equivalent to such a standard system $(\mathcal{N}, K'\mu, (K'\mu)')$ in Proposition 2.4.8?

An O*-algebra $\mathcal{M}$ is said to be *semifinite* if $(\mathcal{M}'_{\mathrm{w}})'$ is a semifinite von Neumann algebra.

Proposition 2.4.10. Let $\mathcal{M}$ be a closed semifinite O*-algebra on $\mathcal{D}$ in $\mathcal{H}$ such that $\mathcal{M}'_{\mathrm{w}}\mathcal{D} \subset \mathcal{D}$. Suppose λ is a generalized vector for $\mathcal{M}$ such that

(i) $\lambda((\mathcal{D}(\lambda)^{\dagger} \cap \mathcal{D}(\lambda))^2)$ is total in $\mathcal{H}$;

(ii) $\lambda'((\mathcal{D}(\lambda')^{*} \cap \mathcal{D}(\lambda'))^2)$ is total in $\mathcal{H}$;

(iii) $S_\lambda = S_{\lambda''}$;

(iv) $\overline{Y} \in L^2(\tau'')$ for each $Y \in \mathcal{D}(\lambda)$, where τ'' is a faithful normal semifinite trace on $(\mathcal{M}'_{\mathrm{w}})'$. Then there exist a standard tracial generalized vector μ for a closed O*-algebra $\mathcal{N}$ in $L^2(\tau'')$ and a non-singular positive self-adjoint operator K' in $L^2(\tau'')$ affiliated with $\mathcal{N}'_{\mathrm{w}}$ such that (μ, K') satisfies all of the conditions in Proposition 2.4.8 and λ is unitarily equivalent to the generalized vector $K'\mu$, that is, there exists a unitary operator U of $L^2(\tau'')$ onto $\mathcal{H}$ such that $U^*\mathcal{M}U = \mathcal{N}, U^*\mathcal{D}(\lambda)U = \mathcal{D}(\mu)$ and $\lambda(Y) = U(K'\mu)(U^*YU)$ for each $Y \in \mathcal{D}(\lambda)$.

Proof. By the assumption for λ, $\lambda''(\mathcal{D}(\lambda'')^* \cap \mathcal{D}(\lambda''))$ is an achieved left Hilbert algebra in $\mathcal{H}$ whose left von Neumann algebra equals the semifinite von Neumann algebra $(\mathcal{M}'_{\mathrm{w}})'$, so that the following results have been shown by Takesaki [1]:

(2.4.13) We put $\Pi_0\lambda''(B) = B$, $B \in \mathcal{D}(\lambda'') \cap \mathfrak{N}_{\tau''}$. Then Π_0 is a closable operator of the dense subspace $\lambda''(\mathcal{D}(\lambda'') \cap \mathfrak{N}_{\tau''})$ onto the dense subspace $\mathcal{D}(\lambda'') \cap \mathfrak{N}_{\tau''}$ in $L^2(\tau'')$ whose closure Π is non-singular.

(2.4.14) Let $\Pi = VT'$ be the polar decomposition of Π. Then V is a unitary operator of $\mathcal{H}$ onto $L^2(\tau'')$ and T' is a non-singular positive self-adjoint operator in $\mathcal{H}$ affiliated with $\mathcal{M}'_{\mathrm{w}}$ such that $\Delta_{\lambda''}^{\frac{1}{2}} = T^{-1} \cdot T'$, where $T = J_{\lambda''}T'J_{\lambda''}$.

(2.4.15) Let ρ_0 be the left regular representation of $(\mathcal{M}'_{\mathrm{w}})'$ on $L^2(\tau'')$ defined by $\rho_0(A)B = AB$, $A \in (\mathcal{M}'_{\mathrm{w}})'$, $B \in \mathfrak{N}_{\tau''}$. Then the unitary operator V implements a spatial isomorphism between $(\mathcal{M}'_{\mathrm{w}})'$ and $\rho_0((\mathcal{M}'_{\mathrm{w}})')$ such that $VBV^* = \rho_0(B)$ for each $B \in \mathcal{D}(\lambda'') \cap \mathfrak{N}_{\tau''}$.

(2.4.16) $\lambda''(\mathcal{D}(\lambda'')^* \cap \mathcal{D}(\lambda'') \cap \mathfrak{N}_{\tau''})$ is dense in the Hilbert space $\mathcal{D}(T')$.

Let Λ be the inverse of Π and $\Lambda = UK'$ be the polar decomposition of Λ. Then we have $U = V^*$ and $K' = U^*T'^{-1}U$. It follows from (2.4.14) that U is a unitary operator of $L^2(\tau'')$ onto $\mathcal{H}$ and K' is a non-singular positive self-adjoint operator in $L^2(\tau'')$ affiliated with the von Neumann algebra $\rho_0((\mathcal{M}'_{\mathrm{w}})')'$. We put

$$\mathcal{N} = U^*\mathcal{M}U,$$
$$\mathcal{D}(\mu) = U^*\mathcal{D}(\lambda)U \text{ and } \mu(U^*YU) = \overline{Y} \in L^2(\tau''), \quad Y \in \mathcal{D}(\lambda).$$

Then $\mathcal{N}$ is a closed O*-algebra on $U^*\mathcal{D}$ in $L^2(\tau'')$ such that $\mathcal{N}'_{\mathrm{w}} = U^*\mathcal{M}'_{\mathrm{w}}U$ and $(\mathcal{N}'_{\mathrm{w}})' = U^*(\mathcal{M}'_{\mathrm{w}})'U = \rho_0((\mathcal{M}'_{\mathrm{w}})')$, and by (2.4.15) μ is a tracial generalized vector for $\mathcal{N}$. We show

$$\lambda(Y) \in \mathcal{D}(\Pi) \text{ and } \Pi\lambda(Y) = \mu(U^*YU), Y \in \mathcal{D}(\lambda).$$

In fact, let $X \in \mathcal{M}$ and let $\overline{X} = U_X|\overline{X}|$ be the polar decomposition of $\overline{X}$ and $|\overline{X}| = \int_0^\infty t dE_X(t)$ the spectral resolution of $|\overline{X}|$. Take an arbitrary $Y \in \mathcal{D}(\lambda)$. Then it is shown that $\overline{Y}E_Y(n) \in \mathcal{D}(\lambda'') \cap \mathfrak{N}_{\tau''}$ and $\lambda''(\overline{Y}E_Y(n)) = E_{Y^\dagger}(n)\lambda(Y)$, $n \in \mathbb{N}$. Hence we have

$$\lim_{n\to\infty} \|E_{Y^\dagger}(n)\lambda(Y) - \lambda(Y)\| = 0,$$
$$\lim_{n\to\infty} \|\Pi E_{Y^\dagger}(n)\lambda(Y) - \mu(U^*YU)\| = \lim_{n\to\infty} \tau''((I - E_Y(n))|\overline{Y}|^2) = 0,$$

which implies $\lambda(Y) \in \mathcal{D}(\Pi)$ and $\Pi\lambda(Y) = \mu(U^*YU)$. Hence we have $\mu(\mathcal{D}(\mu)) \subset \mathcal{D}(K')$. Since $K'\mu(U^*YU) = U^*\lambda(Y)$ for each $Y \in \mathcal{D}(\lambda)$ and $\lambda((\mathcal{D}(\lambda)^\dagger \cap \mathcal{D}(\lambda))^2)$ is total in $\mathcal{H}$, it follows that $(K'\mu)((\mathcal{D}(K'\mu)^\dagger \cap \mathcal{D}(K'\mu))^2)$ is total in $L^2(\tau'')$. Further, since we have

$$(K'\mu)(U^*YU) = U^*\lambda(Y),$$
$$(K \cdot K'^{-1})(K'\mu)(U^*YU) = U^*T^{-1} \cdot T'\lambda(Y) = U^*J_\lambda\lambda(Y^\dagger)$$

for each $Y \in \mathcal{D}(\lambda)^\dagger \cap \mathcal{D}(\lambda)$, and further $S_\lambda = S_{\lambda''} = J_\lambda T^{-1} \cdot T'$ by (2.4.14) and $\lambda(\mathcal{D}(\lambda)^\dagger \cap \mathcal{D}(\lambda))$ is dense in the Hilbert space $\mathcal{D}(S_{\lambda''}) = \mathcal{D}(T^{-1} \cdot T')$, it follows that $K'\mu(\mathcal{D}(\mu)^\dagger \cap \mathcal{D}(\mu))$ is dense in the Hilbert space $\mathcal{D}(K \cdot K'^{-1})$. Thus the pair (μ, K') satisfies all of the conditions in Proposition 2.4.8. It is clear that $\mathcal{D}(\mu) = U^*\mathcal{D}(\lambda)U$ and $\lambda(Y) = U(K'\mu)(UYU^*)$ for each $Y \in \mathcal{D}(\lambda)$. This copmletes the proof.

By Propositions 2.4.8, 2.4.10 we have the following

Theorem 2.4.11. Let $\mathcal{M}$ be a closed semifinite O*-algebra on $\mathcal{D}$ in $\mathcal{H}$ such that $\mathcal{M}'_{\rm w}\mathcal{D} \subset \mathcal{D}$, and let λ be a generalized vector for $\mathcal{M}$. The following statements are equivalent:

(i) $(\mathcal{M}, \lambda, \lambda')$ is a standard system such that $\overline{Y} \in L^2(\tau'')$ for each $Y \in \mathcal{D}(\lambda)$, where τ'' is a faithful normal semifinite trace on $(\mathcal{M}'_{\rm w})'_+$.

(ii) There exists a closed O*-algebra $\mathcal{N}$ on $\mathcal{E}$ in $\mathcal{K}$, a standard tracial generalized vector μ for $\mathcal{N}$ and a non-singular positive self-adjoint operator K' in $\mathcal{K}$ affiliated with $\mathcal{N}'_{\rm w}$ such that

(ii)$_1$ $\mu(\mathcal{D}(\mu)) \subset \mathcal{D}(K')$ and $K'\mu((\mathcal{D}(\mu)^\dagger \cap \mathcal{D}(\mu))^2)$ is total in $\mathcal{K}$;

(ii)$_2$ $K'\mu(\mathcal{D}(\mu)^\dagger \cap \mathcal{D}(\mu))$ is dense in the Hilbert space $\mathcal{D}(K \cdot K'^{-1})$, where $K \equiv J_\mu K' J_\mu$;

(ii)$_3$ $K^{it}\mathcal{E} \subset \mathcal{E}$ and $K^{it}YK^{-it}\lceil\mathcal{E} \in \mathcal{D}(\mu)$ for all $Y \in \mathcal{D}(\mu)$ and $t \in \mathbb{R}$;

(ii)$_4$ λ is unitarily equivalent to the generalized vector $K'\mu$, that is, there exists a unitary operator U of $\mathcal{K}$ onto $\mathcal{H}$ such that $U^*\mathcal{M}U = \mathcal{N}$, $U^*\mathcal{D}(\lambda)U = \mathcal{D}(\mu)$ and $\lambda(Y) = U(K'\mu)(U^*YU)$ for each $Y \in \mathcal{D}(\lambda)$.

Corollary 2.4.12. Let $\mathcal{M}$ be an integrable O*-algebra on $\mathcal{D}$ in $\mathcal{H}$ with a standard tracial vector ξ_0, and let $\xi \in \mathcal{D}$. Then ξ is a standard vector for $\mathcal{M}$ if and only if there exists a non-singular positive self-adjoint operator K' in $\mathcal{H}$ affiliated with $\mathcal{M}'_{\mathrm{w}}$ such that (a) $\xi_0 \in \mathcal{D}(K')$ and $K'\xi_0 \in \mathcal{D}$; (b) $K'\mathcal{M}\xi_0$ is dense in the Hilbert space $\mathcal{D}(K \cdot K'^{-1})$, where $K \equiv J_{\xi_0}K'J_{\xi_0}$; (c) $K^{it}\lceil\mathcal{D} \in \mathcal{M}$ for each $t \in \mathbb{R}$; (d) ξ is unitarily equivalent to $K'\xi_0$.

D. Standard generalized vectors in the Hilbert space of Hilbert-Schmidt operators

For physical applications, it is customary to study the Hilbert space $\mathcal{H}\otimes\overline{\mathcal{H}}$ of all Hilbert-Schmidt operators on $\mathcal{H}$, together with the natural representation π of an O*-algebra $\mathcal{M}$ on $\mathcal{D}$ in $\mathcal{H}$ on $\mathcal{H}\otimes\overline{\mathcal{H}}$. In this section we show how every positive Hilbert-Schmidt operator Ω determines a generalized vector λ_Ω for $\pi(\mathcal{M})$, and we investigate under which conditions λ_Ω is standard or modular. Further, we prove that a positive self-adjoint unbounded operator Ω also defines a generalized vector λ_Ω, and we ask the above questions as before. We denote by $\mathcal{H}\otimes\overline{\mathcal{H}}$ the Hilbert space of all Hilbert-Schmidt operators on a separable Hilbert space $\mathcal{H}$ with the inner product $< S|T > \equiv \operatorname{tr} T^*S$, $S,T \in \mathcal{H}\otimes\overline{\mathcal{H}}$, and the norm $\|T\|_2 \equiv < T|T >^{1/2}$, $T \in \mathcal{H}\otimes\overline{\mathcal{H}}$. We define some operators on $\mathcal{H}\otimes\overline{\mathcal{H}}$: Let H and K be closed operators in $\mathcal{H}$ and put

$$\begin{cases} \mathcal{D}(\pi''(H)) = \{T \in \mathcal{H}\otimes\overline{\mathcal{H}}; T\mathcal{H} \subset \mathcal{D}(H) \text{ and } HT \in \mathcal{H}\otimes\overline{\mathcal{H}}\} \\ \pi''(H)T = HT, \quad T \in \mathcal{D}(\pi''(H)), \end{cases}$$

$$\begin{cases} \mathcal{D}(\pi'(K)) = \{T \in \mathcal{H}\otimes\overline{\mathcal{H}}; TK \text{ is closable and } \overline{TK} \in \mathcal{H}\otimes\overline{\mathcal{H}}\} \\ \pi'(K)T = \overline{TK}, \quad T \in \mathcal{D}(\pi'(H)). \end{cases}$$

Then we have the following

Lemma 2.4.13. (1) π'' is a $*$-homomorphism of $\mathcal{B}(\mathcal{H})$ onto a von Neumann algebra on $\mathcal{H}\otimes\overline{\mathcal{H}}$ and π' is an anti $*$-homomorphism of $\mathcal{B}(\mathcal{H})$ onto the commutant of the von Neumann algebra $\pi''(\mathcal{B}(\mathcal{H}))$, and they have the relation: $\pi''(\mathcal{B}(\mathcal{H})) = J\pi'(\mathcal{B}(\mathcal{H}))J$, where J denotes the isometry on $\mathcal{H}\otimes\overline{\mathcal{H}}$ defined by $JT = T^*$, $T \in \mathcal{H}\otimes\overline{\mathcal{H}}$.

(2) $\pi''(H)$ and $\pi'(H)$ are closed operators in $\mathcal{H}\otimes\overline{\mathcal{H}}$ affiliated with $\pi''(\mathcal{B}(\mathcal{H}))$ and $\pi'(\mathcal{B}(\mathcal{H}))$, respectively, and $J\mathcal{D}(\pi''(H)) = \mathcal{D}(\pi'(H))$ and $\pi''(H) = J\pi'(H)J$.

(3) Suppose H and K are (positive) self-adjoint operators in $\mathcal{H}$. Then, $\pi''(H)$ and $\pi'(K)$ are (positive) self-adjoint operators in $\mathcal{H}\otimes\overline{\mathcal{H}}$, and further $\pi''(H)\pi'(K)$ and $\pi'(K)\pi''(H)$ are (positive) essentially self-adjiont operators in $\mathcal{H}\otimes\overline{\mathcal{H}}$ and $\overline{\pi''(H)\pi'(K)} = \overline{\pi'(K)\pi''(H)}$.

Let $\mathcal{M}$ be a closed O*-algebra on $\mathcal{D}$ in $\mathcal{H}$ such that $\mathcal{M}'_{\mathrm{w}}\mathcal{D} \subset \mathcal{D}$. We put

$$\begin{cases} \mathcal{D}(\pi) = \bigcap_{X \in \mathcal{M}} \mathcal{D}(\pi''(\overline{X})), \\ \pi(X) = \pi''(\overline{X}) \lceil_{\mathcal{D}(\pi)}. \end{cases}$$

Then we have the following

Lemma 2.4.14. Let $\mathcal{M}$ be a closed (resp. self-adjoint) O*-algebra on $\mathcal{D}$ in $\mathcal{H}$ such that $\mathcal{M}'_{\rm w}\mathcal{D} \subset \mathcal{D}$. Then π is a closed (resp. self-adjoint) $*$-representation of $\mathcal{M}$ on $\mathcal{H} \otimes \overline{\mathcal{H}}$ with the domain

$$\mathcal{D}(\pi) = \mathfrak{S}_2(\mathcal{M}) \equiv \{T \in \mathcal{H} \otimes \overline{\mathcal{H}}; T\mathcal{H} \subset \mathcal{D} \text{ and } XT \in \mathcal{H} \otimes \overline{\mathcal{H}} \text{ for all } X \in \mathcal{M}\}.$$

Suppose $\mathcal{M}'_{\rm w} = \mathbb{C}I$. Then, $\pi(\mathcal{M})'_{\rm w} = \pi'(\mathcal{B}(\mathcal{H}))$ and $(\pi(\mathcal{M})'_{\rm w})' = \pi''(\mathcal{B}(\mathcal{H}))$.

Proof. It is not difficult to show the first part. We show the last part. Suppose $\mathcal{M}'_{\rm w} = \mathbb{C}I$. It is clear that $\pi'(\mathcal{B}(\mathcal{H})) \subset \pi(\mathcal{M})'_{\rm w}$. Conversely take an arbitrary $\delta \in \pi(\mathcal{M})'_{\rm w}$. Let ξ and η be any elements of $\mathcal{D}$. Since the continous sesquilinear form on $\mathcal{H} \times \mathcal{H}$ is defined by

$$(x, y) \in \mathcal{H} \times \mathcal{H} \longrightarrow < \delta(x \otimes \overline{\eta}) | y \otimes \overline{\xi} >,$$

it follows from the Riesz theorem that there exists a bounded linear operator $\Gamma(\xi, \eta)$ on $\mathcal{H}$ such that

$$< \delta(x \otimes \overline{\eta}) | y \otimes \overline{\xi} > = (\Gamma(\xi, \eta)x | y)$$

for each $x, y \in \mathcal{H}$. Further, we have

$$\begin{aligned} (\Gamma(\xi, \eta)X\zeta_1 | \zeta_2) &= < \delta\pi(X)(\zeta_1 \otimes \overline{\eta}) | \zeta_2 \otimes \overline{\xi} > \\ &= < \delta(\zeta_1 \otimes \overline{\eta}) | X^\dagger \zeta_2 \otimes \overline{\xi} > \\ &= (\Gamma(\xi, \eta)\zeta_1 | X^\dagger \zeta_2) \end{aligned}$$

for each $X \in \mathcal{M}$ and $\zeta_1, \zeta_2 \in \mathcal{D}$. Hence it follows from $\mathcal{M}'_{\rm w} = \mathbb{C}I$ that $\Gamma(\xi, \eta) = \alpha(\xi, \eta)I$ for some $\alpha(\xi, \eta) \in \mathbb{C}$. Further, since α can be extended to a continous sesquilinear form on $\mathcal{H} \times \mathcal{H}$, there exists an element A of $\mathcal{B}(\mathcal{H})$ such that $\alpha(\xi, \eta) = (A\xi | \eta)$ for each $\xi, \eta \in \mathcal{D}$, which implies

$$\begin{aligned} < \delta(x \otimes \overline{\eta}) | y \otimes \overline{\xi} > &= (A\xi | \eta)(x | y) \\ &= < \pi'(A)(x \otimes \overline{\eta}) | y \otimes \overline{\xi} > \end{aligned}$$

for each $x, y \in \mathcal{H}$ and $\xi, \eta \in \mathcal{D}$. Hence we have $\delta = \pi'(A) \in \pi'(\mathcal{B}(\mathcal{H}))$. Thus we have $\pi'(\mathcal{B}(\mathcal{H})) = \pi(\mathcal{M})'_{\rm w}$. This completes the proof.

Thoughout the rest of this section let $\mathcal{M}$ be a closed O*-algebra on $\mathcal{D}$ in $\mathcal{H}$ such that $\mathcal{M}'_{\rm w} = \mathbb{C}I$. We now show that every positive Hilbert-Schmidt operator Ω on $\mathcal{H}$ determines a generalized vector λ_Ω for $\pi(\mathcal{M})$. Indeed, for $\Omega \geq 0 \in \mathcal{H} \otimes \overline{\mathcal{H}}$ let us put

$$\begin{cases} \mathcal{D}(\lambda_\Omega) = \{\pi(X); X \in \mathcal{M}, \Omega \in \mathcal{D}(\pi(X^\dagger)^*) \text{ and } \pi(X^\dagger)^*\Omega \in \mathfrak{S}_2(\mathcal{M})\}, \\ \lambda_\Omega(\pi(X)) = \pi(X^\dagger)^*\Omega, \quad \pi(X) \in \mathcal{D}(\lambda_\Omega). \end{cases}$$

Then λ_Ω is a generalized vector for $\pi(\mathcal{M})$ and

$$\begin{cases} \mathcal{D}(\lambda_\Omega) = \{\pi(X); X \in \mathcal{M}, \Omega\mathcal{H} \subset \mathcal{D}(X^{\dagger *}) \text{ and } X^{\dagger *}\Omega \in \mathfrak{S}_2(\mathcal{M})\}, \\ \lambda_\Omega(\pi(X)) = X^{\dagger *}\Omega, \quad \pi(X) \in \mathcal{D}(\lambda_\Omega). \end{cases} \tag{2.4.17}$$

We search sufficient conditions for λ_Ω to be a standard or a modular generalized vector for $\pi(\mathcal{M})$.

For the condition $(\mathrm{S})_2$ in Definition 2.2.7 we have the following

Lemma 2.4.15. Let $\Omega \geq 0 \in \mathcal{H} \otimes \overline{\mathcal{H}}$ such that $\Omega\mathcal{H}$ is dense in $\mathcal{H}$. Suppose there exists an orthonormal basis $\{\xi_n\}$ in $\mathcal{H}$ such that $\{\xi_n\} \subset \mathcal{D}$ and $\xi_n \otimes \overline{\xi_m} \in \mathcal{M}$ for $n, m \in \mathbb{N}$. Then λ_Ω is a cyclic generalized vector for $\pi(\mathcal{M})$ such that $\lambda_\Omega((\mathcal{D}(\lambda_\Omega)^\dagger \cap \mathcal{D}(\lambda_\Omega))^2)$ is total in $\mathcal{H} \otimes \overline{\mathcal{H}}$. Furthermore, if $\{\xi_n\}$ is total in $\mathcal{D}[t_\mathcal{M}]$, then λ_Ω is strongly cyclic.

Proof. We put

$$\mathcal{E} = \{\sum_{k,l} \alpha_k\xi_k \otimes \overline{\beta_l\xi_l}; \alpha_k, \beta_l \in \mathbb{C}\}.$$

Then it is clear that $\mathcal{E} \subset (\mathcal{D}(\lambda_\Omega)^\dagger \cap \mathcal{D}(\lambda_\Omega))^2$ and

$$\lambda_\Omega(\sum_{k,l} \alpha_k\xi_k \otimes \overline{\beta_l\xi_l}) = \sum_{k,l} \alpha_k\xi_k \otimes \overline{\beta_l\Omega\xi_l},$$

and since $\{\xi_n\}$ is an ONB in $\mathcal{H}$ and $\Omega\mathcal{H}$ is dense in $\mathcal{H}$, it follows that $\lambda_\Omega(\mathcal{E})$ is dense in $\{x \otimes \overline{y}; x, y \in \mathcal{H}\}$, which implies that $\lambda_\Omega((\mathcal{D}(\lambda_\Omega)^\dagger \cap \mathcal{D}(\lambda_\Omega))^2)$ is total in $\mathcal{H} \otimes \overline{\mathcal{H}}$. When $\{\xi_n\}$ is total in $\mathcal{D}[t_\mathcal{M}]$, it is similarly shown that λ_Ω is strongly cyclic.

For the regularity of λ_Ω we have the following

Lemma 2.4.16. Suppose $\Omega \geq 0 \in \mathcal{H} \otimes \overline{\mathcal{H}}$ such that $\lambda_\Omega((\mathcal{D}(\lambda_\Omega)^\dagger \cap \mathcal{D}(\lambda_\Omega))^2)$ is total in $\mathcal{H} \otimes \overline{\mathcal{H}}$. Then the following statements hold:

(1) λ_Ω is regular if and only if there exists a net $\{K_\alpha\}$ in $\mathcal{B}(\mathcal{H})$ such that $0 \leq K_\alpha \leq I$, $K_\alpha \to I$ strongly and $\Omega K_\alpha \in \mathfrak{S}_2(\mathcal{M})$ for every α.

(2) λ_Ω is strongly regular if and only if there exists a net $\{K_\alpha\}$ in $\mathcal{B}(\mathcal{H})$ such that $0 \leq K_\alpha \leq I$, $K_\alpha \uparrow I$ strongly, $\Omega K_\alpha \in \mathfrak{S}_2(\mathcal{M})$ for every α and $K_\alpha K_\beta = K_\beta K_\alpha$ for every α, β.

(3) Suppose that $\Omega\mathcal{E} \subset \mathcal{D}$ for some dense subspace $\mathcal{E}$ of $\mathcal{H}$. Then λ_Ω is strongly regular.

Proof. (1) Suppose λ_Ω is regular. Then there exists a net $\{\pi'(K_\alpha)\}$ in $\pi'(\mathcal{B}(\mathcal{H}))$ such that $O \leq \pi'(K_\alpha) \leq I$, $\pi'(K_\alpha) \to I$ strongly and $\pi'(K_\alpha)\lambda_\Omega(\pi(X)) = \pi(X^\dagger)^*\lambda_\Omega^C(\pi'(K_\alpha))$ for all $X \in \mathcal{D}(\lambda_\Omega)$. It is clear that $O \leq K_\alpha \leq I$ and $K_\alpha \to I$ strongly. Since

$$
\begin{aligned}
< \lambda_\Omega^C(\pi'(K_\alpha))|\lambda_\Omega(\pi(X^\dagger Y)) > &= < \pi(X)\lambda_\Omega^C(\pi'(K_\alpha))|\lambda_\Omega(\pi(Y)) > \\
&= < \pi'(K_\alpha)\pi(X^\dagger)^*\Omega|\lambda_\Omega(\pi(Y)) > \\
&= < \pi(X^\dagger)^*\Omega|\pi(Y)\lambda_\Omega^C(\pi'(K_\alpha)) > \\
&= < \Omega|\pi(X^\dagger Y)\lambda_\Omega^C(\pi'(K_\alpha)) > \\
&= < \Omega|\pi'(K_\alpha)\lambda_\Omega(\pi(X^\dagger Y)) > \\
&= < \pi'(K_\alpha)\Omega|\lambda_\Omega(\pi(X^\dagger Y)) >
\end{aligned}
$$

for all $X, Y \in \mathcal{D}(\lambda_\Omega)^\dagger \cap \mathcal{D}(\lambda_\Omega)$ and $\lambda_\Omega((\mathcal{D}(\lambda_\Omega)^\dagger \cap \mathcal{D}(\lambda_\Omega))^2)$ is total in $\mathcal{H}\otimes\overline{\mathcal{H}}$, it follows that $\Omega K_\alpha = \pi'(K_\alpha)\Omega = \lambda_\Omega^C(\pi'(K_\alpha)) \in \mathfrak{S}_2(\mathcal{M})$. The converse is trivial.

(2) This is shown in the same way as (1).

(3) Since $\mathcal{H}$ is a separable Hilbert space and $\mathcal{E}$ is dense in $\mathcal{H}$, there exists an ONB $\{\xi_n\}$ in $\mathcal{H}$ contained in $\mathcal{E}$. Since $\Omega\mathcal{E} \subset \mathcal{D}$, the sequence $\{\sum_{k=1}^{n} \xi_k \otimes \overline{\xi_k}; n \in \mathbb{N}\}$ satisfies the conditions in (2). Hence λ_Ω is strongly regular.

Lemma 2.4.17. Let $\Omega \geq 0 \in \mathcal{H} \otimes \overline{\mathcal{H}}$ such that Ω^{-1} is densely defined. Suppose $\lambda_\Omega((\mathcal{D}(\lambda_\Omega)^\dagger \cap \mathcal{D}(\lambda_\Omega))^2)$ is total in $\mathcal{H} \otimes \overline{\mathcal{H}}$, and λ_Ω is regular. Then $\lambda_\Omega^C((\mathcal{D}(\lambda_\Omega^C)^* \cap \mathcal{D}(\lambda_\Omega^C))^2)$ is total in $\mathcal{H} \otimes \overline{\mathcal{H}}$ and $\lambda_\Omega^{CC}(\mathcal{D}(\lambda_\Omega^{CC})^* \cap \mathcal{D}(\lambda_\Omega^{CC}))$ is an achieved left Hilbert algebra in $\mathcal{H}\otimes\overline{\mathcal{H}}$, which equals $\pi''(\mathcal{B}(\mathcal{H}))\Omega$. Its modular conjugation operator $J_{\lambda_\Omega^{CC}}$ coincides with the anti-isometry $J : T \to T^*$, $T \in \mathcal{H} \otimes \overline{\mathcal{H}}$ and its modular operator $\Delta_{\lambda_\Omega^{CC}}$ coincides with the positive self-adjoint operator $\pi'(\Omega^{-2})\pi''(\Omega^2)$.

Proof. By Lemma 2.4.16 there exists a net $\{K_\alpha\}$ in $\mathcal{B}(\mathcal{H})$ such that $O \leq K_\alpha \leq I$, $K_\alpha \to I$ strongly and $\Omega K_\alpha \in \mathfrak{S}_2(\mathcal{M})$ for each α. Then we have

$$\pi'(K_\alpha A K_\beta) \in \mathcal{D}(\lambda_\Omega^C)^* \cap \mathcal{D}(\lambda_\Omega^C). \tag{2.4.18}$$

Since

$$
\begin{aligned}
&\|\pi'(K_\alpha A K_\beta)\Omega - \pi'(A)\Omega\|_2 \\
&\leq \|\pi'(K_\alpha)\pi'(A)\pi'(K_\beta)\Omega - \pi'(K_\alpha)\pi'(A)\Omega\|_2 \\
&\qquad + \|\pi'(K_\alpha)\pi'(A)\Omega - \pi'(A)\Omega\|_2 \\
&\leq \|\pi'(A)(\pi'(K_\beta) - I)\Omega\|_2 + \|(\pi'(K_\alpha) - I)\pi'(A)\Omega\|_2
\end{aligned}
$$

for all α, β and $A \in \mathcal{B}(\mathcal{H})$, we have

$$\lim_{\alpha,\beta}\|\pi'(K_\alpha AK_\beta)\Omega-\pi'(A)\Omega\|_2=0,$$
$$\lim_{\alpha,\beta}\|\pi'(K_\beta A^*K_\alpha)\Omega-\pi'(A^*)\Omega\|_2=0 \tag{2.4.19}$$

for all $A\in\mathcal{B}(\mathcal{H})$. Since $\pi'(\mathcal{B}(\mathcal{H}))\Omega$ is dense in $\mathcal{H}\otimes\overline{\mathcal{H}}$, it follows from (2.4.18) and (2.4.19) that $\lambda_\Omega^{\rm C}(\mathcal{D}(\lambda_\Omega^{\rm C})^*\cap\mathcal{D}(\lambda_\Omega^{\rm C}))$ is dense in $\mathcal{H}\otimes\overline{\mathcal{H}}$. Furthermore, since $\pi'(K_\gamma)\pi'(K_\alpha AK_\beta)\in(\mathcal{D}(\lambda_\Omega^{\rm C})^*\cap\mathcal{D}(\lambda_\Omega^{\rm C}))^2$ for every α,β,γ and $A\in\mathcal{B}(\mathcal{H})$, it follows that $\lambda_\Omega^{\rm C}((\mathcal{D}(\lambda_\Omega^{\rm C})^*\cap\mathcal{D}(\lambda_\Omega^{\rm C}))^2)$ is total in $\mathcal{H}\otimes\overline{\mathcal{H}}$. By (2.4.18) we have

$$\{\pi'(K_\alpha AK_\beta)\Omega;A\in\mathcal{B}(\mathcal{H}),\alpha,\beta\}\subset\lambda_\Omega^{\rm C}(\mathcal{D}(\lambda_\Omega^{\rm C})^*\cap\mathcal{D}(\lambda_\Omega^{\rm C}))\subset\pi'(\mathcal{B}(\mathcal{H}))\Omega,$$

and so it follows from (2.4.19) that $\lambda_\Omega^{\rm CC}(\mathcal{D}(\lambda_\Omega^{\rm CC})^*\cap\mathcal{D}(\lambda_\Omega^{\rm CC}))$ is an achieved left Hilbert algebra in $\mathcal{H}\otimes\overline{\mathcal{H}}$ and it equals the achieved left Hilbert algebra $\pi''(\mathcal{B}(\mathcal{H}))\Omega$. For any $T\in\mathcal{H}\otimes\overline{\mathcal{H}}$ we have

$$J\pi'(\Omega^{-1})\pi''(\Omega)\pi''(T)\Omega=J\pi'(\Omega)T=S_{\lambda_\Omega^{\rm CC}}\pi''(T)\Omega. \tag{2.4.20}$$

Let $\{\xi_n\}$ be the ONB in $\mathcal{H}$ consisting of eigenvectors of non-zero eigenvalues $\{\omega_n\}$ of Ω. We now put $P_n=\sum_{k=1}^{n}\xi_k\otimes\overline{\xi_k}$, $n\in\mathbb{N}$. Since

$$\begin{aligned}\|\pi''(P_nAP_n)\Omega-\pi''(A)\Omega\|_2&\\ \le\|\pi''(P_nAP_n)\Omega-\pi''(P_nA)\Omega\|_2&+\|\pi''(P_nA)\Omega-\pi''(A)\Omega\|_2\\ \le\|A\|\|P_n\Omega-\Omega\|_2&+\|\Omega A^*P_n-\Omega A^*\|_2\\ =\|A\|\left(\sum_{k=n+1}^{\infty}\|\Omega\xi_k\|^2\right)^{1/2}&+\left(\sum_{k=n+1}^{\infty}\|\Omega A^*\xi_k\|^2\right)^{1/2}\end{aligned}$$

for each $A\in\mathcal{B}(\mathcal{H})$, we have

$$\lim_{n\to\infty}\pi''(P_nAP_n)\Omega=\pi''(A)\Omega \tag{2.4.21}$$

for each $A\in\mathcal{B}(\mathcal{H})$. Since $P_nAP_n\in\mathcal{H}\otimes\overline{\mathcal{H}}$ for each $A\in\mathcal{B}(\mathcal{H})$, it follows that $\pi''(\mathcal{H}\otimes\overline{\mathcal{H}})\Omega$ is dense in the Hilbert space $\mathcal{D}(S_{\lambda_\Omega^{\rm CC}})$. By (2.4.20) we have

$$S_{\lambda_\Omega^{\rm CC}}\subset J\pi'(\Omega^{-1})\pi''(\Omega). \tag{2.4.22}$$

Since $\pi'(\Omega^{-1})\pi''(\Omega)$ is a positive self-adjoint operator in $\mathcal{H}\otimes\overline{\mathcal{H}}$ by Lemma 2.4.13, we have

$$\begin{aligned}<\Omega|\pi''(A)J\pi''(A)J\Omega>&=<S_{\lambda_\Omega^{\rm CC}}\pi''(A)\Omega|J\pi''(A)\Omega>\\&=<J\pi'(\Omega^{-1})\pi''(\Omega)\pi''(A)\Omega|J\pi''(A)\Omega>\\&=<\pi''(A)\Omega|\pi'(\Omega^{-1})\pi''(\Omega)\pi''(A)\Omega>\\&\ge 0\end{aligned}$$

for each $A \in \mathcal{B}(\mathcal{H})$. It hence follows from Theorem 1 in Araki [3] that $J = J_{\lambda_\Omega^{CC}}$. By (2.4.22) we have $\Delta^{\frac{1}{2}}_{\lambda_\Omega^{CC}} \subset \pi'(\Omega^{-1})\pi''(\Omega)$. By the maximality of self-adjoint operators, we have $\Delta^{\frac{1}{2}}_{\lambda_\Omega^{CC}} = \pi'(\Omega^{-1})\pi''(\Omega)$. This completes the proof.

Theorem 2.4.18. Let $\mathcal{M}$ be a closed O*-algebra on $\mathcal{D}$ in $\mathcal{H}$ such that $\mathcal{M}'_{\mathrm{w}} = \mathbb{C}I$ and let $\Omega \geq 0 \in \mathcal{H} \otimes \overline{\mathcal{H}}$. Suppose
(i) Ω^{-1} is densely defined;
(ii) $\lambda_\Omega((\mathcal{D}(\lambda_\Omega)^\dagger \cap \mathcal{D}(\lambda_\Omega))^2)$ is total in $\mathcal{H} \otimes \overline{\mathcal{H}}$;
(iii) λ_Ω is regular;
(iv) $\Omega^{it}\mathcal{D} \subset \mathcal{D}$ and $\Omega^{it}\mathcal{M}\Omega^{-it} = \mathcal{M}, \quad {}^\forall t \in \mathbb{R}$.
Then λ_Ω is a standard generalized vector for $\pi(\mathcal{M})$ such that $J_{\lambda_\Omega} = J$ and $\Delta_{\lambda_\Omega} = \pi'(\Omega^{-2})\pi''(\Omega^2)$.

Proof. By (iv) we have

$$X\Omega^{it}T = \Omega^{it}(\Omega^{-it}X\Omega^{it})T \in \mathcal{H} \otimes \overline{\mathcal{H}}$$

for all $X \in \mathcal{M}, T \in \mathfrak{S}_2(\mathcal{M})$ and $t \in \mathbb{R}$. Hence it follows that $\pi''(\Omega^{it})\mathfrak{S}_2(\mathcal{M}) \subset \mathfrak{S}_2(\mathcal{M})$ for all $t \in \mathbb{R}$, which implies by Lemma 2.4.17 that

$$\Delta^{it}_{\lambda_\Omega^{CC}}\mathfrak{S}_2(\mathcal{M}) = \pi'(\Omega^{-2it})\pi''(\Omega^{2it})\mathfrak{S}_2(\mathcal{M}) \subset \mathfrak{S}_2(\mathcal{M}),$$
$$\Delta^{it}_{\lambda_\Omega^{CC}}\pi(\mathcal{M})\Delta^{-it}_{\lambda_\Omega^{CC}} = \pi(\Omega^{2it}\mathcal{M}\Omega^{-2it}) = \pi(\mathcal{M})$$

for all $t \in \mathbb{R}$. Further, for any $\pi(X) \in \mathcal{D}(\lambda_\Omega)^\dagger \cap \mathcal{D}(\lambda_\Omega)$ and $t \in \mathbb{R}$ we have

$$\Omega\mathcal{H} \subset \mathcal{D}(\Omega^{2it}X^{\dagger *}\Omega^{-2it}) \text{ and } \Omega^{2it}X^{\dagger *}\Omega^{-2it}\Omega \in \mathfrak{S}_2(\mathcal{M}).$$

Hence it follows from (2.4.17) that $\Delta^{it}_{\lambda_\Omega^{CC}}(\mathcal{D}(\lambda_\Omega)^\dagger \cap \mathcal{D}(\lambda_\Omega))\Delta^{-it}_{\lambda_\Omega^{CC}} = \mathcal{D}(\lambda_\Omega)^\dagger \cap \mathcal{D}(\lambda_\Omega)$ for all $t \in \mathbb{R}$. Therefore λ_Ω is standard, and by Lemma 2.4.17 $J_{\lambda_\Omega} = J$ and $\Delta_{\lambda_\Omega} = \pi'(\Omega^{-2})\pi''(\Omega^2)$. This completes the proof.

Corollary 2.4.19. Let H be a positive self-adjoint operator in $\mathcal{H}$, $\mathcal{D} = \bigcap_{n=1}^{\infty} \mathcal{D}(H^n)$ and $\Omega \geq 0 \in \mathcal{H} \otimes \overline{\mathcal{H}}$. Suppose Ω^{-1} is densely defined and $\Omega H \subset H\Omega$. Then λ_Ω is a standard generalized vector for $\pi(\mathcal{L}^\dagger(\mathcal{D}))$.

Proof. Let us take an ONB $\{\xi_n\}$ in $\mathcal{H}$ contained in $\mathcal{D}$. Since $\xi_n \otimes \overline{\xi_m} \in \mathcal{L}^\dagger(\mathcal{D})$ for $n, m \in \mathbb{N}$, it follows from Lemma 2.4.15 that $\lambda_\Omega((\mathcal{D}(\lambda_\Omega)^\dagger \cap \mathcal{D}(\lambda_\Omega))^2)$ is total in $\mathcal{H} \otimes \overline{\mathcal{H}}$. Since $\Omega H \subset H\Omega$, it follows that $\Omega\lceil_\mathcal{D}$, $\Omega^{it}\lceil_\mathcal{D} \in \mathcal{L}^\dagger(\mathcal{D})$ for all $t \in \mathbb{R}$, so that λ_Ω is standard by Lemma 2.4.17 and Theorem 2.4.18.

We now look for sufficient conditions for λ_Ω to be a modular generalized vector.

Theorem 2.4.20. Let $\mathcal{M}$ be a closed O*-algebra on $\mathcal{D}$ in $\mathcal{H}$ such that $\mathcal{M}'_{\rm w} = \mathbb{C}I$ and let $\Omega \geq 0 \in \mathcal{H} \otimes \overline{\mathcal{H}}$. Suppose
(i) Ω^{-1} is densely defined;
(ii) $\lambda_\Omega((\mathcal{D}(\lambda_\Omega)^\dagger \cap \mathcal{D}(\lambda_\Omega))^2)$ is total in $\mathcal{H} \otimes \overline{\mathcal{H}}$;
(iii) there exists a dense subspace $\mathcal{E}$ of $\mathcal{D}[t_\mathcal{M}]$ such that
(iii)$_1$ $\mathcal{M}\mathcal{E} \subset \mathcal{E}$.
(iii)$_2$ $\Omega\mathcal{E} \subset \mathcal{E}$.
(iii)$_3$ $\Omega^{it}\mathcal{E} \subset \mathcal{E}$ for each $t \in \mathbb{R}$.
Then λ_Ω is a modular generalized vector for $\pi(\mathcal{M})$.

Proof. It follows from (ii), (iii)$_2$ and Lemma 2.4.16 that λ_Ω is regular, which implies by Lemma 2.4.17 that $\lambda_\Omega^{cc}(\mathcal{D}(\lambda_\Omega^{cc})^* \cap \mathcal{D}(\lambda_\Omega^{cc}))$ is an achieved left Hilbert algebra in $\mathcal{H}\otimes\overline{\mathcal{H}}$, and it equals $\pi''(\mathcal{B}(\mathcal{H}))\Omega$ and $\Delta_{\lambda_\Omega^{cc}} = \pi'(\Omega^{-2})\pi''(\Omega^2)$. We denote by $\mathcal{K}$ the linear span of $\{\xi \otimes \overline{y}; \xi \in \mathcal{E}, y \in \mathcal{H}\}$. Since $\mathcal{E}$ is dense in $\mathcal{D}[t_\mathcal{M}]$, it follows that $\mathcal{K}$ is dense in $\mathfrak{S}_2(\mathcal{M})[t_{\pi(\mathcal{M})}]$. We next show that

$$\{\lambda_\Omega^c(K); K \in \mathcal{D}(\lambda_\Omega^c)^* \cap \mathcal{D}(\lambda_\Omega^c) \text{ s.t. } \lambda_\Omega^c(K), \lambda_\Omega^c(K^*) \in \mathcal{K}\}^2$$
$$\text{is total in the Hilbert space } \mathcal{D}(S^*_{\lambda_\Omega^{cc}}). \quad (2.4.23)$$

In fact, let us take an ONB $\{\xi_n\}$ in $\mathcal{H}$ contained in $\mathcal{E}$ and put $P_n = \sum_{k=1}^{n} \xi_k \otimes \overline{\xi_k}$, $n \in \mathbb{N}$. Then we have

$$\pi'(P_n A P_n) \in (\mathcal{D}(\lambda_\Omega^c)^* \cap \mathcal{D}(\lambda_\Omega^c))^2,$$
$$\lambda_\Omega^c(\pi'(P_n A P_n)) = \sum_{k=1}^{n} \sum_{j=1}^{m} (A\xi_j|\xi_k)\Omega\xi_k \otimes \overline{\xi_k} \in \mathcal{K} \quad (\text{ by (iii)}_2)$$

for all $A \in \mathcal{B}(\mathcal{H})$. Furthermore, we have

$$\lim_{n,m\to\infty} \|\pi'(P_n A P_m)\Omega - \pi'(A)\Omega\|_2 = 0,$$
$$\lim_{n,m\to\infty} \|\pi'(P_n A P_m)^*\Omega - \pi'(A)^*\Omega\|_2 = 0.$$

Thus the statement (2.4.23) holds. By (iii)$_2$ we have $\pi(\mathcal{M})\mathcal{K} \subset \mathcal{K}$, and by (iii)$_3$, $\Delta^{it}_{\lambda_\Omega^{cc}}\mathcal{K} \subset \mathcal{K}$ for all $t \in \mathbb{R}$. Thus, λ_Ω is modular. This completes the proof.

For the standardness and the modularity of a vector $\Omega \in \mathfrak{S}_2(\mathcal{M})$ we have the following

Corollary 2.4.21. Let $\mathcal{M}$ be a closed O*-algebra on $\mathcal{D}$ in $\mathcal{H}$ such that $\mathcal{M}'_w = \mathbb{C}I$ and let $\Omega \geq 0 \in \mathfrak{S}_2(\mathcal{M})$. Suppose Ω^{-1} is densely defined and $\pi(\mathcal{M})\Omega$ is dense in $\mathcal{H} \otimes \overline{\mathcal{H}}$. Then the following statements hold:

(1) Suppose $\Omega \in \mathfrak{S}_2(\mathcal{L}^\dagger(\mathcal{D}))$ and $\Omega^{it}\mathcal{D} \subset \mathcal{D}$ for all $t \in \mathbb{R}$. Then Ω is a modular vector for $\pi(\mathcal{M})$.

(2) Suppose that there exists an element N of $\mathcal{L}^\dagger(\mathcal{D})$ such that $\overline{N^{-1}} \in \mathcal{H} \otimes \overline{\mathcal{H}}$, and $\Omega^{it}\mathcal{D} \subset \mathcal{D}$ for all $t \in \mathbb{R}$. Then Ω is a quasi-standard vector for $\pi(\mathcal{M})$. Further, if $\Omega^{it}\mathcal{M}\Omega^{-it} = \mathcal{M}$ for all $t \in \mathbb{R}$, then Ω is a standard vector for $\pi(\mathcal{M})$.

Proof. (1) Since $\pi(\mathcal{M})\Omega$ is dense in $\mathcal{H} \otimes \overline{\mathcal{H}}$ and $\Omega\mathcal{D} \subset \mathcal{D}$, it follows from Lemma 2.4.16 that λ_Ω is regular, so that by Lemma 2.4.17 $\lambda_\Omega^{cc}(\mathcal{D}(\lambda_\Omega^{cc})^* \cap \mathcal{D}(\lambda_\Omega^{cc})) = \pi''(\mathcal{B}(\mathcal{H}))\Omega$ and it is an achieved left Hilbert algebra in $\mathcal{H} \otimes \overline{\mathcal{H}}$ such that $\Delta_{\lambda_\Omega^{cc}} = \pi'(\Omega^{-2})\pi''(\Omega^2)$. Let $\{\xi_n\}$ be an ONB in $\mathcal{H}$ consisting of eigenvectors of non-zero eigenvalues $\{\omega_n\}$ of Ω. Then $\{\xi_n\} \subset \mathcal{D}$ and $\Omega = \sum_{n=1}^{\infty} \omega_n \xi_n \otimes \overline{\xi_n}$. We denote by $\mathcal{E}$ the linear span of $\{\xi_n \otimes \overline{\xi_m}; n, m \in \mathbb{N}\}$. Then since $\mathcal{E} \subset \mathfrak{S}_2(\mathcal{L}^\dagger(\mathcal{D})) \subset \mathfrak{S}_2(\mathcal{M}) \subset \mathcal{D}\otimes\overline{\mathcal{H}}$ and $\mathcal{E}$ is dense in $\mathfrak{S}_2(\mathcal{M})[t_{\pi(\mathcal{M})}]$, it follows that $\mathfrak{S}_2(\mathcal{L}^\dagger(\mathcal{D}))$ is dense in $\mathfrak{S}_2(\mathcal{M})[t_{\pi(\mathcal{M})}]$. Since $\Omega^{2it}\lceil_{\mathcal{D}} \in \mathcal{L}^\dagger(\mathcal{D})$ for all $t \in \mathbb{R}$, it follows that $\Delta^{it}_{\lambda_\Omega^{cc}}\mathfrak{S}_2(\mathcal{L}^\dagger(\mathcal{D})) \subset \mathfrak{S}_2(\mathcal{L}^\dagger(\mathcal{D}))$ for all $t \in \mathbb{R}$. It is clear that $\pi(\mathcal{M})\mathfrak{S}_2(\mathcal{L}^\dagger(\mathcal{D})) \subset \mathfrak{S}_2(\mathcal{L}^\dagger(\mathcal{D}))$. Thus Ω is a modular generalized vector for $\pi(\mathcal{M})$.

(2) Since

$$XT = N^{-1}(NXT) \in (\mathcal{H} \otimes \overline{\mathcal{H}})\mathcal{B}(\mathcal{H}) = \mathcal{H} \otimes \overline{\mathcal{H}}$$

for all $X \in \mathcal{L}^\dagger(\mathcal{D})$ and $T \in \mathcal{D} \otimes \overline{\mathcal{H}}$, it follows that $\mathfrak{S}_2(\mathcal{L}^\dagger(\mathcal{D})) = \mathfrak{S}_2(\mathcal{M}) = \mathcal{D} \otimes \overline{\mathcal{H}}$. Hence statement (2) follows from (1). This completes the proof.

We next investigate the standardness and the modularity of a generalized vectors λ_Ω defined by a positive self-adjoint unbounded operator Ω. We put

$$\begin{cases} \mathcal{D}(\lambda_\Omega) = \{\pi(X); X \in \mathcal{M} \text{ and } \overline{X^{\dagger *}\Omega} \in \mathfrak{S}_2(\mathcal{M})\}, \\ \lambda_\Omega(\pi(X)) = \overline{X^{\dagger *}\Omega}, \quad \pi(X) \in \mathcal{D}(\lambda_\Omega). \end{cases}$$

Then λ_Ω is a generalized vector for $\pi(\mathcal{M})$.

Lemma 2.4.22. Let $\mathcal{M}$ be a closed O*-algebra on $\mathcal{D}$ in $\mathcal{H}$ and Ω a positive self-adjoint operator in $\mathcal{H}$ such that Ω^{-1} is densely defined. Suppose there exists a subspace $\mathcal{E}$ of $\mathcal{D} \cap \mathcal{D}(\Omega)$ such that

(i) $\{\xi \otimes \overline{\eta}; \xi, \eta \in \mathcal{E}\} \subset \mathcal{M}$,

(ii) $\mathcal{E}$ is a core for Ω.

Then $\lambda_\Omega((\mathcal{D}(\lambda_\Omega)^\dagger \cap \mathcal{D}(\lambda_\Omega))^2)$ is total in $\mathcal{H} \otimes \overline{\mathcal{H}}$. Furthermore, if $\mathcal{E}$ is dense in $\mathcal{D}[t_\mathcal{M}]$, then λ_Ω is a strongly cyclic generalized vector for $\pi(\mathcal{M})$.

Proof. Since $\{\xi \otimes \overline{\eta}; \xi, \eta \in \mathcal{E}\} \subset \mathcal{M}$ and $\mathcal{E}$ is dense in $\mathcal{H}$, it follows that $\mathcal{M}'_{\mathrm{w}} = \mathbb{C}I$. It is easily shown that $\{\pi(\xi \otimes \overline{\eta}); \xi, \eta \in \mathcal{E}\} \subset (\mathcal{D}(\lambda_\Omega)^\dagger \cap \mathcal{D}(\lambda_\Omega))^2$ and $\lambda_\Omega(\pi(\xi \otimes \overline{\eta})) = \xi \otimes \overline{\Omega\eta}$ for each $\xi, \eta \in \mathcal{E}$. Since $\Omega\mathcal{E}$ is total in $\mathcal{H}$, it follows that $\{(\xi \otimes \eta)\Omega; \xi, \eta \in \mathcal{E}\}$ is dense in $\{\xi \otimes \overline{\eta}; \xi, \eta \in \mathcal{D}\}$, and further, since $\{\xi \otimes \overline{\eta}; \xi, \eta \in \mathcal{D}\}$ is total in $\mathfrak{S}_2(\mathcal{M})$, it follows that $\lambda_\Omega((\mathcal{D}(\lambda_\Omega)^\dagger \cap \mathcal{D}(\lambda_\Omega))^2)$ is total in $\mathcal{H} \otimes \overline{\mathcal{H}}$. When $\mathcal{E}$ is dense in $\mathcal{D}[t_\mathcal{M}]$, we can similarly show that λ_Ω is strongly cyclic. This completes the proof.

Theorem 2.4.23. Let $\mathcal{M}$ be a closed O*-algebra on $\mathcal{D}$ in $\mathcal{H}$ such that $\mathcal{M}'_{\mathrm{w}} = \mathbb{C}I$ and Ω a positive self-adjoint operator in $\mathcal{H}$. Suppose

(i) Ω^{-1} is densely defined and $\mathcal{D} \cap \mathcal{D}(\Omega^{-1})$ is a core for Ω^{-1};

(ii) there exists a subspace $\mathcal{N}$ of $\mathcal{M}$ such that $\pi(\mathcal{N}) \subset \mathcal{D}(\lambda_\Omega)$, $\mathcal{N}^\dagger\mathcal{D} \subset \mathcal{D}(\Omega)$ and the linear span of $\mathcal{N}^\dagger\mathcal{D}$ is a core for Ω;

(iii) $\lambda_\Omega((\mathcal{D}(\lambda_\Omega)^\dagger \cap \mathcal{D}(\lambda_\Omega))^2)$ is total in $\mathcal{H} \otimes \overline{\mathcal{H}}$.

Then the following statements hold:

$$(1) \begin{cases} \mathcal{D}(\lambda_\Omega^c) = \{\pi'(A); A \in \mathcal{B}(\mathcal{H}) \text{ s.t. } A\mathcal{H} \subset \mathcal{D}(\Omega) \text{ and } \Omega A \in \mathfrak{S}_2(\mathcal{M})\}, \\ \lambda_\Omega^c(\pi'(A)) = \Omega A, \quad \pi'(A) \in \mathcal{D}(\lambda_\Omega^c) \end{cases}$$

and $\lambda_\Omega^c((\mathcal{D}(\lambda_\Omega^c)^* \cap \mathcal{D}(\lambda_\Omega^c))^2)$ is total in $\mathcal{H} \otimes \overline{\mathcal{H}}$.

$$(2) \begin{cases} \mathcal{D}(\lambda_\Omega^{cc}) = \{\pi''(A); A \in \mathcal{B}(\mathcal{H}) \text{ and } \overline{A\Omega} \in \mathcal{H} \otimes \overline{\mathcal{H}}\}, \\ \lambda_\Omega^{cc}(\pi''(A)) = \overline{A\Omega}, \quad \pi''(A) \in \mathcal{D}(\lambda_\Omega^{cc}) \end{cases}$$

and $\lambda_\Omega^{cc}(\mathcal{D}(\lambda_\Omega^{cc})^* \cap \mathcal{D}(\lambda_\Omega^{cc}))$ is an achieved left Hilbert algebra in $\mathcal{H} \otimes \overline{\mathcal{H}}$.

(3) $S_{\lambda_\Omega^{cc}} = J\overline{\pi''(\Omega)\pi'(\Omega^{-1})} = J\overline{\pi'(\Omega^{-1})\pi''(\Omega)}$, and so $J_{\lambda_\Omega^{cc}} = J$ and $\Delta_{\lambda_\Omega^{cc}} = \overline{\pi''(\Omega)\pi'(\Omega^{-1})} = \overline{\pi'(\Omega^{-1})\pi''(\Omega)}$.

(4) Suppose $\Omega^{it}\mathcal{D} \subset \mathcal{D}$ for all $t \in \mathbb{R}$. Then λ_Ω is a quasi-standard generalized vector for $\pi(\mathcal{M})$.

(5) Suppose $\Omega^{it}\mathcal{D} \subset \mathcal{D}$ and $\Omega^{it}\mathcal{M}\Omega^{-it} = \mathcal{M}$ for all $t \in \mathbb{R}$. Then λ_Ω is a standard generalized vector for $\pi(\mathcal{M})$.

Proof. (1) Take an arbitrary $\pi'(A) \in \mathcal{D}(\lambda_\Omega^c)$.Then there exists an element T of $\mathfrak{S}_2(\mathcal{M})$ such that

$$\pi(X)T = \pi'(A)\lambda_\Omega(\pi(X)) = \overline{X^{\dagger *}\Omega A}$$

for all $\pi(X) \in \mathcal{D}(\lambda_\Omega)$. For each $X \in \mathcal{N}$, $x \in \mathcal{H}$ and $\xi \in \mathcal{D}$ we have

$$(Ax|\Omega X^\dagger\xi) = (x|A^*\Omega X^\dagger\xi) = (\overline{X^{\dagger *}\Omega A}x|\xi) = (XTx|\xi),$$

and since the linear span of $\mathcal{N}^\dagger\mathcal{D}$ is a core for Ω, it follows that $Ax \in \mathcal{D}(\Omega)$ and $\Omega A = T \in \mathfrak{S}_2(\mathcal{M})$. Conversely, suppose that $A \in \mathcal{B}(\mathcal{H})$, $A\mathcal{H} \subset \mathcal{D}(\Omega)$ and $\Omega A \in \mathfrak{S}_2(\mathcal{M})$. Then we have

$$\pi(X)\Omega A = X(\Omega A) = \overline{X^{\dagger *}\Omega A} = \pi'(A)\lambda_\Omega(\pi(X))$$

for all $\pi(X) \in \mathcal{D}(\lambda_\Omega)$, and so $\pi'(A) \in \mathcal{D}(\lambda_\Omega^c)$ and $\lambda_\Omega^c(\pi'(A)) = \Omega A$. Thus we have

$$\begin{cases} \mathcal{D}(\lambda_\Omega^C) = \{\pi'(A); A \in \mathcal{B}(\mathcal{H}) \text{ s.t. } A\mathcal{H} \subset \mathcal{D}(\Omega) \text{ and } \Omega A \in \mathfrak{S}_2(\mathcal{M})\}, \\ \lambda_\Omega^C(\pi'(A)) = \Omega A, \quad \pi'(A) \in \mathcal{D}(\lambda_\Omega^C). \end{cases} \tag{2.4.24}$$

We next show that $\lambda_\Omega^C((\mathcal{D}(\lambda_\Omega^C)^* \cap \mathcal{D}(\lambda_\Omega^C))^2)$ is total in $\mathcal{H} \otimes \overline{\mathcal{H}}$. In fact, it follows from (2.4.24) that

$$\pi'(\Omega^{-1}\xi \otimes \overline{\Omega^{-1}\eta}) \in (\mathcal{D}(\lambda_\Omega^C)^* \cap \mathcal{D}(\lambda_\Omega^C))^2, \quad \lambda_\Omega^C(\pi'(\Omega^{-1}\xi \otimes \overline{\Omega^{-1}\eta})) = \xi \otimes \overline{\Omega^{-1}\eta} \tag{2.4.25}$$

for all $\xi, \eta \in \mathcal{D} \cap \mathcal{D}(\Omega^{-1})$. Since $\Omega^{-1}(\mathcal{D} \cap \mathcal{D}(\Omega^{-1}))$ is dense in $\mathcal{H}$, it follows from (2.4.24) and (2.4.25) that $\lambda_\Omega^C((\mathcal{D}(\lambda_\Omega^C)^* \cap \mathcal{D}(\lambda_\Omega^C))^2)$ is total in $\mathcal{H} \otimes \overline{\mathcal{H}}$.

(2) Take an arbitrary $\pi''(A) \in \mathcal{D}(\lambda_\Omega^{CC})$. Then there exists an element T of $\mathcal{H} \otimes \overline{\mathcal{H}}$ such that $\pi''(A)\lambda_\Omega^C(\pi'(B)) = \pi'(B)T$ for all $\pi'(B) \in \mathcal{D}(\lambda_\Omega^C)$. By (2.4.24) we have

$$A(\Omega B) = TB, \quad {}^\forall \pi'(B) \in \mathcal{D}(\lambda_\Omega^C). \tag{2.4.26}$$

By (2.4.25) and (2.4.26) we have

$$A\Omega(\Omega^{-1}\xi \otimes \overline{\Omega^{-1}\eta}) = T(\Omega^{-1}\xi \otimes \overline{\Omega^{-1}\eta})$$

for all $\xi, \eta \in \mathcal{D} \cap \mathcal{D}(\Omega^{-1})$, and so $A\xi = T\Omega^{-1}\xi$ for all $\xi \in \mathcal{D} \cap \mathcal{D}(\Omega^{-1})$. Since $\mathcal{D} \cap \mathcal{D}(\Omega^{-1})$ is a core for Ω^{-1}, it follows that $A\xi = T\Omega^{-1}\xi$ for all $\xi \in \mathcal{D}(\Omega^{-1})$, and so $A\Omega\xi = T\xi$ for all $\xi \in \mathcal{D}(\Omega)$. Hence, $A\Omega$ is closable and $\overline{A\Omega} = T \in \mathcal{H} \otimes \overline{\mathcal{H}}$. Conversely, take an arbitrary $A \in \mathcal{B}(\mathcal{H})$ such that $\overline{A\Omega} \in \mathcal{H} \otimes \overline{\mathcal{H}}$. Then it follows from (2.4.24) that

$$\pi''(A)\lambda_\Omega^C(\pi'(B)) = A(\Omega B) = \overline{A\Omega}B = \pi'(B)\overline{A\Omega}$$

for all $\pi'(B) \in \mathcal{D}(\lambda_\Omega^C)$, so that $\pi''(A) \in \mathcal{D}(\lambda_\Omega^{CC})$ and $\lambda_\Omega^{CC}(\pi''(A)) = \overline{A\Omega}$.

(3) By Lemma 2.4.13 $\pi''(\Omega)$ and $\pi'(\Omega^{-1})$ are positive self-adjoint operators in $\mathcal{H} \otimes \overline{\mathcal{H}}$ affiliated with the von Neumann algebras $\pi''(\mathcal{B}(\mathcal{H}))$ and $\pi'(\mathcal{B}(\mathcal{H}))$, respectively, and $\pi''(\Omega)\pi'(\Omega^{-1})$ and $\pi'(\Omega^{-1})\pi''(\Omega)$ are positive, essentially self-adjoint operators in $\mathcal{H} \otimes \overline{\mathcal{H}}$ and $\pi''(A) \in \mathcal{D}(\lambda_\Omega^{CC})^* \cap \mathcal{D}(\lambda_\Omega^{CC})$. Let $\Omega = \int_0^\infty t dE(t)$ be the spectral decomposition of Ω and put $E_n = \int_0^n dE(t)$, $n \in \mathbb{N}$. Then we have

$$\lim_{n,m\to\infty} \pi'(E_n)\pi''(E_m)\overline{A\Omega} = \overline{A\Omega},$$

$$\begin{aligned} &\lim_{n,m\to\infty} \pi''(\Omega)\pi'(\Omega^{-1})\pi'(E_n)\pi''(E_m)\overline{A\Omega} \\ &\qquad = \lim_{n,m\to\infty} \pi''(E_m)\pi'(E_n)\Omega A = \Omega A. \end{aligned}$$

Hence, it follows that $\overline{A\Omega} \in \mathcal{D}(\overline{\pi''(\Omega)\pi'(\Omega^{-1})})$ and $\overline{\pi''(\Omega)\pi'(\Omega^{-1})}\,\overline{A\Omega} = \Omega A$, which implies by (2) that

$$J\overline{\pi''(\Omega)\pi'(\Omega^{-1})\pi''(A)\Omega} = J\overline{\pi''(\Omega)\pi'(\Omega^{-1})}\,\overline{A\Omega}$$
$$= (\Omega A)^* = \overline{A^*\Omega} = S_{\lambda_\Omega^{CC}}\pi''(A)\Omega.$$

Hence we have $S_{\lambda_\Omega^{CC}} \subset J\overline{\pi''(\Omega)\pi'(\Omega^{-1})}$. Conversely, take an arbitrary $T \in \mathcal{D}(\pi''(\Omega)\pi'(\Omega^{-1}))$. Then, $T = \overline{A\Omega}$ for some $A \in \mathcal{D}(\pi''(\Omega))$. Hence we have

$$\pi''(\Omega)\pi'(\Omega^{-1})T = \pi''(\Omega)A = \Omega A \in \mathcal{H} \otimes \overline{\mathcal{H}},$$

and so $(\Omega A)^* = \overline{A^*\Omega} \in \mathcal{H} \otimes \overline{\mathcal{H}}$. Hence, $\pi''(A) \in \mathcal{D}(\lambda_\Omega^{CC})^* \cap \mathcal{D}(\lambda_\Omega^{CC})$. Thus we have

$$T = \pi''(A)\Omega \in \mathcal{D}(S_{\lambda_\Omega^{CC}}) \text{ and } S_{\lambda_\Omega^{CC}}T = J\pi''(\Omega)\pi'(\Omega^{-1})T,$$

and hence $J\overline{\pi''(\Omega)\pi'(\Omega^{-1})} \subset S_{\lambda_\Omega^{CC}}$. Thus we have $S_{\lambda_\Omega^{CC}} = J\overline{\pi''(\Omega)\pi'(\Omega^{-1})}$. The statements (4) and (5) follow from (3). This completes the proof.

We give examples of standard generalized vectors for O*-algebras on the Schwartz space $\mathcal{S}(\mathbb{R})$.

Example 2.4.24. Let $\mathcal{S}(\mathbb{R})$ be the Schwartz space of infinitely differentiable rapidly decreasing functions and $\{f_n\}_{n=0,1,\cdots} \subset \mathcal{S}(\mathbb{R})$ the othonormal basis in the Hilbert space $L^2(\mathbb{R})$ of normalized Hermite functions. Let $\mathcal{A}$ be the unbounded CCR-algebra, π_0 the Schrödinger representation of $\mathcal{A}$ and $\mathcal{M}$ an O*-algebra on $\mathcal{S}(\mathbb{R})$ containing $\pi_0(\mathcal{A})$. Then $\pi(\mathcal{M})$ is a self-adjoint O*-algebra on $\mathcal{S} \otimes \overline{L^2} (\equiv \mathcal{S}(\mathbb{R}) \otimes \overline{L^2(\mathbb{R})})$ in $L^2 \otimes \overline{L^2} (\equiv L^2(\mathbb{R}) \otimes \overline{L^2(\mathbb{R})})$ such that $\pi(\mathcal{M})'_{\mathrm{w}} = \pi'(\mathcal{B}(L^2(\mathbb{R})))$ and $(\pi(\mathcal{M})'_{\mathrm{w}})' = \pi''(\mathcal{B}(L^2(\mathbb{R})))$. In fact, as well-known, $\mathcal{S}(\mathbb{R}) = \bigcap_{k=1}^{\infty} \mathcal{D}(N^k)$, where $N = \sum_{n=0}^{\infty}(n+1) f_n \otimes \overline{f_n}$ in $\mathcal{M}$, and hence it follows that

$$XT = N^{-1}(NXT) \in (L^2 \otimes \overline{L^2})\mathcal{B}(L^2(\mathbb{R})) = L^2 \otimes \overline{L^2}$$

for each $X \in \mathcal{M}$ and $T \in \mathcal{S} \otimes \overline{L^2}$, which implies $\mathfrak{S}_2(\mathcal{M}) = \mathcal{S} \otimes \overline{L^2}$. Since $\pi_0(\mathcal{A})$ is a self-adjoint O*-algebra on $\mathcal{S}(\mathbb{R})$ such that $\pi_0(\mathcal{A})'_{\mathrm{w}} = \mathbb{C}I$, it follows by Lemma 2.4.14 that $\pi(\mathcal{M})$ is a self-adjoint O*-algebra on $\mathcal{S} \otimes \overline{L^2}$ such that $\pi(\mathcal{M})'_{\mathrm{w}} = \pi'(\mathcal{B}(L^2(\mathbb{R})))$ and $(\pi(\mathcal{M})'_{\mathrm{w}})' = \pi''(\mathcal{B}(L^2(\mathbb{R})))$. We put

$$l^2_+ = \{\{\alpha_n\} \in l^2; \alpha_n > 0,\ n = 0, 1, 2, \cdots\},$$
$$s_+ = \{\{\alpha_n\} \in l^2_+; \sup_n n^k \alpha_n < \infty \text{ for each } k \in \mathbb{N}\},$$

$$\Omega_{\{\alpha_n\}} = \sum_{n=0}^{\infty} \alpha_n f_n \otimes \overline{f_n}, \{\alpha_n\} \in l^2_+.$$

Then $\Omega_{\{\alpha_n\}} \in \mathcal{S} \otimes \overline{L^2}$ for each $\{\alpha_n\} \in s_+$. But, for a general $\{\alpha_n\} \in l^2_+$ $\Omega_{\{\alpha_n\}}$ is not a vector for $\pi(\mathcal{M})$, and so we need to consider the generalized vector $\lambda_{\Omega_{\{\alpha_n\}}}$ for $\pi(\mathcal{M})$. Let $\{\alpha_n\} \in l^2_+$. We have the following results for the standardness of $\lambda_{\{\alpha_n\}}$:

(1) Suppose $\{YX^{\dagger *}\Omega_{\{\alpha_n\}}; X, Y \in \mathcal{D}(\lambda_{\Omega_{\{\alpha_n\}}})^\dagger \cap \mathcal{D}(\lambda_{\Omega_{\{\alpha_n\}}})\}$ is total in $L^2 \otimes \overline{L^2}$. Then $\lambda_{\Omega_{\{\alpha_n\}}}$ is a quasi-standard generalized vector for $\pi(\mathcal{M})$.

(2) Suppose $\mathcal{M} \supset \{f_n \otimes \overline{f_m}; n, m \in \mathbb{N} \cup \{0\}\}$. Then $\lambda_{\Omega_{\{\alpha_n\}}}$ is a quasi-standard generalized vector for $\pi(\mathcal{M})$.

(3) $\lambda_{\Omega_{\{\alpha_n\}}}$ is a standard generalized vector for $\pi(\mathcal{L}^\dagger(\mathcal{S}(\mathbb{R})))$.

We first show the statement (1). Since $\Omega_{\{\alpha_n\}}\mathcal{E} \subset \mathcal{S}(\mathbb{R})$, where $\mathcal{E}$ is the dense subspace of $L^2(\mathbb{R})$ generated by $\{f_n\}_{n=0,1,\cdots}$, it follows from Lemma 2.4.16 that $\Omega_{\{\alpha_n\}}$ is strongly regular, so that by Lemma 2.4.17 $\lambda^{CC}_{\Omega_{\{\alpha_n\}}}$ is a standard generalized vector for the von Neumann algebra $\pi''(\mathcal{B}(L^2(\mathbb{R})))$ such that the modular conjugation operator $J_{\lambda^{CC}_{\Omega_{\{\alpha_n\}}}}$ equals the involution $T \in L^2 \otimes \overline{L^2} \to T^* \in L^2 \otimes \overline{L^2}$ and the modular operator $\Delta_{\lambda^{CC}_{\Omega_{\{\alpha_n\}}}}$ equals $\pi'(\Omega^{-2}_{\{\alpha_n\}})\pi''(\Omega^2_{\{\alpha_n\}})$. Further, since $\Omega^{it}_{\{\alpha_n\}}N \subset N\Omega^{it}_{\{\alpha_n\}}$, $^\forall t \in \mathbb{R}$ and $\mathcal{S}(\mathbb{R}) = \bigcap_{k=1}^{\infty} \mathcal{D}(N^k)$, it follows that $\Omega^{it}_{\{\alpha_n\}}\mathcal{S}(\mathbb{R}) \subset \mathcal{S}(\mathbb{R})$ for $^\forall t \in \mathbb{R}$, which implies $\Delta^{it}_{\lambda^{CC}_{\Omega_{\{\alpha_n\}}}} \mathcal{S} \otimes \overline{L^2} \subset \mathcal{S} \otimes \overline{L^2}$ for $^\forall t \in \mathbb{R}$. By Theorem 2.4.18 $\lambda_{\Omega_{\{\alpha_n\}}}$ is quasi-standard. We show the statement (2). Since $\{\pi(f_n \otimes \overline{f_m}); n, m \in \mathbb{N} \cup \{0\}\} \subset \mathcal{D}(\lambda_{\Omega_{\{\alpha_n\}}})^\dagger \cap \mathcal{D}(\lambda_{\Omega_{\{\alpha_n\}}})$ and $\{\pi(Y)\pi(X^{\dagger *})\Omega_{\{\alpha_n\}}; X, Y \in \mathcal{D}(\lambda_{\Omega_{\{\alpha_n\}}})^\dagger \cap \mathcal{D}(\lambda_{\Omega_{\{\alpha_n\}}})\}$ is total in $L^2 \otimes \overline{L^2}$, it follows from (1) that $\lambda_{\Omega_{\{\alpha_n\}}}$ is quasi-standard. The statement (3) follows from (2) and the equality:

$$\Delta^{it}_{\lambda^{CC}_{\Omega}}\pi(\mathcal{L}^\dagger(\mathcal{S}(\mathbb{R})))\Delta^{-it}_{\lambda^{CC}_{\Omega}} = \pi(\Omega^{it}_{\{\alpha_n\}}\mathcal{L}^\dagger(\mathcal{S}(\mathbb{R}))\Omega^{-it}_{\{\alpha_n\}}) = \pi(\mathcal{L}^\dagger(\mathcal{S}(\mathbb{R}))), t \in \mathbb{R}.$$

Let $\{\alpha_n\} \in s_+$. Then the following statements hold:

(4) Suppose $\pi(\mathcal{M})\Omega_{\{\alpha_n\}}$ is dense in $L^2 \otimes \overline{L^2}$. Then $\Omega_{\{\alpha_n\}}$ is a quasi-standard vector for $\pi(\mathcal{M})$.

(5) Suppose $\alpha_n \leq \gamma e^{-n\beta}, {}^\forall n \in \mathbb{N}$ for some $\gamma > 0$ and $\beta > 0$. Then $\Omega_{\{\alpha_n\}}$ is a quasi-standard vector for $\pi(\mathcal{M})$. In particular, for any $\beta > 0$ $\Omega_{\{e^{-n\beta}\}}$ is a standard vector for $\pi(\pi_0(\mathcal{A}))$.

This proof and more results will be given in Chapter IV.

We shall apply the results of this section to physical examples (namely the BCS-Bogolubov model of super conductively, a class of interacting boson models in Fock space and the quantum field theory) in Chapter IV.

E. Standard systems constructed by von Neumann algebras with standard generalized vector

Here we consider the following question:

Let $\mathcal{M}_0$ be a von Neumann algebra on a Hilbert space $\mathcal{H}$ with a full standard generalized vector λ_0. Do there exist a generalized von Neumann algebra

$\mathcal{M}$ on a dense subspace $\mathcal{D}$ of $\mathcal{H}$ over $\mathcal{M}_0$ and a standard generalized vector λ for $\mathcal{M}$ such that $\lambda^{cc} = \lambda_0$?

Let Δ_{λ_0} and J_{λ_0} be the modular operator and the modular conjugation operator for the achieved left Hilbert algebra $\lambda_0(\mathcal{D}(\lambda_0)^* \cap \mathcal{D}(\lambda_0))$, respectively, and let $\mathcal{T}(\lambda_0)$ be the $*$-subalgebra of $\mathcal{M}_0$ such that $\lambda_0(\mathcal{T}(\lambda_0))$ is the maximal Tomita algebra equivalent to the achieved left Hilbert algebra $\lambda_0(\mathcal{D}(\lambda_0)^* \cap \mathcal{D}(\lambda_0))$.

Theorem 2.4.25. Let $\mathcal{M}_0$ be a von Neumann algebra on $\mathcal{H}$ with a full generalized vector λ_0. Suppose there exists a dense subspace $\mathcal{E}$ of $\mathcal{H}$ such that

(i) $\mathcal{M}_0'\mathcal{E} \subset \mathcal{E}$;

(ii) $\Delta_{\lambda_0}^{it}\mathcal{E} \subset \mathcal{E}, \quad {}^\forall t \in \mathbb{R}$;

(iii) $\overline{\mathcal{M}_0^{\mathcal{E}}}^{\tau_s^*} \cap \mathcal{L}^\dagger(\mathcal{D}) \neq \mathcal{M}_0^{\mathcal{E}} \equiv \{A\lceil\mathcal{E}; A \in \mathcal{M}_0 \text{ s.t. } A\mathcal{E} \subset \mathcal{E} \text{ and } A^*\mathcal{E} \subset \mathcal{E}\}$;

(iv) $\lambda_0((\mathcal{D}(\lambda_0^{\mathcal{E}})^* \cap \mathcal{D}(\lambda_0^{\mathcal{E}}))^2)$ is total in $\mathcal{H}$, where $\mathcal{D}(\lambda_0^{\mathcal{E}}) = \{A \in \mathcal{D}(\lambda_0); A\mathcal{E} \subset \mathcal{E} \text{ and } \lambda_0(A) \in \mathcal{E}\}$;

(v) $\lambda_0(\mathcal{D}(\lambda_0^{\mathcal{E}})^* \cap \mathcal{D}(\lambda_0^{\mathcal{E}}))$ is dense in the Hilbert space $\mathcal{D}(S_{\lambda_0})$;

(vi) $\{K \in \mathcal{D}(\lambda_0')^* \cap \mathcal{D}(\lambda_0'); \lambda_0'(K), \lambda_0'(K^*) \in \mathcal{E}\}$ is a nondegenerate $*$-subalgebra of $\mathcal{M}_0'$;

(vii) $\{\lambda_0'(K); K \in \mathcal{D}(\lambda_0')^* \cap \mathcal{D}(\lambda_0') \text{ s.t. } \lambda_0'(K), \lambda_0'(K^*) \in \mathcal{E}\}$ is dense in the Hilbert space $\mathcal{D}(S_{\lambda_0}^*)$.

Then, for any $\{\sigma_t^{\lambda_0}\}_{t\in\mathbb{R}}$-invariant subset $\mathcal{T}$ of $(\overline{\mathcal{M}_0^{\mathcal{E}}}^{\tau_s^*} \cap \mathcal{L}^\dagger(\mathcal{D})) \setminus \mathcal{M}_0^{\mathcal{E}}$ there exist a generalized von Neumann algebra $\mathcal{M}_{(\mathcal{E},\mathcal{T})}$ in $\mathcal{H}$ and a standard generalized vector $\lambda_{(\mathcal{E},\mathcal{T})}$ for $\mathcal{M}_{(\mathcal{E},\mathcal{T})}$ such that $(\mathcal{M}_{(\mathcal{E},\mathcal{T})})_{\mathrm{w}}' = \mathcal{M}_0'$ and $\lambda_{(\mathcal{E},\mathcal{T})}^{cc} = \lambda_0$.

Proof. By (i) $[\mathcal{M}_0^{\mathcal{E}}]^{s^*} \equiv \overline{\mathcal{M}_0^{\mathcal{E}}}^{\tau_s^*} \cap \mathcal{L}^\dagger(\mathcal{D})$ is an O*-algebra on $\mathcal{E}$ and since $\mathcal{D}(\lambda_0^{\mathcal{E}})^* \cap \mathcal{D}(\lambda_0^{\mathcal{E}}) \subset \mathcal{M}_0^{\mathcal{E}}$, it follows from (iv) that $(\overline{\mathcal{M}_0^{\mathcal{E}}})'' = \mathcal{M}_0$. Hence we have by (iii) that

$$[\mathcal{M}_0^{\mathcal{E}}]_b^{s^*} = \mathcal{M}_0^{\mathcal{E}} \subsetneqq [\mathcal{M}_0^{\mathcal{E}}]^{s^*} \subset \{X \in \mathcal{L}^\dagger(\mathcal{E}); \overline{X} \text{ is affiliated with } \mathcal{M}_0\}. \tag{2.4.27}$$

Let $\mathcal{T}$ be any $\{\sigma_t^{\lambda_0}\}$-invariant subset of $[\mathcal{M}_0^{\mathcal{E}}]^{s^*} \setminus \mathcal{M}_0^{\mathcal{E}}$. We denote by $\mathcal{M}_{(\mathcal{E},\mathcal{T})}$ the closere of the O*-algebra $[\mathcal{M}_0^{\mathcal{E}}, \mathcal{T}]$ on $\mathcal{E}$ generated by $\mathcal{M}_0^{\mathcal{E}}$ and $\mathcal{T}$, that is,

$$\mathcal{D}_{(\mathcal{E},\mathcal{T})} = \bigcap_{X \in [\mathcal{M}_0^{\mathcal{E}},\mathcal{T}]} \mathcal{D}(\overline{X}) \text{ and } \mathcal{M}_{[\mathcal{E},\mathcal{T}]} = \{\overline{X}\lceil_{\mathcal{D}_{(\mathcal{E},\mathcal{T})}}; X \in [\mathcal{M}_0^{\mathcal{E}}, \mathcal{T}]\}.$$

We simply write $\mathcal{D}_{(\mathcal{E},\mathcal{T})}$ and $\mathcal{M}_{(\mathcal{E},\mathcal{T})}$ by $\mathcal{D}$ and $\mathcal{M}$, respectively. By (2.4.27) we have $[\mathcal{M}_0^{\mathcal{E}}, \mathcal{T}]_{\mathrm{w}}' = \mathcal{M}_0'$, and by the $\{\sigma_t^{\lambda_0}\}$-invariance of $\mathcal{T}$ and (ii) we have

$$\Delta_{\lambda_0}^{it}[\mathcal{M}_0^{\mathcal{E}}, \mathcal{T}]\Delta_{\lambda_0}^{-it} = [\mathcal{M}_0^{\mathcal{E}}, \mathcal{T}], \quad {}^\forall t \in \mathbb{R},$$

so that

$$\mathcal{M} \subset \{X \in \mathcal{L}^\dagger(\mathcal{D}); \overline{X} \text{ is affiliated with } \mathcal{M}_0\}; \tag{2.4.28}$$
$$\Delta_{\lambda_0}^{it}\mathcal{D} \subset \mathcal{D}, \quad {}^\forall t \in \mathbb{R}; \tag{2.4.29}$$
$$\Delta_{\lambda_0}^{it}\mathcal{M}\Delta_{\lambda_0}^{-it} = \mathcal{M}, \quad {}^\forall t \in \mathbb{R}. \tag{2.4.30}$$

We next define a generalized vector $\lambda_{(\mathcal{E},T)}$ for the O*-algebra $\mathcal{M}$ as follows:

$$\begin{cases} \mathcal{D}(\lambda_{(\mathcal{E},T)}) = \{X \in \mathcal{M}; {}^\exists\xi_X \in \mathcal{D} \text{ s.t.} \\ \qquad\qquad X\lambda^c(K) = K\xi_X \text{ for } {}^\forall K \in \mathcal{D}(\lambda_0') \text{ s.t. } \lambda_0'(K) \in \mathcal{E}\}, \\ \lambda_{(\mathcal{E},T)}(X) = \xi_X, \quad X \in \mathcal{D}(\lambda_{(\mathcal{E},T)}). \end{cases}$$

We simply write $\lambda_{(\mathcal{E},T)}$ by λ. By (iv) and (vi) λ is well-defined and by (2.4.27) and (iv) λ is a generalized vector for $\mathcal{M}$ such that $\lambda((\mathcal{D}(\lambda)^\dagger \cap \mathcal{D}(\lambda))^2)$ is total in $\mathcal{H}$. We next show that

$$\begin{aligned} &\{K \in \mathcal{D}(\lambda_0')^* \cap \mathcal{D}(\lambda_0'); \lambda_0'(K) \in \mathcal{E}\} \subset \mathcal{D}(\lambda^c)^* \cap \mathcal{D}(\lambda^c) \subset \mathcal{D}(\lambda_0')^* \cap \mathcal{D}(\lambda_0'), \\ &\lambda_0'(K) = \lambda^c(K), \quad {}^\forall K \in \mathcal{D}(\lambda_0')^* \cap \mathcal{D}(\lambda_0') \text{ s.t. } \lambda_0'(K) \in \mathcal{E}. \end{aligned} \tag{2.4.31}$$

In fact, it is clear by the definition of λ that for any $K \in \mathcal{D}(\lambda_0')^* \cap \mathcal{D}(\lambda_0')$ s.t. $\lambda_0'(K) \in \mathcal{E}$, $K \in \mathcal{D}(\lambda^c)^* \cap \mathcal{D}(\lambda^c)$ and $\lambda^c(K) = \lambda_0'(K)$. Take arbitrary $K \in \mathcal{D}(\lambda^c)^* \cap \mathcal{D}(\lambda^c)$ and $A \in \mathcal{D}(\lambda_0)^* \cap \mathcal{D}(\lambda_0)$. By (v) there exists a sequence $\{A_n\}$ in $\mathcal{D}(\lambda_0^\mathcal{E})^* \cap \mathcal{D}(\lambda_0^\mathcal{E})$ such that $\lim_{n\to\infty} \lambda_0(A_n) = \lambda_0(A)$ and $\lim_{n\to\infty} \lambda_0(A_n^*) = \lambda_0(A^*)$, and since $(\mathcal{D}(\lambda_0^\mathcal{E})^* \cap \mathcal{D}(\lambda_0^\mathcal{E}))\lceil\mathcal{D} \subset \mathcal{D}(\lambda)^\dagger \cap \mathcal{D}(\lambda)$, we have $K\lambda_0(A_n) = A_n\lambda^c(K)$ for ${}^\forall n \in \mathbb{N}$, so that for each $K_1 \in \mathcal{D}(\lambda_0')^* \cap \mathcal{D}(\lambda_0')$ we have

$$\begin{aligned} (K\lambda_0(A)|\lambda_0'(K_1)) &= \lim_{n\to\infty} (K\lambda_0(A_n)|\lambda_0'(K_1)) \\ &= \lim_{n\to\infty} (A_n\lambda^c(K)|\lambda_0'(K_1)) \\ &= \lim_{n\to\infty} (\lambda^c(K)|A_n^*\lambda_0'(K_1)) \\ &= \lim_{n\to\infty} (\lambda^c(K)|K_1\lambda_0(A_n^*)) \\ &= (\lambda^c(K)|K_1\lambda_0(A^*)) \\ &= (\lambda^c(K)|A^*\lambda_0'(K_1)) \\ &= (A\lambda^c(K)|\lambda_0'(K_1)) \end{aligned}$$

and similarly

$$(K^*\lambda_0(A)|\lambda_0'(K_1)) = (A\lambda^c(K^*)|\lambda_0'(K_1)).$$

Hence we have $K \in \mathcal{D}(\lambda_0')^* \cap \mathcal{D}(\lambda_0')$. Thus the statement (2.4.31) holds. By (vi), (vii) and (2.4.31) we have $\lambda^c((\mathcal{D}(\lambda^c)^* \cap \mathcal{D}(\lambda^c))^2)$ is total in $\mathcal{H}$ and $\lambda^{ccc} = \lambda_0'$. Therefore we have $\lambda^{cc} = \lambda_0$, which implies by (2.4.29) and (2.4.30) that λ is a standard generalized vector for $\mathcal{M}$. This completes the proof.

Corollary 2.4.26. Let $(\mathcal{M}_0, \lambda_0)$ be in Theorem 2.4.24. For any subspace $\mathcal{E}$ of $\mathcal{H}$ satisfying the conditions (i) $\sim$ (vii) in Theorem 2.4.24 there exist a generalized von Neumann algebra $\mathcal{M}_{\mathcal{E}}$ on $\mathcal{D}_{\mathcal{E}}$ over $\mathcal{M}_0$ and a standard generalized vector $\lambda_{\mathcal{E}}$ for $\mathcal{M}_{\mathcal{E}}$ such that $\lambda_{\mathcal{E}}^{cc} = \lambda_0$.

Proof. This follows from Theorem 2.4.24 by taking $(\overline{\mathcal{M}_0^{\mathcal{E}}}^{\tau_s^*} \cap \mathcal{L}^\dagger(\mathcal{D}))\backslash \mathcal{M}_0^{\mathcal{E}}$ as $\mathcal{T}$ in Theorem 2.4.24.

We denote by $\mathcal{S}(\mathcal{M}_0, \lambda_0)$ the set of all subspaces $\mathcal{E}$ of $\mathcal{H}$ satisfying (i) $\sim$ (vii) in Theorem 2.4.24.

Corollary 2.4.27. Let $\mathcal{M}_0$ be a von Neumann algebra on $\mathcal{H}$ with a cyclic and separating vector ξ_0. Suppose there exists a dense subspace $\mathcal{E}$ for $\mathcal{H}$ such that

(i) $\mathcal{M}_0'\mathcal{E} \subset \mathcal{E}$;
(ii) $\Delta_{\xi_0}^{it}\mathcal{E} \subset \mathcal{E}, \quad {}^\forall t \in \mathbb{R}$;
(iii) $\overline{\mathcal{M}_0^{\mathcal{E}}}^{\tau_s^*} \cap \mathcal{L}^\dagger(\mathcal{D}) \neq \mathcal{M}_0^{\mathcal{E}}$;
(iv) $(\overline{\mathcal{M}_0^{\mathcal{E}}})'' = \mathcal{M}_0$;
(v) $\xi_0 \in \mathcal{E}$.

Then $\mathcal{E} \in \mathcal{S}(\mathcal{M}_0, \lambda_{\xi_0})$, and so ξ_0 is a standard vector for the closed O*-algebra $\mathcal{M}_{(\mathcal{E},\mathcal{T})}$ for any $\{\sigma_t^{\xi_0}\}_{t\in\mathbb{R}}$-invariant subset $\mathcal{T}$ of $(\overline{\mathcal{M}_0^{\mathcal{E}}}^{\tau_s^*} \cap \mathcal{L}^\dagger(\mathcal{D})) \setminus \mathcal{M}_0^{\mathcal{E}}$.

Proof. By (v) we have $\mathcal{D}(\lambda_{\xi_0}^{\mathcal{E}})^* \cap \mathcal{D}(\lambda_{\xi_0}^{\mathcal{E}}) = \mathcal{M}_0^{\mathcal{E}}$, which implies by (iv) and (v) that the conditons (iv) and (v) in Theorem 2.4.24 are satisfied. Further, since $\mathcal{D}(\lambda_{\xi_0}') = \mathcal{M}_0'$ and $\lambda_{\xi_0}'(K) = K\xi_0$ for each $K \in \mathcal{M}_0'$, it follows from (i) that the conditions (vi) and (vii) in Theorem 2.4.24 are satisfied. Thus we have $\mathcal{E} \in \mathcal{S}(\mathcal{M}_0, \lambda_{\xi_0})$. This completes the proof.

Example 2.4.28. *Let $\mathcal{M}_0$ be a von Neumann algebra on $\mathcal{H}$ and λ_0 a tracial standard generalized vector for $\mathcal{M}_0$.* Then $\mathcal{H}_{\lambda_0}^{\omega}$ in Example 2.4.5 belongs to $\mathcal{S}(\mathcal{M}_0, \lambda_0)$. In fact, this follows since $\Delta_{\lambda_0} = I$ and $\mathcal{D}(\lambda_0^{\mathcal{H}_{\lambda_0}^{\omega}}) = \mathcal{H}_{\lambda_0}^{\omega}$.

Example 2.4.29. *Let $\mathcal{M}_0$ be a von Neumann algebra on $\mathcal{H}$ with a cyclic and separating vector ξ_0 and H a positive self-adjoint operator* in $\mathcal{H}$ *affiliated with the fixed point algebra $\mathcal{M}_0^{\sigma^{\lambda_{\xi_0}}} \equiv \{A \in \mathcal{M}_0; \sigma_t^{\lambda_{\xi_0}}(A) = A, {}^\forall t \in \mathbb{R}\}$ of $\{\sigma^{\lambda_{\xi_0}}\}$ such that $\xi_0 \in \mathcal{D}^\infty(H)$. Then $\mathcal{D}^\infty(H) \in \mathcal{S}(\mathcal{M}_0, \lambda_{\xi_0})$.* In fact, it is clear that the conditions (i), (ii) and (v) in Corollary 2.4.26 are satisfied. Let $H = \int_0^\infty t dE_H(t)$ be the spectral resolution of H. Take an arbitrary $A \in \mathcal{M}_0$. Since $E_H(n)AE_H(n) \in \mathcal{M}_0^{\mathcal{D}^\infty(H)}$, ${}^\forall n \in \mathbb{N}$ and $\lim_{n\to\infty} E_H(n)AE_H(n) = A$ strongly, it follows that the condition (iv) in Corollary 2.4.26 holds. Further, we have $H^n E_H(m) \in \mathcal{M}_0^{\mathcal{D}^\infty(H)}$ for ${}^\forall n, m \in \mathbb{N}$ and $H^n E_H(m)\lceil_{\mathcal{D}^\infty(H)} \xrightarrow[m\to\infty]{}$

$H^n\lceil_{\mathcal{D}^\infty(H)}$, strongly*, and hence $H^n\lceil_{\mathcal{D}^\infty(H)} \in (\overline{\mathcal{M}_0^{\mathcal{D}^\infty(H)}}^{\tau_s^*} \cap \mathcal{L}^\dagger(\mathcal{D}^\infty(H))) \setminus \mathcal{M}_0^{\mathcal{D}^\infty(H)}$, and so the condition (iii) in Corollary 2.4.26 holds. Thus we have $\mathcal{D}^\infty(H) \in \mathcal{S}(\mathcal{M}_0, \lambda_{\xi_0})$.

Example 2.4.30. Let $\mathcal{M}_0$ be a von Neumann algebra on $\mathcal{H}$ with a cyclic and separating vector ξ_0.

When can we construct such a positive self-adjoint operator H in $\mathcal{H}$ as in Example 2.4.28? The above question is affirmative in cases of the following (i) and (ii):

(i) $\mathcal{M}_0^{\sigma^{\lambda_{\xi_0}}}$ *is infinitely dimensional.*

(ii) $\mathcal{M}_0$ *is semifinite and the spectrum $S_p(\Delta_{\xi_0})$ of Δ_{ξ_0} is an infinite set.*

Suppose $\mathcal{M}_0^{\sigma^{\lambda_{\xi_0}}}$ is infinitely dimensional. Then there is a sequence $\{E_n\}$ of mutually orthogonal projections in $\mathcal{M}_0^{\sigma^{\lambda_{\xi_0}}}$ such that $\|E_n\xi_0\| < 1$ and $\log\|E_n\xi_0\| - \log\|E_{n+1}\xi_0\| > 1$ for $n \in \mathbb{N}$. Then it is easily shown that $H \equiv \sum_{n=1}^{\infty}(-\log\|E_n\xi_0\|)E_n$ is a positive self-adjoint operator in $\mathcal{H}$ affiliated with $\mathcal{M}_0^{\sigma^{\lambda_{\xi_0}}}$ such that $\xi_0 \in \mathcal{D}^\infty(H)$. Suppose the statement (ii) holds. By Theorem 14.2 in Takesaki [1] there exists a positive self-adjoint operator K in $\mathcal{H}$ affiliated with $\mathcal{M}_0$ such that $\Delta_{\xi_0} = K^{-2} \cdot K'^2$ where $K' = J_{\xi_0}KJ_{\xi_0}$. Then K is affiliated with $\mathcal{M}_0^{\sigma^{\lambda_{\xi_0}}}$. Since $S_P(\Delta_{\xi_0})$ is an infinite set, it follows that $S_P(K)$ is an infinite set, which implies that $\mathcal{M}_0^{\sigma^{\lambda_{\xi_0}}}$ is infinitely dimensional. Therefore, the question is affirmative by the above argument.

2.5 Generalized Connes cocycle theorem

In this section we generalize the Connes cocycle theorem for von Neumann algebras to generalized von Neumann algebras. Let $\mathcal{M}$ be a closed O*-algebra on $\mathcal{D}$ in $\mathcal{H}$. Let $\mathcal{K}_4$ be a four-dimensional Hilbert space with an orthonormal basis $\{\eta_{ij}\}_{i,j=1,2}$ and $M_2(\mathbb{C})$ a 2×2-matrix algebra generated by the matrices E_{ij} which are defined by $\mathrm{E}_{ij}\eta_{kl} = \delta_{ik}\eta_{kl}$. Identifying $\zeta = \zeta_1\otimes\eta_{11} + \zeta_2\otimes\eta_{21} + \zeta_3\otimes\eta_{12} + \zeta_4\otimes\eta_{22} \in \mathcal{H}\otimes\mathcal{K}_4$ with $\zeta = (\zeta_1,\zeta_2,\zeta_3,\zeta_4) \in \mathcal{H}^4 \equiv \mathcal{H}\oplus\mathcal{H}\oplus\mathcal{H}\oplus\mathcal{H}$, $\mathcal{M}\otimes M_2(\mathbb{C})$ is regarded as the matrix algebra on $\mathcal{D}^4 \equiv \mathcal{D}\oplus\mathcal{D}\oplus\mathcal{D}\oplus\mathcal{D}$:

$$\left\{\begin{pmatrix} X_{11} & X_{12} & 0 & 0 \\ X_{21} & X_{22} & 0 & 0 \\ 0 & 0 & X_{11} & X_{12} \\ 0 & 0 & X_{21} & X_{22} \end{pmatrix} ; X_{ij} \in \mathcal{M}\right\}.$$

Suppose λ and μ are generalized vectors for $\mathcal{M}$. We put

$$\mathcal{D}(\theta_{\lambda,\mu}) = \left\{ X = \begin{pmatrix} X_{11} & X_{12} & 0 & 0 \\ X_{21} & X_{22} & 0 & 0 \\ 0 & 0 & X_{11} & X_{12} \\ 0 & 0 & X_{21} & X_{22} \end{pmatrix} ; \begin{matrix} X_{11}, X_{21} \in \mathcal{D}(\lambda) \\ X_{12}, X_{22} \in \mathcal{D}(\mu) \end{matrix} \right\},$$

$$\theta_{\lambda,\mu}(X) = \begin{pmatrix} \lambda(X_{11}) \\ \lambda(X_{21}) \\ \mu(X_{12}) \\ \mu(X_{22}) \end{pmatrix}, \quad X \in \mathcal{D}(\theta_{\lambda,\mu}).$$

Then $\theta_{\lambda,\mu}$ is a generalized vector for $\mathcal{M} \otimes M_2(\mathbb{C})$. We consider when $\theta_{\lambda,\mu}$ is a standard generalized vector for $\mathcal{M} \otimes M_2(\mathbb{C})$. We here need the notion of semifiniteness of generalized vectors:

Definition 2.5.1. A generalized vetor λ for $\mathcal{M}$ is said to be semifinite if there exists a net $\{U_\alpha\}$ in $\mathcal{D}(\lambda)^\dagger \cap \mathcal{D}(\lambda)$ such that $\|\overline{U_\alpha}\| \leq 1$ for each α and $\{U_\alpha\}$ converges strongly to I.

Lemma 2.5.2. Suppose λ is a semifinite generalized vector for $\mathcal{M}$. Then the following statements are equivalent:

(i) λ is cyclic.

(ii) $\lambda((\mathcal{D}(\lambda)^\dagger \cap \mathcal{D}(\lambda))^2)$ is total in $\mathcal{H}$.

Proof. (i) $\Rightarrow$ (ii) Take an arbitrary $X \in \mathcal{D}(\lambda)$. Since λ is semifinite, there exists a net $\{U_\alpha\}$ in $\mathcal{D}(\lambda)^\dagger \cap \mathcal{D}(\lambda)$ such that $\|\overline{U_\alpha}\| \leq 1$, $^\forall\alpha$ and $\{U_\alpha\}$ converges strongly to I. Then we have

$$\begin{aligned} &U_\alpha X \in \mathcal{D}(\lambda)^\dagger \cap \mathcal{D}(\lambda), \quad ^\forall\alpha, \\ &\lim_\alpha \|\lambda(U_\alpha X) - \lambda(X)\| = \lim_\alpha \|U_\alpha \lambda(X) - \lambda(X)\| = 0, \end{aligned}$$

and hence it follows since λ is cyclic that $\lambda(\mathcal{D}(\lambda)^\dagger \cap \mathcal{D}(\lambda))$ is dense in $\mathcal{H}$. Repeating the same argument, we can show that $\lambda((\mathcal{D}(\lambda)^\dagger \cap \mathcal{D}(\lambda))^2)$ is total in $\mathcal{H}$.

(ii) $\Rightarrow$ (i) This is trivial.

Lemma 2.5.3. Let $\mathcal{M}$ be a closed O*-algebra on $\mathcal{D}$ in $\mathcal{H}$ such that $\mathcal{M}'_{\mathrm{w}}\mathcal{D} \subset \mathcal{D}$ and let λ and μ are semifinite, cyclic generalized vectors for $\mathcal{M}$. Then the following statements hold:

(1) $\lambda(\mathcal{D}(\lambda) \cap \mathcal{D}(\mu)^\dagger)$ and $\mu(\mathcal{D}(\mu) \cap \mathcal{D}(\lambda)^\dagger)$ are dense in $\mathcal{H}$.

(2) $\theta_{\lambda,\mu}$ is semifinite and cyclic, and so $\theta_{\lambda,\mu}((\mathcal{D}(\theta_{\lambda,\mu})^\dagger \cap \mathcal{D}(\theta_{\lambda,\mu}))^2)$ is total in $\mathcal{H}^4$.

(3)

$$\mathcal{D}(\theta^C_{\lambda,\mu}) = \left\{ K = \begin{pmatrix} K_{11} & 0 & K_{12} & 0 \\ 0 & K_{11} & 0 & K_{12} \\ K_{21} & 0 & K_{22} & 0 \\ 0 & K_{21} & 0 & K_{22} \end{pmatrix} ; \begin{matrix} K_{11}, K_{21} \in \mathcal{D}(\lambda^C) \\ K_{12}, K_{22} \in \mathcal{D}(\mu^C) \end{matrix} \right\},$$

$$\theta^C_{\lambda,\mu}(K) = \begin{pmatrix} \lambda^C(K_{11}) \\ \mu^C(K_{12}) \\ \lambda^C(K_{21}) \\ \mu^C(K_{22}) \end{pmatrix}, \quad K \in \mathcal{D}(\theta^C_{\lambda,\mu}). \tag{2.5.1}$$

(4) Suppose $\lambda^C((\mathcal{D}(\lambda^C)^* \cap \mathcal{D}(\lambda^C))^2)$ and $\mu^C((\mathcal{D}(\mu^C)^* \cap \mathcal{D}(\mu^C))^2)$ are total in $\mathcal{H}$. Then $\theta^C_{\lambda,\mu}((\mathcal{D}(\theta^C_{\lambda,\mu})^\dagger \cap \mathcal{D}(\theta^C_{\lambda,\mu}))^2)$ is total in $\mathcal{H}^4$ and

$$\mathcal{D}(\theta^{cc}_{\lambda,\mu}) = \left\{ A = \begin{pmatrix} A_{11} & A_{12} & 0 & 0 \\ A_{21} & A_{22} & 0 & 0 \\ 0 & 0 & A_{11} & A_{12} \\ 0 & 0 & A_{21} & A_{22} \end{pmatrix} ; \begin{matrix} A_{11}, A_{21} \in \mathcal{D}(\lambda^{cc}) \\ A_{12}, A_{22} \in \mathcal{D}(\mu^{cc}) \end{matrix} \right\},$$

$$\theta^{cc}_{\lambda,\mu}(A) = \begin{pmatrix} \lambda^{cc}(A_{11}) \\ \lambda^{cc}(A_{21}) \\ \mu^{cc}(A_{12}) \\ \mu^{cc}(A_{22}) \end{pmatrix}, A \in \mathcal{D}(\theta^{cc}_{\lambda,\mu}). \tag{2.5.2}$$

Proof. Since λ and μ are semifinite, there exist nets $\{U_\alpha\}$ and $\{V_\beta\}$ in $\mathcal{D}(\lambda)^\dagger \cap \mathcal{D}(\lambda)$ and $\mathcal{D}(\mu)^\dagger \cap \mathcal{D}(\mu)$ such that $\|\overline{U_\alpha}\| \leq 1, {}^\forall\alpha$ and $\|\overline{V_\beta}\| \leq 1, {}^\forall\beta$ and $\{U_\alpha\}$ and $\{V_\beta\}$ converge strongly to I, respectively.

(1) Take an arbitrary $X \in \mathcal{D}(\lambda)$. Then we have $V_\beta X \in \mathcal{D}(\lambda) \cap \mathcal{D}(\mu)^\dagger$ for each β and $\lim_\beta \lambda(V_\beta X) = \lim_\beta V_\beta \lambda(X) = \lambda(X)$, which implies that $\lambda(\mathcal{D}(\lambda) \cap \mathcal{D}(\mu)^\dagger)$ is dense in $\mathcal{H}$. Similarly, $\mu(\mathcal{D}(\mu) \cap \mathcal{D}(\lambda)^\dagger)$ is dense in $\mathcal{H}$.

(2) It is easily shown that

$$\begin{aligned} &\mathcal{D}(\theta_{\lambda,\mu})^\dagger \cap \mathcal{D}(\theta_{\lambda,\mu}) \\ &= \left\{ X = \begin{pmatrix} X_{11} & X_{12} & 0 & 0 \\ X_{21} & X_{22} & 0 & 0 \\ 0 & 0 & X_{11} & X_{12} \\ 0 & 0 & X_{21} & X_{22} \end{pmatrix} ; \begin{matrix} X_{11} \in \mathcal{D}(\lambda)^\dagger \cap \mathcal{D}(\lambda), \\ X_{21} \in \mathcal{D}(\lambda) \cap \mathcal{D}(\mu)^\dagger, \\ X_{12} \in \mathcal{D}(\mu) \cap \mathcal{D}(\lambda)^\dagger, \\ X_{22} \in \mathcal{D}(\mu)^\dagger \cap \mathcal{D}(\mu) \end{matrix} \right\}, \end{aligned} \tag{2.5.3}$$

and

$$\left\{ \begin{pmatrix} U_\alpha & 0 & 0 & 0 \\ 0 & V_\beta & 0 & 0 \\ 0 & 0 & U_\alpha & 0 \\ 0 & 0 & 0 & V_\beta \end{pmatrix} \right\}$$

is a net in $\mathcal{D}(\theta_{\lambda,\mu})^\dagger \cap \mathcal{D}(\theta_{\lambda,\mu})$ which converges strongly to I, which implies that $\theta_{\lambda,\mu}$ is semifinite. By (2.5.3) and (1), $\theta_{\lambda,\mu}$ is cyclic. Hence it follows from Lemma 2.5.2 that $\theta_{\lambda,\mu}((\mathcal{D}(\theta_{\lambda,\mu})^\dagger \cap \mathcal{D}(\theta_{\lambda,\mu}))^2)$ is total in $\mathcal{H}^4$.

(3) This is easily shown.

(4) By the assumption λ^C and μ^C are semifinite and cyclic, and by (2.5.2) it is proved similarly to the proof of (2) that $\theta^C_{\lambda,\mu}$ is semifinite and cyclic. Hence it follows from Lemma 2.5.2 that $\theta^C_{\lambda,\mu}((\mathcal{D}(\theta^C_{\lambda,\mu})^* \cap \mathcal{D}(\theta^C_{\lambda,\mu}))^2)$ is total in $\mathcal{H}^4$. It is not difficult to show the statement (2.5.3). This completes the proof.

Proposition 2.5.4. Let $\mathcal{M}$ be a generalized von Neumann algebra on $\mathcal{D}$ in $\mathcal{H}$ and let λ and μ be semifinite generalized vectors for $\mathcal{M}$. The following statements are equivalent.

(i) λ and μ are (full) standard generalized vectors for $\mathcal{M}$.

(ii) $\theta_{\lambda,\mu}$ is a (full) standard generalized vector for $\mathcal{M} \otimes M_2(\mathbb{C})$.

Proof. (i) $\Rightarrow$ (ii) By (2.5.2) and (2.5.3) we have

$$S_{\theta^{CC}_{\lambda,\mu}} = \begin{pmatrix} S_{\lambda^{cc}} & 0 & 0 & 0 \\ 0 & 0 & S_{\lambda^{CC}\mu^{CC}} & 0 \\ 0 & S_{\mu^{CC}\lambda^{CC}} & 0 & 0 \\ 0 & 0 & 0 & S_{\mu^{CC}} \end{pmatrix},$$

and so

$$\Delta_{\theta^{CC}_{\lambda,\mu}} = \begin{pmatrix} \Delta_{\lambda^{CC}} & 0 & 0 & 0 \\ 0 & \Delta_{\mu^{CC}\lambda^{CC}} & 0 & 0 \\ 0 & 0 & \Delta_{\lambda^{CC}\mu^{CC}} & 0 \\ 0 & 0 & 0 & \Delta_{\mu^{CC}} \end{pmatrix}, \tag{2.5.4}$$

where $S_{\lambda^{CC}\mu^{CC}}$ is the closure of the conjugate linear operator $\mu^{CC}(A) \rightarrow \lambda^{CC}(A^*)$, $A \in D(\mu^{CC}) \cap D(\lambda^{CC})^*$ and $S_{\lambda^{CC}\mu^{CC}} = J_{\lambda^{CC}\mu^{CC}} \Delta^{\frac{1}{2}}_{\lambda^{CC}\mu^{CC}}$ is the polar decomposition of $S_{\lambda^{CC}\mu^{CC}}$, and $S_{\mu^{CC}\lambda^{CC}}, J_{\mu^{CC}\lambda^{CC}}$ and $\Delta_{\mu^{CC}\lambda^{CC}}$ are operators defined similarly. Since

$$\Delta^{it}_{\theta^{CC}_{\lambda,\mu}}((\mathcal{M}'_{\mathrm{w}})' \otimes M_2(\mathbb{C}))\, \Delta^{-it}_{\theta^{CC}_{\lambda,\mu}} = (\mathcal{M}'_{\mathrm{w}})' \otimes M_2(\mathbb{C}), \quad t \in \mathbb{R},$$

it follows from (2.5.4) that

$$\sigma^{\lambda^{CC}}_t(A_{11}) = \Delta^{it}_{\lambda^{CC}\mu^{CC}} A_{11} \Delta^{-it}_{\lambda^{CC}\mu^{CC}}, \tag{2.5.5}$$

$$\Delta^{it}_{\lambda^{CC}} A_{12} \Delta^{-it}_{\mu^{CC}\lambda^{CC}} = \Delta^{it}_{\lambda^{CC}\mu^{CC}} A_{12} \Delta^{-it}_{\mu^{CC}}, \tag{2.5.6}$$

$$\Delta^{it}_{\mu^{CC}\lambda^{CC}} A_{21} \Delta^{-it}_{\lambda^{CC}} = \Delta^{it}_{\mu^{CC}} A_{21} \Delta^{-it}_{\lambda^{CC}\mu^{CC}}, \tag{2.5.7}$$

$$\sigma^{\mu^{CC}}_t(A_{22}) = \Delta^{it}_{\mu^{CC}\lambda^{CC}} A_{22} \Delta^{-it}_{\mu^{CC}\lambda^{CC}} \tag{2.5.8}$$

for all $A_{11}, A_{12}, A_{21}, A_{22} \in (\mathcal{M}'_{\mathrm{w}})'$ and $t \in \mathbb{R}$. We now denote by $[D\mu^{CC} : D\lambda^{CC}]_t$ the Connes cocycle associated with the weight $\varphi_{\mu^{CC}}$ with respect to the weight $\varphi_{\lambda^{CC}}$ (Stratila [1]), that is,

$$[D\mu^{CC} : D\lambda^{CC}]_t = \Delta^{it}_{\mu^{CC}} \Delta^{-it}_{\lambda^{CC}\mu^{CC}}, \quad t \in \mathbb{R}. \tag{2.5.9}$$

By (2.5.7) we have

$$[D\mu^{CC} : D\lambda^{CC}]_t = \Delta^{it}_{\mu^{CC}\lambda^{CC}} \Delta^{-it}_{\lambda^{CC}} \in (\mathcal{M}'_{\mathrm{w}})', \quad t \in \mathbb{R} \tag{2.5.10}$$

and by (2.5.5) and (2.5.8)

$$\Delta^{it}_{\lambda^{CC}} \Delta^{-it}_{\lambda^{CC}\mu^{CC}}, \quad \Delta^{it}_{\mu^{CC}} \Delta^{-it}_{\mu^{CC}\lambda^{CC}} \in \mathcal{M}'_{\mathrm{w}}, \quad t \in \mathbb{R}. \tag{2.5.11}$$

Since $\Delta^{it}_{\lambda^{CC}}\mathcal{D} \subset \mathcal{D}$, $\Delta^{it}_{\mu^{CC}}\mathcal{D} \subset \mathcal{D}$ $(\forall t \in \mathbb{R})$ and $\mathcal{M}'_{\mathrm{w}}\mathcal{D} \subset \mathcal{D}$, it follows from (2.5.11) that

$$\Delta^{it}_{\mu^{CC}\lambda^{CC}}\mathcal{D} = \Delta^{it}_{\mu^{CC}} \Delta^{-it}_{\mu^{CC}} \Delta^{it}_{\mu^{CC}\lambda^{CC}}\mathcal{D} \subset \Delta^{it}_{\mu^{CC}}\mathcal{M}'_{\mathrm{w}}\mathcal{D} \subset \mathcal{D}, \tag{2.5.12}$$

and similarly

$$\Delta^{it}_{\lambda^{CC}\mu^{CC}}\mathcal{D} \subset \mathcal{D}, \tag{2.5.13}$$

which implies by (2.5.9), (2.5.10) and (2.5.4) that

$$[D\mu^{CC} : D\lambda^{CC}]_t \lceil \mathcal{D} \in \mathcal{M} \text{ and } \Delta^{it}_{\theta^{CC}_{\lambda,\mu}} \mathcal{D}^4 \subset \mathcal{D}^4, \quad t \in \mathbb{R}. \tag{2.5.14}$$

Furthermore, it follows from (2.5.5) $\sim$ (2.5.8), (2.5.12) and (2.5.13) that

$$\begin{aligned}
\sigma^{\lambda^{CC}}_t(X_{11}) &= \Delta^{it}_{\lambda^{CC}\mu^{CC}} X_{11} \Delta^{-it}_{\lambda^{CC}\mu^{CC}}, \\
\Delta^{it}_{\lambda^{CC}} X_{12} \Delta^{-it}_{\mu^{CC}\lambda^{CC}} &= \Delta^{it}_{\lambda^{CC}\mu^{CC}} X_{12} \Delta^{-it}_{\mu^{CC}}, \\
\Delta^{it}_{\mu^{CC}\lambda^{CC}} X_{21} \Delta^{-it}_{\lambda^{CC}} &= \Delta^{it}_{\mu^{CC}} X_{21} \Delta^{-it}_{\lambda^{CC}\mu^{CC}}, \\
\sigma^{\mu^{CC}}_t(X_{22}) &= \Delta^{it}_{\mu^{CC}\lambda^{CC}} X_{22} \Delta^{-it}_{\mu^{CC}\lambda^{CC}}
\end{aligned}$$

for each $X_{11}, X_{12}, X_{21}, X_{22} \in \mathcal{M}$ and $t \in \mathbb{R}$, and they belong to $\mathcal{M}$ since $\mathcal{M}$ is a generalized von Neumann algebra. Hence we have

$$\sigma^{\mu^{CC}}_t(X_{11}) = [D\mu^{CC} : D\lambda^{CC}]_t \, \sigma^{\lambda^{CC}}_t(X_{11}) \, [D\mu^{CC} : D\lambda^{CC}]^*_t \\ , X_{11} \in \mathcal{M}, \quad t \in \mathbb{R}, \tag{2.5.15}$$

and so

$$\sigma_t^{\theta_{\lambda,\mu}^{CC}}(X)$$

$$= \begin{pmatrix} \sigma_t^{\lambda^{CC}}(X_{11}) & D_{\mu\lambda}^* \sigma_t^{\mu^{CC}}(X_{12}) & 0 & 0 \\ D_{\mu\lambda}\sigma_t^{\lambda^{CC}}(X_{21}) & \sigma_t^{\mu^{CC}}(X_{22}) & 0 & 0 \\ 0 & 0 & \sigma_t^{\lambda^{CC}}(X_{11}) & \sigma_t^{\lambda^{CC}}(X_{12})D_{\mu\lambda}^* \\ 0 & 0 & \sigma_t^{\mu^{CC}}(X_{21})D_{\mu\lambda} & \sigma_t^{\mu^{CC}}(X_{22}) \end{pmatrix}$$

$$= \begin{pmatrix} \sigma_t^{\lambda^{CC}}(X_{11}) & D_{\mu\lambda}^* \sigma_t^{\mu^{CC}}(X_{12}) & 0 & 0 \\ D_{\mu\lambda}\sigma_t^{\lambda^{CC}}(X_{21}) & \sigma_t^{\mu^{CC}}(X_{22}) & 0 & 0 \\ 0 & 0 & \sigma_t^{\lambda^{CC}}(X_{11}) & D_{\mu\lambda}^*\sigma_t^{\mu^{CC}}(X_{12}) \\ 0 & 0 & D_{\mu\lambda}\sigma_t^{\lambda^{CC}}(X_{21}) & \sigma_t^{\mu^{CC}}(X_{22}) \end{pmatrix}$$

$$\in \mathcal{M} \otimes M_2(\mathbb{C}) \tag{2.5.16}$$

for each $X \in \mathcal{M} \otimes M_2(\mathbb{C})$ and $t \in \mathbb{R}$, where $D_{\mu\lambda} \equiv [D\mu^{CC} : D\lambda^{CC}]_t$. Therefore $\theta_{\lambda,\mu}$ is a standard generalized vector for $\mathcal{M} \otimes M_2(\mathbb{C})$.

(ii) $\Rightarrow$ (i) This is trivial.

It is easily shown by (2.5.2) that $\theta_{\lambda,\mu}$ is full if and only if λ and μ are full. This completes the proof.

Remark 2.5.5. Suppose λ and μ are standard, semifinite generalized vectors for $\mathcal{M}$. We put

$$S_{\mu\lambda}\lambda(X) = \mu(X^\dagger),\ X \in \mathcal{D}(\lambda) \cap \mathcal{D}(\mu)^\dagger.$$

$$S_{\lambda\mu}\mu(X) = \lambda(X^\dagger),\ X \in \mathcal{D}(\lambda) \cap \mathcal{D}(\mu)^\dagger.$$

Then $S_{\mu\lambda}$ and $S_{\lambda\mu}$ are closable operators in $\mathcal{H}$ whose closures are denoted by the same $S_{\mu\lambda}$ and $S_{\lambda\mu}$,respectively. Let $S_{\mu\lambda} = J_{\mu\lambda}\Delta_{\mu\lambda}^{\frac{1}{2}}$ and $S_{\lambda\mu} = J_{\lambda\mu}\Delta_{\lambda\mu}^{\frac{1}{2}}$ be polar decompositions of $S_{\mu\lambda}$ and $S_{\lambda\mu}$, respectively. By Proposition 2.5.4 $\theta_{\lambda,\mu}$ is a standard generalized vector for $\mathcal{M} \otimes M_2(\mathbb{C})$, and so by Theorem 2.2.8 $S_{\theta_{\lambda,\mu}^{CC}} = S_{\theta_{\lambda,\mu}}$. Therefore, we have

$$S_{\lambda^{CC}} = S_\lambda,\quad S_{\mu^{CC}} = S_\mu,\quad S_{\mu^{CC}\lambda^{CC}} = S_{\mu\lambda},\quad S_{\lambda^{CC}\mu^{CC}} = S_{\lambda\mu}$$

and so

$$\Delta_{\lambda^{CC}} = \Delta_\lambda,\quad \Delta_{\mu^{CC}} = \Delta_\mu,\quad \Delta_{\mu^{CC}\lambda^{CC}} = \Delta_{\mu\lambda},\quad \Delta_{\lambda^{CC}\mu^{CC}} = \Delta_{\lambda\mu}.$$

Hence we have

$$[D\mu^{CC} : D\lambda^{CC}]_t = \Delta_\mu^{it}\Delta_{\lambda\mu}^{-it} = \Delta_{\mu\lambda}^{it}\Delta_\lambda^{-it},\quad t \in \mathbb{R}. \tag{2.5.17}$$

Theorem 2.5.6. (Generalized Connes cocycle theorem) Let $\mathcal{M}$ be a generalized von Neumann algebra on $\mathcal{D}$ in $\mathcal{H}$. Suppose λ and μ are full standard, semifinite generalized vectors for $\mathcal{M}$. Then there uniquely exists a strongly continous map $t \in \mathbb{R} \to U_t \in \mathcal{M}$ such that

(i) $\overline{U_t}$ is unitary, $t \in \mathbb{R}$;

(ii) $U_{t+s} = U_t \sigma_t^\lambda(U_s)$, $s, t \in \mathbb{R}$;

(iii) $\sigma_t^\mu(X) = U_t \sigma_t^\lambda(X) U_t^\dagger$, $X \in \mathcal{M}$, $t \in \mathbb{R}$;

(iv) for each $X \in \mathcal{D}(\mu) \cap \mathcal{D}(\lambda)^\dagger$ and $Y \in \mathcal{D}(\lambda) \cap \mathcal{D}(\mu)^\dagger$ there exists an element $F_{X,Y} \in A(0,1)$ such that

$$F_{X,Y}(t) = (\lambda(U_t \sigma_t^\lambda(Y)) \mid \lambda(X^\dagger)), \; F_{X,Y}(t+i) = (\mu(X) \mid \mu(U_t^\dagger \sigma_t^\mu(Y^\dagger))$$

for all $t \in \mathbb{R}$.

Proof. We put

$$U_t = [D\mu^{cc} : D\lambda^{cc}]_t \lceil \mathcal{D}, \quad t \in \mathbb{R}.$$

Then it follows from Proposition 2.5.4, (2.5.15), (2.5.16) and Theorem 3.1 in Stratila [1] that $t \in \mathbb{R} \to U_t \in \mathcal{M}$ is a strongly continous map satisfying (i) $\sim$ (iv). We show the uniqueness of $\{U_t\}_{t \in \mathbb{R}}$. Let $t \in \mathbb{R} \to V_t \in \mathcal{M}$ be a strongly continous map satisfying (i) $\sim$ (iv). We put

$$\delta_t\left(\begin{pmatrix} X_{11} & X_{12} & 0 & 0 \\ X_{21} & X_{22} & 0 & 0 \\ 0 & 0 & X_{11} & X_{12} \\ 0 & 0 & X_{21} & X_{22} \end{pmatrix}\right) = \begin{pmatrix} \sigma_t^\lambda(X_{11}) & V_t^* \sigma_t^\mu(X_{12}) & 0 & 0 \\ V_t \sigma_t^\lambda(X_{21}) & \sigma_t^\mu(X_{22}) & 0 & 0 \\ 0 & 0 & \sigma_t^\lambda(X_{11}) & V_t^* \sigma_t^\mu(X_{12}) \\ 0 & 0 & V_t \sigma_t^\lambda(X_{21}) & \sigma_t^\mu(X_{22}) \end{pmatrix}$$

$$, X \in \mathcal{M} \otimes M_2(\mathbb{C}), \quad t \in \mathbb{R}.$$

Then $\{\delta_t\}$ is a strongly continous one-parameter group of $*$-automorphisms of $\mathcal{M} \otimes M_2(\mathbb{C})$ such that $\delta_t(\mathcal{D}(\theta)^\dagger \cap \mathcal{D}(\theta)) \subset \mathcal{D}(\theta)^\dagger \cap \mathcal{D}(\theta)$ for each $t \in \mathbb{R}$, where $\theta \equiv \theta_{\lambda,\mu}$, and θ satisfies the KMS-condition with respect to $\{\delta_t\}$. Further, we have $\delta_t(X) = W_t X W_t^*$ for each $X \in \mathcal{M} \otimes M_2(\mathbb{C})$ and $t \in \mathbb{R}$, where

$$W_t = \begin{pmatrix} \Delta_t^{it} & & & O \\ & \overline{V_t \Delta_\lambda^{it}} & & \\ & & V_t^* \Delta_\mu^{it} & \\ O & & & \Delta_\mu^{it} \end{pmatrix}, \quad t \in \mathbb{R}.$$

Hence it follows that the one-parameter group $\{W_t\}_{t \in \mathbb{R}}$ of unitary operators satisfies the KMS-condition with respect to $\mathcal{K}$ ($\equiv$ the closure of $\{\theta(X); X^\dagger = X \in \mathcal{D}(\theta)^\dagger \cap \mathcal{D}(\theta)\}$) in the sense of Definition 3.4 in Rieffel-Van Daele [1] and $W_t \mathcal{K} \subset \mathcal{K}$ for each $t \in \mathbb{R}$, so that by Theorem 3.8 in Rieffel-Van Daele [1] $W_t = \Delta_\theta^{it}$ for each $t \in \mathbb{R}$. Hence it follows from (2.5.16) that $V_t = [D\mu^{cc} : D\lambda^{cc}]_t \lceil \mathcal{D} = U_t$ for all $t \in \mathbb{R}$. This completes the proof.

Let λ and μ be *full* standard, semifinite generalized vectors for $\mathcal{M}$. The map $t \in \mathbb{R} \to U_t \in \mathcal{M}$, uniquely determined by the above theorem, is called *the cocycle associated with* μ *with respect to* λ, and is denoted by $[D\mu : D\lambda]$.

For every standard vector ξ_0 for $\mathcal{M}$ λ_{ξ_0} is full and semifinite, and so the Connes cocycle $[D\lambda_{\eta_0} : D\lambda_{\xi_0}]$ associated with λ_{η_0} with respect to λ_{ξ_0} is defined for every standard vectors ξ_0 and η_0. We simply denote it by $[D\eta_0 : D\xi_0]$, and also denote it by $[D\omega_{\eta_0} : D\omega_{\xi_0}]$ according to the usual Connes cocycle associated with the state ω_{η_0} with respect to the state ω_{ξ_0}.

Suppose standard, semifinite generalized vectors λ and μ are not necessarily full, then we put $[D\mu : D\lambda]_t = [D\mu_e : D\lambda_e]_t$, $t \in \mathbb{R}$. Then $t \to [D\mu : D\lambda]_t$ is a strongly continuous map satisfying the conditions (i) $\sim$ (iv) in Theorem 2.5.6 and it is called the cocycle associated with μ with respect to λ. This equals the Connes cocycle $[D\mu^{cc} : D\lambda^{cc}]$ associated with μ^{cc} with respect to λ^{cc}.

Lemma 2.5.7. Let λ, μ_1 and μ_2 be full standard, semifinite generalized vectors for $\mathcal{M}$. Suppose $[D\mu_1 : D\lambda]_t = [D\mu_2 : D\lambda]_t$ for all $t \in \mathbb{R}$. Then $\mu_1 = \mu_2$.

Proof. By Corollary 3.6 in Stratila [1] we have $\mu_1^{cc} = \mu_2^{cc}$, and so $\mu_1^c = \mu_2^c$. Take an arbitrary $X \in \mathcal{D}(\mu_1)$. By there exists a sequence $\{X_n\}$ in $\mathcal{D}(\mu_1^{cc}) = \mathcal{D}(\mu_2^{cc})$ such that $\lim_{n\to\infty} X_n\xi = X\xi$ for each $\xi \in \mathcal{D}$ and $\lim_{n\to\infty} \mu_2^{cc}(X_n) = \lim_{n\to\infty} \mu_1^{cc}(X_n) = \mu_1(X)$. Hence we have

$$\begin{aligned} K\mu_1(X) = \lim_{n\to\infty} K\mu_2^{cc}(X_n) &= \lim_{n\to\infty} X_n\mu_2^c(K) \\ &= X\mu_2^c(K) \end{aligned}$$

for all $K \in \mathcal{D}(\mu_2^c)^* \cap \mathcal{D}(\mu_2^c)$, which implies by the fullness of μ_2 that $\mu_1 \subset \mu_2$. Similarly we can show $\mu_2 \subset \mu_1$.

We define the notion of relative modular pair (λ_1, λ_2) of semifinite, modular generalized vectors λ_1 and λ_2 to apply the generalized Connes cocycle theorem to more examples.

Definition 2.5.8. Let $\mathcal{M}$ be a closed O*-algebra on $\mathcal{D}$ in $\mathcal{H}$ such that $\mathcal{M}'_{\mathrm{w}}\mathcal{D} \subset \mathcal{D}$. A pair (λ_1, λ_2) of semifinite generalized vectors for $\mathcal{M}$ is said to be *relative modular* if the conditions $(\mathrm{S})_2$ and $(\mathrm{S})_3^{\mathrm{c}}$ in Definition 2.3.1 and the following condition (RM) hold:

(RM) There exists a dense subspace $\mathcal{E}$ of $\mathcal{D}[t_{\mathcal{M}}]$ such that

$(\mathrm{RM})_1$ $\lambda_i(\mathcal{D}(\lambda_i)^\dagger \cap \mathcal{D}(\lambda_i)) \subset \mathcal{E}$, $\quad i = 1, 2$;

$(\mathrm{RM})_2$ $\{\lambda_i^c(K_1K_2); K_1, K_2 \in \mathcal{D}(\lambda_i^c)^* \cap \mathcal{D}(\lambda_i^c)$ s.t. $\lambda^c(K_1), \lambda^c(K_1^*), \lambda^c(K_2), \lambda^c(K_2^*) \in \mathcal{E}\}$

is total in the Hilbert space $\mathcal{D}(S^*_{\lambda_i^{cc}})$, $i = 1, 2$;

$(\mathrm{RM})_3$ $\mathcal{M}\mathcal{E} \subset \mathcal{E}$;

$(\mathrm{RM})_4$ $\Delta^{it}_{\lambda_1^{cc}}\mathcal{E} \subset \mathcal{E}$ and $\Delta^{it}_{\lambda_2^{cc}}\mathcal{E} \subset \mathcal{E}$, $^\forall t \in \mathbb{R}$.

Similarly to Section 2.3 we can show the following

Lemma 2.5.9. Let $\mathcal{M}$ be a closed O*-algebra on $\mathcal{D}$ in $\mathcal{H}$ such that $\mathcal{M}'_{\mathrm{w}}\mathcal{D} \subset \mathcal{D}$, and λ_1 and λ_2 semifinite generalized vectors for $\mathcal{M}$. Suppose (λ_1, λ_2) is *relative modular.* We denote by $\mathcal{D}^{\mathrm{M}}_{(\lambda_1,\lambda_2)}$ the subspace of $\mathcal{D}$ generated by $\bigcup_{\mathcal{E}\in\mathcal{F}} \mathcal{E}$, where $\mathcal{F}$ is the set of all subspaces $\mathcal{E}$ of $\mathcal{D}$ satisfying the conditions $(\mathrm{RM})_1 \sim (\mathrm{RM})_4$ in Definition 2.5.8. The following statements hold:

(1) $\mathcal{M}'_{\mathrm{w}}\mathcal{D}^{\mathrm{M}}_{(\lambda_1,\lambda_2)} \subset \mathcal{D}^{\mathrm{M}}_{(\lambda_1,\lambda_2)}$, and so $(\mathcal{M}\lceil\mathcal{D}^{\mathrm{M}}_{(\lambda_1,\lambda_2)})''_{\mathrm{wc}}$ is a generalized von Neumann algebra on $\mathcal{D}^{\mathrm{M}}_{(\lambda_1,\lambda_2)}$ in $\mathcal{H}$ over $(\mathcal{M}'_{\mathrm{w}})'$.

(2) λ_1 and λ_2 are modular generalized vectors for $\mathcal{M}$ such that $\mathcal{D}^{\mathrm{M}}_{(\lambda_1,\lambda_2)} \subset \mathcal{D}^{\mathrm{M}}_{\lambda_1} \cap \mathcal{D}^{\mathrm{M}}_{\lambda_2}$.

(3) $(\lambda_1)_{RS} \equiv \overline{(\lambda_1)_{\mathcal{D}^{M}_{(\lambda_1,\lambda_2)}}}$ and $(\lambda_2)_{RS} \equiv \overline{(\lambda_2)_{\mathcal{D}^{M}_{(\lambda_1,\lambda_2)}}}$ are full standard, semifinite generalized vectors for the generalized von Neumann algebra $(\mathcal{M}\lceil\mathcal{D}^{\mathrm{M}}_{(\lambda_1,\lambda_2)})''_{\mathrm{wc}}$.

By Lemma 2.5.9 and Theorem 2.5.6 we have the following

Corollary 2.5.10. Suppose a pair (λ_1, λ_2) of semifinite generalized vectors is relative modular. Then

$$[D(\lambda_2)_{RS} : D(\lambda_1)_{RS}]_t \mathcal{D}^{\mathrm{M}}_{(\lambda_1,\lambda_2)} \subset \mathcal{D}^{\mathrm{M}}_{(\lambda_1,\lambda_2)},$$
$$\sigma_t^{\lambda_2^{cc}}(X)\xi = [D(\lambda_2)_{RS} : D(\lambda_1)_{RS}]_t \sigma_t^{\lambda_1^{cc}}(X)[D(\lambda_2)_{RS} : D(\lambda_1)_{RS}]_t^* \xi$$

for each $X \in \mathcal{M}$, $\xi \in \mathcal{D}^{\mathrm{M}}_{(\lambda_1,\lambda_2)}$ and $t \in \mathbb{R}$.

As seen above, for any relative modular pair (λ_1, λ_2) of semifinite generalized vectors we can apply all results obtained for the full standard, semifinite generalized vectors $(\lambda_1)_{RS}$ and $(\lambda_2)_{RS}$ for the generalized von Neumann algebra $(\mathcal{M}\lceil\mathcal{D}^{\mathrm{M}}_{(\lambda_1,\lambda_2)})''_{\mathrm{wc}}$ to it.

2.6 Generalized Pedersen and Takesaki Radon-Nikodym theorem

In this section we construct the standard, semifinite generalized vector λ_A associated with a given full standard, semifinite generalized vector λ and a given positive self-adjoint operator A affiliated with the centralizer of λ, and consider when a full standard, semifinite generalized vector μ is represented as the full extension of such a λ_A.

Let $\mathcal{M}$ be a generalized von Neumann algebra on $\mathcal{D}$ in a Hilbert space $\mathcal{H}$ and λ a standard generalized vector for $\mathcal{M}$. We put

$$\mathcal{M}^{\sigma^\lambda} = \{A \in \mathcal{M}; A\Delta_\lambda^{it} \supset \Delta_\lambda^{it} A, \ \ \forall t \in \mathbb{R}\}, \ \ \mathcal{M}_b^{\sigma^\lambda} = \mathcal{M}_b \cap \mathcal{M}^{\sigma^\lambda}.$$

Then $\mathcal{M}^{\sigma^\lambda}$ and $\mathcal{M}_b^{\sigma^\lambda}$ are O^*-subalgebras of $\mathcal{M}$.

Lemma 2.6.1. Let λ be a full standard generalized vector for $\mathcal{M}$. Then the following statements hold.

(1) Suppose $A \in \mathcal{M}_b$ such that $\Delta_\lambda^{\frac{1}{2}} A^\dagger \Delta_\lambda^{-\frac{1}{2}}$ is bounded. Then $XA \in \mathcal{D}(\lambda)^\dagger \cap \mathcal{D}(\lambda)$ and $\lambda(XA) = J_\lambda \Delta_\lambda^{\frac{1}{2}} A^\dagger \Delta_\lambda^{-\frac{1}{2}} J_\lambda \lambda(X)$ for each $X \in \mathcal{D}(\lambda)^\dagger \cap \mathcal{D}(\lambda)$.

(2) Suppose $X \in \mathcal{D}(\lambda)^\dagger \cap \mathcal{D}(\lambda)$ and $A \in \mathcal{M}$ such that $XA \in \mathcal{D}(\lambda)^\dagger \cap \mathcal{D}(\lambda)$. Then $\lambda(XA) = J_\lambda \Delta_\lambda^{\frac{1}{2}} A^\dagger \Delta_\lambda^{-\frac{1}{2}} J_\lambda \lambda(X)$.

(3) Suppose $A \in \mathcal{M}_b^{\sigma^\lambda}$. Then $XA \in \mathcal{D}(\lambda)$ and $\lambda(XA) = J_\lambda A^* J_\lambda \lambda(X)$ for each $X \in \mathcal{D}(\lambda)$.

Proof. (1) Since $\Delta_\lambda^{\frac{1}{2}} A^\dagger \Delta_\lambda^{-\frac{1}{2}}$ is bounded, it follows that

$$A^\dagger \lambda(X^\dagger) \in \mathcal{D}(S_\lambda) \text{ and } S_\lambda A^\dagger \lambda(X^\dagger) = J_\lambda \Delta_\lambda^{\frac{1}{2}} A^\dagger \Delta_\lambda^{-\frac{1}{2}} J_\lambda \lambda(X),$$

which implies

$$\begin{aligned}
(XA\lambda^c(K) \mid \lambda^c(K_1)) &= (\lambda^c(K) \mid A^\dagger X^\dagger \lambda^c(K_1)) \\
&= (\lambda^c(K) \mid K_1 \lambda(A^\dagger X^\dagger)) \\
&= (\lambda^c(K_1^* K) \mid \lambda(A^\dagger X^\dagger)) \\
&= (S_\lambda^* \lambda^c(K^* K_1) \mid \lambda(A^\dagger X^\dagger)) \\
&= (S_\lambda \lambda(A^\dagger X^\dagger) \mid \lambda^c(K^* K_1)) \\
&= (K J_\lambda \Delta_\lambda^{\frac{1}{2}} A^\dagger \Delta_\lambda^{-\frac{1}{2}} J_\lambda \lambda(X) \mid \lambda^c(K_1))
\end{aligned}$$

for each $K, K_1 \in \mathcal{D}(\lambda^c)^* \cap \mathcal{D}(\lambda^c)$. Hence we have

$$XA\lambda^c(K) = K J_\lambda \Delta_\lambda^{\frac{1}{2}} A^\dagger \Delta_\lambda^{-\frac{1}{2}} J_\lambda \lambda(X)$$

for each $K \in \mathcal{D}(\lambda^c)^* \cap \mathcal{D}(\lambda^c)$. Since $J_\lambda \Delta_\lambda^{\frac{1}{2}} A^* \Delta_\lambda^{-\frac{1}{2}} J_\lambda \in \mathcal{M}'_{\mathrm{w}}$ and λ is full, it follows that $XA \in \mathcal{D}(\lambda)$ and $\lambda(XA) = J_\lambda \Delta_\lambda^{\frac{1}{2}} A^\dagger \Delta_\lambda^{-\frac{1}{2}} J_\lambda \lambda(X)$.

(2) This follows from

$$\begin{aligned}
(\lambda(XA) \mid \lambda^c(K^* K_1)) &= (XA\lambda^c(K) \mid \lambda^c(K_1)) \\
&= (\lambda^c(K_1^* K) \mid A^\dagger \lambda(X^\dagger)) \\
&= (S_\lambda^* \lambda^c(K^* K_1) \mid A^\dagger \lambda(X^\dagger)) \\
&= (S_\lambda A^\dagger \lambda(X^\dagger) \mid \lambda^c(K^* K_1)) \\
&= (J_\lambda \Delta_\lambda^{\frac{1}{2}} A^\dagger \Delta_\lambda^{-\frac{1}{2}} J_\lambda \lambda(X) \mid \lambda^c(K^* K_1))
\end{aligned}$$

for each $X \in \mathcal{D}(\lambda)^\dagger \cap \mathcal{D}(\lambda)$ and $K, K_1 \in \mathcal{D}(\lambda^C)^* \cap \mathcal{D}(\lambda^C)$.

(3) We first show

$$A\lambda^C(K) \in \mathcal{D}(S_\lambda^*) \text{ and } S_\lambda^* A\lambda^C(K) = J_\lambda \overline{A} \lambda_\lambda \lambda^C(K^*) \tag{2.6.1}$$

for each $K \in \mathcal{D}(\lambda^C)^* \cap \mathcal{D}(\lambda^C)$. This follows from

$$\begin{aligned}(S_\lambda\lambda(Y) \mid A\lambda^C(K)) &= (A^* \Delta_\lambda^{-\frac{1}{2}} J_\lambda \lambda(Y) \mid \lambda^C(K)) \\ &= (\Delta_\lambda^{-\frac{1}{2}} A^* J_\lambda \lambda(Y) \mid \lambda^C(K)) \\ &= (\lambda^C(K^*) \mid J_\lambda A^* J_\lambda \lambda(Y)) \\ &= (J_\lambda \overline{A} J_\lambda \lambda^C(K^*) \mid \lambda(Y))\end{aligned}$$

for each $Y \in \mathcal{D}(\lambda)^\dagger \cap \mathcal{D}(\lambda)$. By (2.6.1) we have

$$\begin{aligned}(XA\lambda^C(K) \mid \lambda(Y)) &= (A\lambda^C(K) \mid \lambda(X^\dagger Y)) \\ &= (\lambda(Y^\dagger X) \mid S_\lambda^* A\lambda^C(K)) \\ &= (\lambda(Y^\dagger X) \mid J_\lambda \overline{A} J_\lambda \lambda^C(K^*)) \\ &= (\lambda(X) \mid J_\lambda \overline{A} J_\lambda K^* \lambda(Y)) \\ &= (K J_\lambda A^* J_\lambda \lambda(X) \mid \lambda(Y))\end{aligned}$$

for each $K \in \mathcal{D}(\lambda^C)^* \cap \mathcal{D}(\lambda^C)$ and $Y \in \mathcal{D}(\lambda)^\dagger \cap \mathcal{D}(\lambda)$, which implies by the fullness of λ that $XA \in \mathcal{D}(\lambda)$ and $\lambda(XA) = J_\lambda A^* J_\lambda \lambda(X)$.

Theorem 2.6.2. Let $\mathcal{M}$ be a generalized von Neumann algebra on $\mathcal{D}$ in $\mathcal{H}$ and λ and μ full standard, semifinite generalized vectors for $\mathcal{M}$. Then the following statements are equivalent.

(i) $\mathcal{D}(\mu)$ is $\{\sigma_t^\lambda\}$-invariant and $\|\mu(\sigma_t^\lambda(X))\| = \|\mu(X)\|$ for all $X \in D(\mu)$.

(i)′ $\mathcal{D}(\lambda)$ is $\{\sigma_t^\mu\}$-invariant and $\|\lambda(\sigma_t^\mu(X))\| = \|\lambda(X)\|$ for all $X \in \mathcal{D}(\lambda)$.

(ii) $[D\mu : D\lambda]_t \in \mathcal{M}^{\sigma^\mu}, \quad \forall t \in \mathbb{R}$.

(ii)′ $[D\mu : D\lambda]_t \in \mathcal{M}^{\sigma^\lambda}, \quad \forall t \in \mathbb{R}$.

(iii) $\{[D\mu : D\lambda]_t\}_{t\in\mathbb{R}}$ is a strongly continous one-parameter group of unitary elements of $\mathcal{M}$.

Proof. The equivalence of (ii), (ii)′ and (iii) follows from Theorem 2.5.6.

(i) $\Rightarrow$ (ii) We now put $U_t = [D\mu : D\lambda]_t, \quad t \in \mathbb{R}$. Take an arbitrary $t \in \mathbb{R}$ and put $A = \sigma_{-t}^\mu(U_t)$. By Theorem 2.5.6 we have

$$\begin{aligned}XA = X\sigma_{-t}^\mu(U_t) &= \sigma_{-t}^\mu(\sigma_t^\mu(X)U_t) \\ &= \sigma_{-t}^\mu(U_t\sigma_t^\lambda(X))\end{aligned}$$

for all $X \in \mathcal{D}(\mu)^\dagger \cap \mathcal{D}(\mu)$, so that by the assumption (i) that

$$XA, YA \in \mathcal{D}(\mu)^\dagger \cap \mathcal{D}(\mu) \text{ and } (\mu(X) \mid \mu(Y)) = (\mu(XA) \mid \mu(YA)) \quad (2.6.2)$$

for all $X, Y \in \mathcal{D}(\mu)^\dagger \cap \mathcal{D}(\mu)$, and further by Lemma 2.6.2, (2)

$$\begin{aligned}\|\mu(X)\| = \|\mu(\sigma_t^\lambda(X))\| &= \|\mu(U_t^* \sigma_t^\mu(X) U_t)\| \\ &= \|\mu(XA)\| \\ &= \|J_\mu \Delta_\mu^{\frac{1}{2}} A^\dagger \Delta_\mu^{-\frac{1}{2}} J_\mu \mu(X)\|\end{aligned}$$

for all $X \in \mathcal{D}(\mu)^\dagger \cap \mathcal{D}(\mu)$. Hence, $J_\mu \Delta_\mu^{\frac{1}{2}} A^\dagger \Delta_\mu^{-\frac{1}{2}} J_\mu$ is bounded. Furthermore, since $\mathcal{D}(\mu)^\dagger \cap \mathcal{D}(\mu)$ is $\{\sigma_t^\mu\}$-invariant and $\{\sigma_t^\lambda\}$-invariant, it follows from Theorem 2.5.6 that

$$XU_s^* = U_s^* \sigma_s^\mu(\sigma_{-s}^\lambda(X)) \in \mathcal{D}(\mu)^\dagger \cap \mathcal{D}(\mu)$$

for all $X \in \mathcal{D}(\mu)^\dagger \cap \mathcal{D}(\mu)$ and $s \in \mathbb{R}$, which implies

$$X\sigma_s^\mu(U_s^*) \in \mathcal{D}(\mu)^\dagger \cap \mathcal{D}(\mu)$$

for all $X \in \mathcal{D}(\mu)^\dagger \cap \mathcal{D}(\mu)$ and $s \in \mathbb{R}$. Hence, by (2.6.2) we have

$$XA^\dagger, YA^\dagger \in \mathcal{D}(\mu)^\dagger \cap \mathcal{D}(\mu) \text{ and } (\mu(X) \mid \mu(YA)) = (\mu(XA^\dagger) \mid \mu(Y))$$

for all $X, Y \in \mathcal{D}(\mu)^\dagger \cap \mathcal{D}(\mu)$, which implies by Lemma 2.6.1, (2) that

$$\begin{aligned}(\mu(X) \mid J_\mu \Delta_\mu^{\frac{1}{2}} A^\dagger \Delta_\mu^{-\frac{1}{2}} J_\mu \mu(Y)) &= (\mu(X) \mid \mu(YA)) \\ &= (\mu(XA^\dagger) \mid \mu(Y)) \\ &= (J_\mu \Delta_\mu^{\frac{1}{2}} A \Delta_\mu^{-\frac{1}{2}} J_\mu \mu(X) \mid \mu(Y)) \\ &= (\mu(X) \mid (J_\mu \Delta_\mu^{\frac{1}{2}} A \Delta_\mu^{-\frac{1}{2}} J_\mu)^* \mu(Y))\end{aligned}$$

for each $X, Y \in \mathcal{D}(\mu)^\dagger \cap \mathcal{D}(\mu)$. Hence we have

$$\overline{J_\mu \Delta_\mu^{\frac{1}{2}} A^\dagger \Delta_\mu^{-\frac{1}{2}} J_\mu} = (J_\mu \Delta_\mu^{\frac{1}{2}} A \Delta_\mu^{-\frac{1}{2}} J_\mu)^*,$$

which implies $\overline{A}\Delta_\mu \subset \Delta_\mu \overline{A}$. Therefore it follows that $U_t \in \mathcal{M}^{\sigma^\mu}$ for all $t \in \mathbb{R}$.

(ii) $\Rightarrow$ (i) It follows from Theorem 2.5.6 and Lemma 2.6.1, (3) that

$$\sigma_t^\lambda(X) = U_t^* \sigma_t^\mu(X) U_t \in \mathcal{D}(\mu)$$

and

$$\begin{aligned}\|\mu(\sigma_t^\lambda(X))\| &= \|\mu(U_t^* \sigma_t^\mu(X) U_t)\| \\ &= \|J_\mu U_t^* J_\mu \mu(\sigma_t^\mu(X))\| \\ &= \|\mu(X)\|\end{aligned}$$

for each $X \in \mathcal{D}(\mu)$ and $t \in \mathbb{R}$.

(i)′ $\Leftrightarrow$ (ii) This is proved similarly to the proof of the equivalence of (i) and (ii).

This completes the proof.

If the equivalent conditions in Theorem 2.6.2 are satisfied, we say that μ *commutes with* λ. If μ commutes with λ, then

$$\sigma_t^\lambda \circ \sigma_t^\mu = \sigma_t^\mu \circ \sigma_t^\lambda, \quad t \in \mathbb{R}.$$

But, the converse is not necessarily true even in the bounded case (4.15 in Stratila [1]).

We next present the canonical construction and the properties of the generalized vector λ_A associated with a given full standard, semifinite generalized vector λ and a given positive self-adjoint operator A affiliated with the centralizer of λ. We investigate when a full standard, semifinite generalized vector μ for $\mathcal{M}$ which commutes with λ is represented as $(\lambda_A)_e$.

Let λ be a full standard, semifinite generalized vector for $\mathcal{M}$ and $\mathcal{M}_\eta^{\sigma^\lambda}$ the set of all non-singular positive self-adjoint operators A in $\mathcal{H}$ satisfying $\{E_A(t); -\infty < t < \infty\}''\lceil\mathcal{D} \subset \mathcal{M}_b^{\sigma^\lambda}$, where $\{E_A(t)\}$ is the spectral resolution of A. Let $A \in \mathcal{M}_\eta^{\sigma^\lambda}$ and put

$$\begin{cases} \mathcal{D}(\lambda_A) = \{X \in \mathcal{D}(\lambda); \lambda(YX) \in \mathcal{D}(J_\lambda A J_\lambda) \text{ for all } Y \in \mathcal{M}\}, \\ \lambda_A(X) = J_\lambda A J_\lambda \lambda(X), \quad X \in \mathcal{D}(\lambda_A). \end{cases}$$

Then we have the following

Lemma 2.6.3. λ_A is a standard, semifinite generalized vector for $\mathcal{M}$ satisfying

$$\sigma_t^{\lambda_A}(X) = A^{2it}\sigma_t^\lambda(X)A^{-2it},$$

$$[D\lambda_A : D\lambda]_t \equiv [D(\lambda_A)_e : D\lambda]_t = A^{2it}\lceil\mathcal{D}, \; X \in \mathcal{M}, \quad t \in \mathbb{R}.$$

Proof. It is clear that λ_A is a generalized vector for $\mathcal{M}$. By Lemma 2.6.1, (3) we have

$$E_A(n)XE_A(m) \in \mathcal{D}(\lambda)^\dagger \cap \mathcal{D}(\lambda),$$
$$\lambda(E_A(n)XE_A(m)) = J_\lambda E_A(m) J_\lambda E_A(n)\lambda(X)$$

for each $n, m \in \mathbb{N}$ and $X \in \mathcal{D}(\lambda)^\dagger \cap \mathcal{D}(\lambda)$, which implies $E_A(n)XE_A(m) \in \mathcal{D}(\lambda_A)^\dagger \cap \mathcal{D}(\lambda_A)$. Further, since λ is semifinite, it follows that λ_A is semifinite. Since

$$\{E_A(n)XA^{-1}E_A(m); X \in \mathcal{D}(\lambda)^\dagger \cap \mathcal{D}(\lambda), m, n, \in \mathbb{N}\} \subset \mathcal{D}(\lambda_A)^\dagger \cap \mathcal{D}(\lambda_A)$$

and

$$\lambda_A(E_A(n)XA^{-1}E_A(m)) = E_A(n)J_\lambda E_A(m)J_\lambda\lambda(X) \longrightarrow \lambda(X) \ (m,n\to\infty),$$

it follows that λ_A is cyclic, so that by Lemma 2.5.2

$$\lambda_A((\mathcal{D}(\lambda_A)^\dagger \cap \mathcal{D}(\lambda_A))^2) \text{ is dense in } \mathcal{H}. \tag{2.6.3}$$

We put

$$\mathcal{K} = \{K \in \mathcal{D}(\lambda^C); \lambda^C(K) \in \mathcal{D}(A) \cap \mathcal{D}(J_\lambda A^{-1} J_\lambda) \text{ and } A\lambda^C(K) \in \mathcal{D}\}.$$

Then we have

$$\mathcal{K} \subset \mathcal{D}(\lambda_A^C) \text{ and } \lambda_A^C(K) = A\lambda^C(K), \ \ \forall K \in \mathcal{K}. \tag{2.6.4}$$

In fact, we have by Lemma 2.6.1, (3)

$$\lim_{n\to\infty} AE_A(n)\lambda^C(K) = A\lambda^C(K),$$

$$\begin{aligned}\lim_{n\to\infty} XAE_A(n)\lambda^C(K) &= \lim_{n\to\infty} K\lambda(XAE_A(n)) \\ &= \lim_{n\to\infty} KJ_\lambda AE_A(n)J_\lambda\lambda(X) \\ &= KJ_\lambda AJ_\lambda\lambda(X) \\ &= K\lambda_A(X)\end{aligned}$$

for each $X \in \mathcal{D}(\lambda_A)$ and $K \in \mathcal{K}$, which implies the statement (2.6.4) is true. We put

$$K_{mn} = J_\lambda E_A(m)J_\lambda KJ_\lambda E_A(n)J_\lambda$$

for $K \in \mathcal{D}(\lambda^C)^* \cap \mathcal{D}(\lambda^C)$ and $m,n \in \mathbb{N}$. Then we have

$$\begin{aligned}K_{mn}\lambda(X) &= (J_\lambda E_A(m)J_\lambda)K(J_\lambda E_A(n)J_\lambda)\lambda(X) \\ &= (J_\lambda E_A(m)J_\lambda)K\lambda(XE_A(n)) \\ &= (J_\lambda E_A(m)J_\lambda)XE_A(n)\lambda^C(K) \\ &= X(J_\lambda E_A(m)J_\lambda)E_A(n)\lambda^C(K)\end{aligned}$$

and

$$K_{mn}^*\lambda(X) = X(J_\lambda E_A(n)J_\lambda)E_A(m)\lambda^C(K^*)$$

for each $X \in \mathcal{D}(\lambda)$, so that

$$\begin{aligned}&K_{mn} \in \mathcal{K} \cap \mathcal{K}^*, \\ &\lambda^C(K_{mn}) = (J_\lambda E_A(m)J_\lambda)E_A(n)\lambda^C(K), \\ &\lambda^C(K_{mn}^*) = (J_\lambda E_A(n)J_\lambda)E_A(m)\lambda^C(K^*).\end{aligned} \tag{2.6.5}$$

Hence we have

$$\begin{aligned}
&C_{mn}K_{mn} \in (\mathcal{K}\cap\mathcal{K}^*)^2,\\
&\lim_{m,n\to\infty} \lambda^c(C_{mn}K_{mn}) = \lim_{m,n\to\infty} C_{mn}\lambda^c(K_{mn})\\
&\qquad = \lim_{m,n\to\infty} C_{mn}(J_\lambda E_A(m)J_\lambda)E_A(n)\lambda^c(K)\\
&\qquad = C\lambda^c(K) = \lambda^c(CK),\\
&\lim_{m,n\to\infty} \lambda^c((C_{mn}K_{mn})^*) = \lambda^c((CK)^*)
\end{aligned}$$

for each $C, K \in \mathcal{D}(\lambda^c)^* \cap \mathcal{D}(\lambda^c)$, which implies that

$$\lambda^c((\mathcal{K}\cap\mathcal{K}^*)^2) \text{ is total in the Hilbert space } \mathcal{D}(S_\lambda^*). \tag{2.6.6}$$

For each $K \in \mathcal{K}\cap\mathcal{K}^*$ and $n \in \mathbb{N}$ we have

$$\begin{aligned}
K_n &\equiv KJ_\lambda A^{-1}E_A(n)J_\lambda \in \mathcal{K}\cap\mathcal{K}^*,\\
\lambda^c(K_n) &= A^{-1}E_A(n)\lambda^c(K),\\
\lambda^c(K_n^*) &= J_\lambda A^{-1}E_A(n)J_\lambda\lambda^c(K^*)
\end{aligned}$$

and so by (2.6.4)

$$\begin{aligned}
\lim_{n\to\infty}\lambda_A^c(CK_n) = \lim_{n\to\infty} C\lambda_A^c(K_n) &= \lim_{n\to\infty} CE_A(n)\lambda^c(K)\\
&= C\lambda^c(K)
\end{aligned}$$

for each $C \in \mathcal{K}\cap\mathcal{K}^*$. Hence it follows from (2.6.6) that $\lambda_A^c((\mathcal{K}\cap\mathcal{K}^*)^2)$ is total in $\mathcal{H}$, which implies by (2.6.4) that

$$\lambda_A^c((\mathcal{D}(\lambda_A^c)^*\cap\mathcal{D}(\lambda_A^c))^2) \text{ is total in } \mathcal{H}. \tag{2.6.7}$$

For each $K \in \mathcal{K}\cap\mathcal{K}^*$ we have by (2.6.4) and (2.6.5)

$$\begin{aligned}
\lim_{m,n\to\infty}\lambda_A^c(K_{mn}) &= \lim_{m,n\to\infty} AE_A(n)J_\lambda E_A(m)J_\lambda\lambda^c(K)\\
&= A\lambda^c(K) = \lambda_A^c(K),\\
\lim_{m,n\to\infty}\lambda_A^c(K_{mn}^*) &= \lim_{m,n\to\infty} AE_A(m)J_\lambda E_A(n)J_\lambda\lambda^c(K^*)\\
&= \lambda_A^c(K^*).
\end{aligned}$$

Furthermore, for each $C \in \mathcal{D}(\lambda_A^c)^*\cap\mathcal{D}(\lambda_A^c)$ and $m, n \in \mathbb{N}$ we put

$$C_{mn} = (J_\lambda E_A(m)J_\lambda)C(J_\lambda E_A(n)J_\lambda).$$

Then we have

$$\begin{aligned}
C_{mn}\lambda(X) &= (J_\lambda E_A(m)J_\lambda)C\lambda_A(XA^{-1}E_A(n))\\
&= (J_\lambda E_A(m)J_\lambda)XA^{-1}E_A(n)\lambda_A^c(C)\\
&= X(J_\lambda E_A(m)J_\lambda)A^{-1}E_A(n)\lambda_A^c(C),\\
C_{mn}^*\lambda(X) &= X(J_\lambda E_A(n)J_\lambda)A^{-1}E_A(m)\lambda_A^c(C^*),
\end{aligned}$$

and so by (2.6.4) and (2.6.5)

$$\begin{aligned} &C_{mn} \in \mathcal{K} \cap \mathcal{K}^*, \\ &\lim_{m,n\to\infty} \lambda_A^C(C_{mn}) = \lim_{m,n\to\infty} (J_\lambda E_A(m) J_\lambda) E_A(n) \lambda_A^C(C) \\ &\qquad\qquad\qquad\quad = \lambda_A^C(C), \\ &\lim_{m,n\to\infty} \lambda_A^C(C_{mn}^*) = \lambda_A^C(C^*). \end{aligned}$$

Therefore it follows that

$$\{\lambda_A^C(K_{mn})\ ;\ K \in \mathcal{K} \cap \mathcal{K}^*,\ \ m,n \in \mathbb{N}\} \quad \text{is total in the Hilbert space } \mathcal{D}(S^*_{\lambda_A^C C}). \tag{2.6.8}$$

For each $K \in \mathcal{K} \cap \mathcal{K}^*$ and $m,n \in \mathbb{N}$ we have by (2.6.4) and (2.6.5)

$$\begin{aligned} S^*_{\lambda_A^C C} \lambda_A^C(K_{mn}) &= A E_A(n) J_\lambda E_A(m) J_\lambda \lambda^C(K^*) \\ &= A E_A(n) J_\lambda E_A(m) J_\lambda S_\lambda^* \lambda^C(K) \\ &= S_\lambda^* J_\lambda A E_A(n) J_\lambda E_A(m) \lambda^C(K) \\ &= S_\lambda^* J_\lambda A E_A(n) J_\lambda A^{-1} E_A(m) \lambda_A^C(K), \end{aligned}$$

and so

$$\lambda_A^C(K) \in \mathcal{D}(\overline{S_\lambda^* J_\lambda A J_\lambda A^{-1}}) \text{ and } \overline{S_\lambda^* J_\lambda A J_\lambda A^{-1}} \lambda_A^C(K) = S^*_{\lambda_A^C C} \lambda_A^C(K)$$

for each $K \in \mathcal{K} \cap \mathcal{K}^*$. By (2.6.8) we have

$$S^*_{\lambda_A^C C} \subset \overline{S_\lambda^* J_\lambda A J_\lambda A^{-1}}. \tag{2.6.9}$$

Similarly we have

$$S_\lambda^* \subset \overline{S^*_{\lambda_A^C C} J_\lambda A^{-1} J_\lambda A}. \tag{2.6.10}$$

By (2.6.9) and (2.6.10) we have

$$S^*_{\lambda_A^C C} = \overline{\bar{S}_\lambda^* J_\lambda A J_\lambda A^{-1}} = J_\lambda \overline{\Delta_\lambda^{-\frac{1}{2}} J_\lambda A J_\lambda A^{-1}}. \tag{2.6.11}$$

Since A is affiliated with $(\mathcal{M}_w'')^{\sigma^\lambda}$, it follows that the two self-adjoint operators $\Delta_\lambda^{-\frac{1}{2}}$ and $\overline{J_\lambda A J_\lambda A^{-1}}$ are strongly commuting, that is, the spectral projections of the two self-adjoint operators are mutually commuting, and so $\overline{\Delta_\lambda^{-\frac{1}{2}} J_\lambda A J_\lambda A^{-1}}$ is self-adjoint and it equals $\overline{J_\lambda A J_\lambda A^{-1} \Delta_\lambda^{-\frac{1}{2}}}$. Hence, it follows from (2.6.11) and the uniqueness of the polar decomposition of $S^*_{\lambda_A^C C}$, it follows that

$$J_{\lambda_A^{CC}} = J_\lambda \text{ and } \Delta_{\lambda_A^{CC}}^{-\frac{1}{2}} = \overline{\Delta_\lambda^{-\frac{1}{2}} J_\lambda A J_\lambda A^{-1}} = \overline{J_\lambda A J_\lambda A^{-1} \Delta_\lambda^{-\frac{1}{2}}},$$

which implies

$$\Delta_{\lambda_A^{CC}}^{it} = J_\lambda A^{-2it} J_\lambda A^{2it} \Delta_\lambda^{it} \text{ and } \sigma_t^{\lambda_A^{CC}}(X) = A^{2it}\sigma_t^\lambda(X)A^{-2it}$$

for $X \in \mathcal{M}$ and $t \in \mathbb{R}$. Hence it follows from Lemma 2.6.1,(3) that

$$\sigma_t^{\lambda_A^{cc}}(\mathcal{D}(\lambda_A)^\dagger \cap \mathcal{D}(\lambda_A)) \subset \mathcal{D}(\lambda_A)^\dagger \cap \mathcal{D}(\lambda_A), \quad t \in \mathbb{R}.$$

Therefore λ_A is a standard generalized vector for $\mathcal{M}$. Further, it follows from Theorem 2.5.6 that $[D\lambda_A : D\lambda]_t \equiv [D(\lambda_A)_e : D\lambda]_t = A^{2it}\lceil\mathcal{D}$ for $t \in \mathbb{R}$. This completes the proof.

Let λ and μ be full standard, semifinite generalized vectors for $\mathcal{M}$. Suppose μ commutes with λ. Then it follows from Theorem 2.6.2 that $\{[D\mu : D\lambda]_t\}_{t\in\mathbf{R}}$ is a strongly continuous one-parameter group of unitary operators in $(\mathcal{M}'_{\mathrm{w}})'^{\sigma^\lambda}$, and so by the Stone theorem there exists a unique non-singular positive self-adjoint operator $A_{\lambda,\mu}$ affiliated with $(\mathcal{M}'_{\mathrm{w}})'^{\sigma^\lambda}$ such that $[D\mu; D\lambda]_t = A_{\lambda,\mu}^{2it}\lceil\mathcal{D}$ for all $t \in \mathbb{R}$. By Lemma 2.5.7, 2.6.3 we have the following generalized Pedersen-Takesaki Radon Nikodym theorem :

Theorem 2.6.4. Let $\mathcal{M}$ be a generalized von Neumann algebra on $\mathcal{D}$ in $\mathcal{H}$ and λ and μ full standard, semifinite generalized vectors for $\mathcal{M}$. Suppose $A_{\lambda,\mu} \in \mathcal{M}_\eta^{\sigma^\lambda}$. Then $\mu = (\lambda_{A_{\lambda,\mu}})_e$.

Corollary 2.6.5. Let $\mathcal{M}$ be an EW^*-algebra on $\mathcal{D}$ in $\mathcal{H}$ and λ and μ full standard generalized vectors for $\mathcal{M}$. Then μ commutes with λ if and only if $\mu = (\lambda_A)_e$ for some non-singular positive self-adjoint operator A affiliated with $(\mathcal{M}'_{\mathrm{w}})'^{\sigma^\lambda}$.

Proof. Since $\mathcal{M}$ is an EW*-algebra on $\mathcal{D}$ in $\mathcal{H}$, it follows that for every full standard generalized vector λ for $\mathcal{M}$, $\{A\lceil\mathcal{D}; A \in \mathcal{D}(\lambda^{cc})\} \subset \mathcal{D}(\lambda)$, which implies that λ is semifinite. Suppose μ commutes with λ. Since $\mathcal{M}$ is an EW^*-algebra on $\mathcal{D}$ in $\mathcal{H}$, we have $A_{\lambda,\mu} \in \mathcal{M}_\eta^{\sigma^\lambda}$, and so $\mu = (\lambda_{A_{\lambda,\mu}})_e$ by Theorem 2.6.4. The converse follows from Lemma 2.6.3.

Theorem 2.6.6. Let $\mathcal{M}$ be a generalized von Neumann algebra on $\mathcal{D}$ in $\mathcal{H}$ and λ and μ full standard, semifinite generalized vectors for $\mathcal{M}$. Then the following statements are equivalent.

(i) μ satisfies the KMS-condition with respect to $\{\sigma_t^\lambda\}$.

(ii) $\sigma_t^\mu = \sigma_t^\lambda$ for each $t \in \mathbb{R}$.

(iii) There exists a non-singular positive self-adjoint operator A affiliated with the center of $(\mathcal{M}'_{\mathrm{w}})'$ such that $\mu = (\lambda_A)_e$.

Proof. (i) ⇒ (ii) This follows from Corollary 4.11 in Stratila [1].
(ii) ⇒ (i) This is trivial.
(ii) ⇒ (iii) By Theorem 2.5.6 we have

$$\sigma_t^{\lambda}(X) = \sigma_t^{\mu}(X) = A_{\lambda,\mu}^{2it}\sigma_t^{\lambda}(X)A_{\lambda,\mu}^{-2it}, \quad X \in \mathcal{M}, \quad t \in \mathbb{R},$$

which implies that $A \equiv A_{\lambda,\mu}$ is affiliated with the center of $(\mathcal{M}'_{\mathrm{w}})'$, so that we can show similarly to the proof of Lemma 2.6.3 that λ_A is a standard, semifinite generalized vector for $\mathcal{M}$ such that $[D\lambda_A : D\lambda]_t = A^{2it}\lceil\mathcal{D}$ for all $t \in \mathbb{R}$. Hence it follows from Lemma 2.5.7 that $\mu = (\lambda_A)_e$.

(iii) ⇒ (ii) Since A is affiliated with the center of $(\mathcal{M}'_{\mathrm{w}})'$, it follows from Lemma 2.6.3 that $A \in \mathcal{M}_\eta^{\sigma^\lambda}$, λ_A is a standard, semifinite generalized vector for $\mathcal{M}$ and

$$\sigma_t^{\mu}(X) = \sigma_t^{(\lambda_A)_e}(X) = A^{2it}\sigma_t^{\lambda}(X)A^{-2it} = \sigma_t^{\lambda}(X)$$

for all $X \in \mathcal{M}$ and $t \in \mathbb{R}$. This completes the proof.

A generalized von Neumann algebra $\mathcal{M}$ is said to be *spatially semifinite* if there exists a standard, semifinite, tracial generalized vector for $\mathcal{M}$.

Proposition 2.6.7. Let $\mathcal{M}$ be a generalized von Neumann algebra on $\mathcal{D}$ in $\mathcal{H}$. The following statements hold.

(1) Suppose $\mathcal{M}$ is spatially semifinite. Then, for each full standard, semifinite generalized vector λ for $\mathcal{M}$ there exists a non-singular positive self-adjoint operator A affiliated with $(\mathcal{M}'_{\mathrm{w}})'^{\sigma^\lambda}$ such that $\sigma_t^{\lambda}(X) = A^{2it}XA^{-2it}$ for all $X \in \mathcal{M}$ and $t \in \mathbb{R}$.

(2) Conversely suppose there exist a full standard, semifinite generalized vector λ for $\mathcal{M}$ and a non-singular positive self-adjoint operator $A \in \mathcal{M}_\eta^{\sigma^\lambda}$ such that $\sigma_t^{\lambda}(X) = A^{2it}XA^{-2it}$ for all $X \in \mathcal{M}$ and $t \in \mathbb{R}$. Then $\mathcal{M}$ is spatially semifinite.

Proof. (1) Since $\mathcal{M}$ is spatially semifinite, there exists a full standard, semifinite generalized vector μ for $\mathcal{M}$ such that $\Delta_\mu = 1$. Hence it follows from Lemma 2.6.3 and Theorem 2.6.4 that $\sigma_t^{\lambda}(X) = A_{\mu,\lambda}^{2it}\sigma_t^{\mu}(X)A_{\mu,\lambda}^{-2it} = A_{\mu,\lambda}^{2it}XA_{\mu,\lambda}^{-2it}$ for all $X \in \mathcal{M}$ and $t \in \mathbb{R}$.

(2) Since $A \in \mathcal{M}_\eta^{\sigma^\lambda}$, it follows from Lemma 2.6.3 that $\mu \equiv \lambda_A$ is well-defined and $\sigma_t^{\mu}(X) = A^{2it}\sigma_t^{\lambda}(X)A^{-2it} = X$ for all $X \in \mathcal{M}$ and $t \in \mathbb{R}$. Therefore, μ is tracial, and so $\mathcal{M}$ is spatially semifinite.

Corollary 2.6.8. An EW^*-algebra $\mathcal{M}$ is spatially semifinite if and only if there exist a full standard generalized vector λ for $\mathcal{M}$ and a non-singular positive self-adjoint operator A affiliated with $(\mathcal{M}'_{\mathrm{w}})'$ such that $\sigma_t^{\lambda}(X) = A^{2it}XA^{-2it}$ for all $X \in \mathcal{M}$ and $t \in \mathbb{R}$. If this is true, then for any full standard generalized vector ν for $\mathcal{M}$ there exists a non-singular positive self-adjoint operator A_ν affiliated with $(\mathcal{M}'_w)'$ such that $\sigma_t^\nu(X) = A_\nu^{2it} X A_\nu^{-2it}$ for all $X \in \mathcal{M}$ and $t \in \mathbb{R}$.

Proof. This follows from Corollary 2.6.5 and Proposition 2.6.7.

We finally note some results for the case of standard vectors:

Lemma 2.6.9. Let $\mathcal{M}$ be a generalized von Neumann algebra on $\mathcal{D}$ in $\mathcal{H}$ and ξ_0 and η_0 standard vectors for $\mathcal{M}$. Suppose $A \equiv A_{\lambda_{\xi_0},\lambda_{\eta_0}} \in \mathcal{M}_\eta^{\sigma^{\xi_0}}$. Then $\xi_0 \in \mathcal{D}(A)$ and $\eta_0 = A\xi_0$.

Proof. For each $X \in \mathcal{M}$ and $n \in \mathbb{N}$ we have

$$J_{\xi_0} A J_{\xi_0} X E_A(n)\xi_0 = X A E_A(n)\xi_0,$$

and so

$$E_A(n) \in \mathcal{D}((\lambda_{\xi_0})_A) \text{ and } (\lambda_{\xi_0})_A(E_A(n)) = AE_A(n)\xi_0.$$

Since $\lambda_{\eta_0} = ((\lambda_{\xi_0})_A)_e$, it follows that

$$\lim_{n\to\infty} E_A(n)\xi_0 = \xi_0 \text{ and } \lim_{n\to\infty} AE_A(n)\xi_0 = \lim_{n\to\infty} E_A(n)\eta_0 = \eta_0,$$

which implies $\xi_0 \in \mathcal{D}(A)$ and $\eta_0 = A\xi_0$.

Corollary 2.6.10. Let $\mathcal{M}$ be a generalized von Neumann algebra on $\mathcal{D}$ in $\mathcal{H}$ and ξ_0 and η_0 standard vectors for $\mathcal{M}$.

(1) The following statements are equivalent:

(i) ω_{η_0} is $\{\sigma_t^{\xi_0}\}$-invariant.

(i)′ ω_{ξ_0} is $\{\sigma_t^{\eta_0}\}$-invariant.

(ii) $[D\omega_{\eta_0} : D\omega_{\xi_0}]_t \in \mathcal{M}^{\sigma^{\eta_0}}, \quad {}^\forall t \in \mathbb{R}$.

(ii)′ $[D\omega_{\xi_0} : D\omega_{\eta_0}]_t \in \mathcal{M}^{\sigma^{\xi_0}}, \quad {}^\forall t \in \mathbb{R}$.

(iii) $\{[D\omega_{\eta_0} : D\omega_{\xi_0}]_t\}_{t\in\mathbb{R}}$ is a strongly continous one-parameter group of unitary elements of $\mathcal{M}$.

(2) The following statements are equivalent:

(i) ω_{η_0} satisfies the KMS-condition with respect to $\{\sigma_t^{\xi_0}\}$.

(ii) $\sigma_t^{\eta_0} = \sigma_t^{\xi_0}, \quad {}^\forall t \in \mathbb{R}$.

(iii) There exists a non-singular positive self-adjoint operator A affiliated with the center of $(\mathcal{M}'_w)'$ such that $\eta_0 = A\xi_0$.

Proof. (1) This follows from Theorem 2.6.2.
(2) This follows from Theorem 2.6.6 and Lemma 2.6.9.

For the case of EW*-algebras we have the following result by Corollary 2.6.5 and Lemma 2.6.9:

Corollary 2.6.11. Let $\mathcal{M}$ be an EW*-algebra on $\mathcal{D}$ in $\mathcal{H}$ and ξ_0 and η_0 standard vectors for $\mathcal{M}$. Then λ_{η_0} commutes with λ_{ξ_0} if and only if $\eta_0 = A\xi_0$ for some non-singular positive self-adjoint operator A affiliated with $(\mathcal{M}'_{\mathrm{w}})'^{\sigma^{\xi_0}}$.

2.7 Generalized standard systems

In Section 2.1 ∼ 2.6 we have treated with the standard systems and modular systems under the assumption $(\mathrm{S})_1$: The weak commutant $\mathcal{M}'_{\mathrm{w}}$ of an O*-algebra $\mathcal{M}$ on $\mathcal{D}$ satisfies always the condition $\mathcal{M}'_{\mathrm{w}}\mathcal{D} \subset \mathcal{D}$. In the Wightman quantum field theory in Chapter IV O*-algebras $\mathcal{M}$ whose commutants $\mathcal{M}'_{\mathrm{w}}$ are not even von Neumann algebras have appeared. So, we consider to generalize the notion of standard systems.

Let $(\mathcal{M}, \lambda, \mathcal{A})$ be a triple of a closed O*-algebra $\mathcal{M}$ on a dense subspace $\mathcal{D}$ in a Hilbert space $\mathcal{H}$, a generalized vector λ for $\mathcal{M}$ and a von Neumann algebra $\mathcal{A}$ on $\mathcal{H}$. Suppose

$(\mathrm{GS})_1$ $\lambda((\mathcal{D}(\lambda)^\dagger \cap \mathcal{D}(\lambda))^2)$ is total in $\mathcal{H}$;

$(\mathrm{GS})_2$ $\mathcal{A}' \subset \mathcal{M}'_{\mathrm{w}}$.

Then we put

$$\mathcal{D}(\lambda_{\mathcal{A}'}) = \{K \in \mathcal{A}'; {}^\exists\xi_K \in \mathcal{D}(e_{\mathcal{A}'}(\mathcal{M})) \text{ s.t. } K\lambda(X) = e_{\mathcal{A}'}(X)\xi_K \text{ for all } X \in \mathcal{D}(\lambda)\},$$
$$\lambda_{\mathcal{A}'}(K) = \xi_K, \qquad K \in \mathcal{D}(\lambda_{\mathcal{A}'}),$$

where $e_{\mathcal{A}'}(\mathcal{M})$ is the induced extension of $\mathcal{M}$ defined in Section 1.4. Then $\lambda_{\mathcal{A}'}$ is a generalized vector for the von Neumann algebra $\mathcal{A}'$. Further, suppose

$(\mathrm{GS})_3$ $\lambda_{\mathcal{A}'}((\mathcal{D}(\lambda_{\mathcal{A}'})^* \cap \mathcal{D}(\lambda_{\mathcal{A}'}))^2)$ is total in $\mathcal{H}$.

Here we put

$$\mathcal{D}(\lambda_{\mathcal{A}}) = \{A \in \mathcal{A}; {}^\exists\xi_A \in \mathcal{H} \text{ s.t. } A\lambda_{\mathcal{A}'}(K) = K\xi_A \text{ for all } K \in \mathcal{D}(\lambda_{\mathcal{A}'})\}$$
$$\lambda_{\mathcal{A}}(A) = \xi_A, \qquad A \in \mathcal{D}(\lambda_{\mathcal{A}}).$$

Then we have the following

Lemma 2.7.1. $\lambda_{\mathcal{A}}$ is a generalized vector for the von Neumann algebra $\mathcal{A}$ and $\lambda_{\mathcal{A}}(\mathcal{D}(\lambda_{\mathcal{A}})^* \cap \mathcal{D}(\lambda_{\mathcal{A}}))$ is a full left Hilbert algebra in $\mathcal{H}$ equipped with the multiplication $\lambda_{\mathcal{A}}(A)\lambda_{\mathcal{A}}(B) = \lambda_{\mathcal{A}}(AB)$ and the involution $\lambda_{\mathcal{A}}(A) \to \lambda_{\mathcal{A}}(A^*)$.

Proof. It is easily shown that $\lambda_{\mathcal{A}}$ is a generalized vector for the von

Neumann algebra $\mathcal{A}$. Take any element X of $\mathcal{D}(\lambda)^\dagger \cap \mathcal{D}(\lambda)$. Let $\overline{e_{\mathcal{A}'}(X)} = U|\overline{e_{\mathcal{A}'}(X)}|$ be the polar decomposition of $\overline{e_{\mathcal{A}'}(X)}$, $|\overline{e_{\mathcal{A}'}(X)}| = \int_0^\infty t\,dE(t)$ the spectral resolution and $E_n = \int_0^n dE(t)$, $n \in \mathbb{N}$. Since $\overline{e_{\mathcal{A}'}(X)}$ is affiliated with $\mathcal{A}$, it is shown similarly to the proof of Lemma 2.2.2 that

$$X_n \equiv \overline{e_{\mathcal{A}'}(X)}E_n \in \mathcal{D}(\lambda_{\mathcal{A}})^* \cap \mathcal{D}(\lambda_{\mathcal{A}}), \tag{2.7.1}$$

$$\lambda_{\mathcal{A}}(X_n) = UE_nU^*\lambda(X), \tag{2.7.2}$$

$$\lambda_{\mathcal{A}}(X_n^*) = E_n\lambda(X^\dagger), \tag{2.7.3}$$

$$\lim_{n\to\infty}\lambda_{\mathcal{A}}(X_n) = \lambda(X), \quad \lim_{n\to\infty}\lambda_{\mathcal{A}}(X_n^*) = \lambda(X^\dagger). \tag{2.7.4}$$

Take arbitrary $X, Y \in \mathcal{D}(\lambda)^\dagger \cap \mathcal{D}(\lambda)$. By (2.7.1) $\sim$ (2.7.4) we have

$$\{Y_m X_m\} \subset (\mathcal{D}(\lambda_{\mathcal{A}})^* \cap \mathcal{D}(\lambda_{\mathcal{A}}))^2, \; m, n \in \mathbb{N},$$

$$\begin{aligned}\lim_{m\to\infty}\lim_{n\to\infty}\lambda_{\mathcal{A}}(Y_m X_n) &= \lim_{m\to\infty}\lim_{n\to\infty} Y_m\lambda_{\mathcal{A}}(X_n)\\ &= \lim_{m\to\infty} Y_m\lambda(X)\\ &= Y\lambda(X)\\ &= \lambda(YX).\end{aligned}$$

Hence it follows from $(\mathrm{GS})_1$ that $\lambda_{\mathcal{A}}((\mathcal{D}(\lambda_{\mathcal{A}})^* \cap \mathcal{D}(\lambda_{\mathcal{A}}))^2)$ is total in $\mathcal{H}$, which implies that $\lambda_{\mathcal{A}}(\mathcal{D}(\lambda_{\mathcal{A}})^* \cap \mathcal{D}(\lambda_{\mathcal{A}}))$ is a full left Hilbert algebra in $\mathcal{H}$.

Let $S_{\lambda_{\mathcal{A}}} = J_{\lambda_{\mathcal{A}}}\Delta_{\lambda_{\mathcal{A}}}^{\frac{1}{2}}$ be the polar decomposition of the involution $\lambda_{\mathcal{A}}(A) \to \lambda_{\mathcal{A}}(A^*)$. By the Tomita fundamental theorem we have

$$J_{\lambda_{\mathcal{A}}}\mathcal{A}J_{\lambda_{\mathcal{A}}} = \mathcal{A}' \tag{2.7.5}$$

$$\Delta_{\lambda_{\mathcal{A}}}^{it}\mathcal{A}\Delta_{\lambda_{\mathcal{A}}}^{-it} = \mathcal{A}, \; t \in \mathbb{R}. \tag{2.7.6}$$

Further, it follows from $(\mathrm{GS})_3$ that the involution $\lambda(X) \to \lambda(X^\dagger)$, $X \in \mathcal{D}(\lambda)^\dagger \cap \mathcal{D}(\lambda)$, is closable and its closure is denoted by S_λ. Let $S_\lambda = J_\lambda\Delta_\lambda^{\frac{1}{2}}$ be the polar decomposition of S_λ. By (2.7.1) $\sim$ (2.7.4) we have the following

Lemma 2.7.2. $S_\lambda \subset S_{\lambda_{\mathcal{A}}}$.

Definition 2.7.3. A triple $(\mathcal{M}, \lambda, \mathcal{A})$ is said to be a *standard system* if it satisfies the above conditions $(\mathrm{GS})_1$, $(\mathrm{GS})_2$ and $(\mathrm{GS})_3$ and the following conditions $(\mathrm{GS})_4$, $(\mathrm{GS})_5$ and $(\mathrm{GS})_6$:

$(\mathrm{GS})_4$ $\Delta_{\lambda_{\mathcal{A}}}^{it}\mathcal{D} = \mathcal{D}$ for all $t \in \mathbb{R}$.

$(\mathrm{GS})_5$ $\Delta_{\lambda_{\mathcal{A}}}^{it}\mathcal{M}\Delta_{\lambda_{\mathcal{A}}}^{-it} = \mathcal{M}$ for all $t \in \mathbb{R}$.

$(\mathrm{GS})_6$ $\Delta_{\lambda_{\mathcal{A}}}^{it}(\mathcal{D}(\lambda)^\dagger \cap \mathcal{D}(\lambda))\Delta_{\lambda_{\mathcal{A}}}^{-it} = \mathcal{D}(\lambda)^\dagger \cap \mathcal{D}(\lambda)$ for all $t \in \mathbb{R}$.

Theorem 2.7.4. Suppose $(\mathcal{M}, \lambda, \mathcal{A})$ is a standard system. Then the following statements hold:

(1) $S_\lambda = S_{\lambda_\mathcal{A}}$.

(2) $\{\sigma_t^\lambda\}_{t\in\mathbb{R}}$ is a one-parameter group of $*$-automorphisms of $\mathcal{M}$, where $\sigma_t^\lambda(X) \equiv \Delta_\lambda^{it} X \Delta_\lambda^{-it}$ for $X \in \mathcal{M}$ and $t \in \mathbb{R}$.

(3) λ satisfies the KMS-condition with respect to $\{\sigma_t^\lambda\}_{t\in\mathbb{R}}$.

Proof. Let $X, Y \in \mathcal{D}(\lambda)^\dagger \cap \mathcal{D}(\lambda)$. Using $\lambda_\mathcal{A}$ satisfies the KMS-condition with respect to the one-parameter group $\{\sigma_t^{\lambda_\mathcal{A}}\}_{t\in\mathbb{R}}$ of $\mathcal{A}$, where $\sigma_t^{\lambda_\mathcal{A}}(A) \equiv \Delta_{\lambda_\mathcal{A}}^{it} A \Delta_{\lambda_\mathcal{A}}^{-it}$ for $A \in \mathcal{A}$ and $t \in \mathbb{R}$, and (2.7.1) $\sim$ (2.7.4), it is shown similarly to the proof of Theorem 2.2.4 that there exists a function $f_{X,Y}$ in $A(0,1)$ such that

$$f_{X,Y}(t) = (\lambda(Y^\dagger)|\lambda(\sigma_t^{\lambda_\mathcal{A}}(Y^\dagger))),$$
$$f_{X,Y}(t+i) = (\lambda(\sigma_t^{\lambda_\mathcal{A}}(X))|\lambda(Y))$$

for all $t \in \mathbb{R}$. Hence it is shown similarly to the proof of Theorem 2.2.4 that $\Delta_\lambda^{it} = \Delta_{\lambda_\mathcal{A}}^{it}$ for all $t \in \mathbb{R}$, which implies all of our assertions. This completes the proof.

Let $(\mathcal{M}, \xi, \mathcal{A})$ be a triple of a closed O*-algebra $\mathcal{M}$ on $\mathcal{D}$ in $\mathcal{H}$, $\xi \in \mathcal{D}$ and a von Neumann algebra $\mathcal{A}$ on $\mathcal{H}$. Suppose $\mathcal{M}\xi$ is dense in $\mathcal{H}$ and $\mathcal{A}' \subset \mathcal{M}'_{\mathrm{w}}$. Then, since $\mathcal{D}(\lambda_\xi) = \mathcal{M}$ and $\lambda_\xi(X) = X\xi$ for all $X \in \mathcal{M}$, it follows that $\mathcal{D}((\lambda_\xi)_{\mathcal{A}'}) = \mathcal{A}'$ and $(\lambda_\xi)_{\mathcal{A}'}(K) = K\xi$ for all $K \in \mathcal{A}'$. Hence, the conditions $(\mathrm{GS})_3$,$(\mathrm{GS})_4$ and $(\mathrm{GS})_5$ in Definition 2.7.3 become as follows:

$(\mathrm{GS})_3$ $\mathcal{A}'\xi$ is dense in $\mathcal{H}$;

$(\mathrm{GS})_4$ $\Delta_{\mathcal{A}\xi}^{it}\mathcal{D} \subset \mathcal{D}$ for all $t \in \mathbb{R}$, where $\Delta_{\mathcal{A}\xi}$ is the modular operator of the full left Hilbert algebra $\mathcal{A}\xi$ in $\mathcal{H}$ equipped with the multiplication $(A\xi)(B\xi) = AB\xi$ and the involution $A\xi \to A^*\xi$;

$(\mathrm{GS})_5$ $\Delta_{\mathcal{A}\xi}^{it}\mathcal{M}\Delta_{\mathcal{A}\xi}^{-it} = \mathcal{M}$ for all $t \in \mathbb{R}$ and $(\mathrm{GS})_6$ holds always.

Hence we have the following

Lemma 2.7.5. Let $(\mathcal{M}, \xi, \mathcal{A})$ be a triple of a closed O*-algebra $\mathcal{M}$ on $\mathcal{D}$ in $\mathcal{H}$, $\xi \in \mathcal{D}$ and a von Neumann algebra $\mathcal{A}$ on $\mathcal{H}$. Then $(\mathcal{M}, \lambda_\xi, \mathcal{A})$ is a standard system if and only if the following statements $(\mathrm{GS})_1 \sim (\mathrm{GS})_5$ hold:

$(\mathrm{GS})_1$ $\mathcal{M}\xi$ is dense in $\mathcal{H}$;

$(\mathrm{GS})_2$ $\mathcal{A}' \subset \mathcal{M}'_{\mathrm{w}}$;

$(\mathrm{GS})_3$ $\mathcal{A}'\xi$ is dense in $\mathcal{H}$;

$(\mathrm{GS})_4$ $\Delta_{\mathcal{A}\xi}^{it}\mathcal{D} = \mathcal{D}$ for all $t \in \mathbb{R}$;

$(\mathrm{GS})_5$ $\Delta_{\mathcal{A}\xi}^{it}\mathcal{M}\Delta_{\mathcal{A}\xi}^{-it} = \mathcal{M}$ for all $t \in \mathbb{R}$.

Definition 2.7.6. A triple $(\mathcal{M}, \xi, \mathcal{A})$ of a closed O*-algebra $\mathcal{M}$ on $\mathcal{D}$ in $\mathcal{H}$, $\xi \in \mathcal{D}$ and a von Neumann algebra $\mathcal{A}$ on $\mathcal{H}$ is said to be a *standard system* if the conditions $(\mathrm{GS})_1 \sim (\mathrm{GS})_5$ of Lemma 2.7.5 hold.

To apply the unbounded Tomita-Takesaki theory to more examples, we weaken the conditions $(\mathrm{GS})_4$ and $(\mathrm{GS})_5$ in Lemma 2.7.5 and define the notion of modular systems.

Definition 2.7.7. A triple $(\mathcal{M}, \xi, \mathcal{A})$ of a closed O*-algebra $\mathcal{M}$ on $\mathcal{D}$ in $\mathcal{H}$, $\xi \in \mathcal{D}$ and a von Neumann algebra $\mathcal{A}$ on $\mathcal{H}$ is said to be a *modular system* if the condition $(\mathrm{GS})_1$, $(\mathrm{GS})_2$ and $(\mathrm{GS})_3$ in Lemma 2.7.5 and the following condition (GM) holds:

(GM) There exists a subspace $\mathcal{E}$ of $\mathcal{D}(e_{\mathcal{A}'}(\mathcal{M}))$ such that

$(\mathrm{GM})_1$ $\xi \in \mathcal{E}$

$(\mathrm{GM})_2$ $e_{\mathcal{A}'}(\mathcal{M})\mathcal{E} = \mathcal{E}$;

$(\mathrm{GM})_3$ $\Delta_{\mathcal{A}\xi}^{it}\mathcal{E} = \mathcal{E}$ for all $t \in \mathbb{R}$.

Let $(\mathcal{M}, \xi, \mathcal{A})$ be a modular system. We put $\mathcal{D}_M = \bigcup_{\mathcal{E} \in \mathcal{F}} \mathcal{E}$, where $\mathcal{F}$ is the set of all subspaces $\mathcal{E}$ of $\mathcal{D}(e_{\mathcal{A}'}(\mathcal{M}))$ satisfying $(\mathrm{GM})_1$, $(\mathrm{GM})_2$ and $(\mathrm{GM})_3$ in Definition 2.7.7. Then $\mathcal{D}_M$ is a subspace of $\mathcal{D}(e_{\mathcal{A}'}(\mathcal{M}))$ containing $\mathcal{M}\xi$, and so it is dense in $\mathcal{H}$. It is easily shown that the linear span of $\mathcal{A}'\mathcal{D}_M$ belongs to $\mathcal{F}$. Since $\mathcal{D}_M$ is maximum in $\mathcal{F}$, we have $\mathcal{A}'\mathcal{D}_M = \mathcal{D}_M$, which implies that

$$\mathcal{U}(\mathcal{M}, \xi, \mathcal{A}) \equiv \{X \in \mathcal{L}^\dagger(\mathcal{D}_M); \overline{X} \text{ is affiliated with } \mathcal{A}\}$$

is an O*-algebra on $\mathcal{D}_M$ in $\mathcal{H}$. Since $\widetilde{\mathcal{D}_M} \equiv \bigcap_{X \in \mathcal{U}(\mathcal{M}, \xi, \mathcal{A})} \mathcal{D}(\overline{X}) \in \mathcal{F}$, it follows from the maximum of $\mathcal{D}_M$ that $\mathcal{U}(\mathcal{M}, \xi, \mathcal{A})$ is closed. Hence $\mathcal{U}(\mathcal{M}, \xi, \mathcal{A})$ is a generalized von Neumann algebra on $\mathcal{D}_M$ over $\mathcal{A}$ and it is called a *left generalized von Neumann algebra* of the modular system $(\mathcal{M}, \xi, \mathcal{A})$. We have the following

Theorem 2.7.8. Suppose $(\mathcal{M}, \xi, \mathcal{A})$ is a modular system. Then a system $(\mathcal{U}(\mathcal{M}, \xi, \mathcal{A}), \xi, \mathcal{A})$ is standard.

We may define the notion of modular systems $(\mathcal{M}, \lambda, \mathcal{A})$ for generalized vectors λ, but we omit it in this note.

Notes

It is well known that the Tomita-Takesaki theory plays an important rule for the study of von Neumann algebras and for the applications of quantum physics (Bratteli-Robinson [1,2]). Inoue [4, 5, 9, 10, 14] tried to develop the Tomita-Takesaki theory in O*-algebras, mainly in case of O*-algebras with cyclic and separating vector. To treat with such a study more systematically,

Inoue-Karwowski [1] defined the notion of generalized vectors for O*-algebras which is a generalization of cyclic vectors, and using it they have developed such a study in (Inoue[15, 17, 18, 19], Antoine-Inoue-Ogi-Trapani [1] and Inoue-Karwowski [1]). Here we have introduced these studies.

The works in Section 2.1, 2.2, 2.3 are due to Inoue-Karwowski [1]. The works A (resp. B and C, D) in Section 2.4 are due to Inoue-Karwowski [1] (resp. Inoue [19], Antoine-Inoue-Ogi-Trapani [1]). The results in Section 2.4, E are generalizations of those obtained in Inoue [9] for von Neumann algebras with cyclic and separating vector. The works in Section 2.5, 2.6 are due to Inoue [17]. The standard system $(\mathcal{M}, \xi, \mathcal{A})$ consisting of an O*-algebra, a cyclic vector ξ and a von Neumann algebra $\mathcal{A}$ in Section 2.7 was introdcued in Inoue [14], and the general standard system $(\mathcal{M}, \lambda, \mathcal{A})$ has been introduced here.

A Tomita-Takesaki theory in partial O*-algebras have been studied in Antoine-Inoue-Ogi [1], Ekhaguere [1] and Inoue [13]

3. Standard weights on O*-algebras

Weights on O*-algebras (that is, linear functionals that take positive, but not necessarily finite valued) appear naturally in the studies of the unbounded Tomita-Takesaki theory. and the quantum physics. Thus it is significant to study weights on O*-algebras for the structure of O*-algebras and the physical applications. Further, the weights on O*-algebras occasion some pathological phenomena which don't occur for weights on C^*- and W^*-algebras. From this viewpoint we should study systematically weights on O*-algebras.

In Section 3.1 we define quasi-weights and weights on O*-algebras and give the fundamental examples. Let $\mathcal{M}$ be a closed O*-algebra on a dense subspace $\mathcal{D}$ in a Hilbert space $\mathcal{H}$. The algebraic positive cone $\mathcal{P}(\mathcal{M})$ and the operational positive cone $\mathcal{M}_+$ are defined and the corresponding weights are defined. The phenomenon arises for the GNS-construction of φ which is important for such a study : $\mathfrak{N}_\varphi^0 \equiv \{X \in \mathcal{M}; \varphi(X^\dagger X) < \infty\}$ is a left ideal of $\mathcal{M}$ in the bounded case, but it is not necessarily a left ideal of $\mathcal{M}$. For example, the condition $\varphi(I) < \infty$ doesn't necessarily imply $\varphi(X^\dagger X) < \infty$ for all $X \in \mathcal{M}$. So, using the left ideal $\mathfrak{N}_\varphi \equiv \{X \in \mathcal{M}\ ;\ \varphi((AX)^\dagger(AX)) < \infty \text{ for all } A \in \mathcal{M}\}$ of $\mathcal{M}$, we construct the GNS-representation π_φ and the vector representation λ_φ on the similar method to positive linear functionals, that is, π_φ is a $*$-homomorphism of $\mathcal{M}$ onto the O*-algebra $\pi_\varphi(\mathcal{M})$ on the dense subspace $\mathcal{D}(\pi_\varphi)$ in the Hilbert space $\mathcal{H}_\varphi$, and λ_φ is a linear map of $\mathfrak{N}_\varphi$ into $\mathcal{D}(\pi_\varphi)$ satisfying $\lambda_\varphi(AX) = \pi_\varphi(A)\lambda_\varphi(X)$ for each $A \in \mathcal{M}$ and $X \in \mathfrak{N}_\varphi$. However, there are non-zero weights φ such that $\mathfrak{N}_\varphi^0$ has many elements but $\mathfrak{N}_\varphi = \{0\}$ (Example 3.6.2) and so the GNS-construction for such a weight is meaningless. We don't treat with such a weight. We give two important examples of weights. For any $\xi \in \mathcal{D}$ the positive linear functional ω_ξ is defined by $\omega_\xi(X) = (X\xi|\xi)$, $X \in \mathcal{M}$, but if $\xi \in \mathcal{H} \setminus \mathcal{D}$ then the definition of the above ω_ξ is impossible. We regard ω_ξ as the map $A \to (A^{\dagger *}\xi|\xi)$ of the positive cone $\mathcal{P}(\mathfrak{N}_{\omega_\xi})$ generated by the left ideal $\mathfrak{N}_{\omega_\xi} \equiv \{X \in \mathcal{M}; \xi \in \mathcal{D}(X^{\dagger *})$ and $X^{\dagger *}\xi \in \mathcal{D}\}$ of $\mathcal{M}$ into $\mathbb{R}_+$ satisfying $\omega_\xi(A+B) = \omega_\xi(A) + \omega_\xi(B)$ and $\omega_\xi(\alpha A) = \alpha\omega_\xi(A)$ for all $A, B \in \mathcal{P}(\mathfrak{N}_{\omega_\xi})$ and $\alpha \geq 0$. So, we need to study such a map (called *quasi-weight*) which is strictly weaker than the notion of weights. We give another important (quasi-)weight constructed from a net $\{f_\alpha\}$ of positive linear functionals on $\mathcal{M}$. It is natural to consider whether $\sup_\alpha f_\alpha$ is a (quasi-)weight on $\mathcal{P}(\mathcal{M})$. We show that if $\{f_\alpha\}$ has a certain net property for $\mathcal{P}(\mathcal{M})$ (resp. $\mathcal{P}(\mathfrak{N}_\varphi)$) then $\sup_\alpha f_\alpha$ is a weight (resp. a quasi-weight) on $\mathcal{P}(\mathcal{M})$. In Section 3.2 we define and study the notions of regularity and singularity for (quasi-)weights φ on $\mathcal{P}(\mathcal{M})$, and give the decomposition theorem of φ into the regular part φ_r and the singular part φ_s. A quasi-weight φ on $\mathcal{P}(\mathcal{M})$ is said to be *regular* if $\varphi = \sup_\alpha f_\alpha$ on $\mathcal{P}(\mathfrak{N}_\varphi)$ for some net $\{f_\alpha\}$ of positive linear functionals on $\mathcal{M}$, and φ is said to be singular if there doesn't exist any positive linear functional f on $\mathcal{M}$ such that $f(X^\dagger X) \leq \varphi(X^\dagger X)$ for all $X \in \mathfrak{N}_\varphi$ and $f \neq 0$ on $\mathcal{P}(\mathfrak{N}_\varphi)$. Using the trio-commutant $T(\varphi)'_C$ for φ, we characterize the regularity and the singularity of φ and show that φ is decomposed into the sum of the regular part φ_r of φ and the singularity part φ_s of φ. In Section 3.3 we define and study an important class in regular (quasi-)weights which is possible to develop the Tomita-Takesaki theory in O*-algebras. Let φ be a faithful (quasi-)weight on $\mathcal{P}(\mathcal{M})$ such that $\pi_\varphi(\mathcal{M})'_{\mathrm{w}}\mathcal{D}(\pi_\varphi) \subset \mathcal{D}(\pi_\varphi)$. Then, the map $\Lambda_\varphi : \pi_\varphi(X) \to \lambda_\varphi(X)$, $X \in \mathfrak{N}_\varphi$ is a generalized vector for the O*-algebra $\pi_\varphi(\mathcal{M})$, that is, it is a linear map of the left ideal $\mathcal{D}(\Lambda_\varphi) \equiv \pi_\varphi(\mathfrak{N}_\varphi)$ into $\mathcal{D}(\pi_\varphi)$ satisfying $\Lambda_\varphi(\pi_\varphi(A)\pi_\varphi(X)) = \pi_\varphi(A)\Lambda_\varphi(\pi_\varphi(X))$ for all $A \in \mathcal{M}$ and $X \in \mathfrak{N}_\varphi$. Using (quasi-)standard generalized vectors defined and studied in Section 2.2, we define the notion of (quasi-)standardness of φ as follows: φ is said to be *standard* (resp. *quasi-standard*) if the generalized vector Λ_φ is standard (resp. quasi-standard). And we obtain that if φ is standard, then the modular automorphism group $\{\sigma_t^\varphi\}_{t\in\mathbb{R}}$ of $\mathfrak{N}_\varphi^\dagger \cap \mathfrak{N}_\varphi$ is defined and φ is a $\{\sigma_t^\varphi\}$-KMS (quasi-)weight, and if φ is quasi-standard, then it is extended to a standard quasi-weight $\overline{\varphi}$ on the positive cone $\mathcal{P}(\pi_\varphi(\mathcal{M})''_{\mathrm{wc}})$ of the generalized von Neumann algebra $\pi_\varphi(\mathcal{M})''_{\mathrm{wc}}$. In Section 3.4 we generalize the Connes cocycle theorem for weights on O*-algebras. In Section 2.5 we generalized the Connes cocycle theorem for standard generalized vectors. As the notion of generalized vectors is spatial, such a generalization is possible to a certain extent, but the notion of (quasi-)weights is purely algebraic and the algebraic properties don't reflect to the topological properties in general, for example, $\pi_\varphi(\mathcal{M})$ is net necessarily a generalized von Neumann algebra when $\mathcal{M}$ is a generalized von Neumann algebra, and so such a generalization for (quasi-)weights have some difficult problems. We here need the notions of semifiniteness and σ-weak continuity of (quasi-)weights. Let φ and ψ be faithful, σ-weakly continuous and semifinite (quasi-)weights on $\mathcal{P}(\mathcal{M})$ such that π_φ and π_ψ are self adjoint. We consider the matrix algebra $\mathcal{M} \otimes M_2(\mathbb{C})$ on $\mathcal{D} \otimes \mathcal{D}$:

$$\left\{ X = \begin{pmatrix} X_{11} & X_{12} \\ X_{21} & X_{22} \end{pmatrix} ; X_{ij} \in \mathcal{M} \right\}$$

and a faithful, σ-weakly continuous semifinite (quasi-)weight θ on $\mathcal{P}(\mathcal{M} \otimes M_2(\mathbb{C}))$ by

$$\theta(X^\dagger X) = \varphi(X_{11}^\dagger X_{11} + X_{21}^\dagger X_{21}) + \psi(X_{12}^\dagger X_{12} + X_{22}^\dagger X_{22}),$$
$$X = (X_{ij}) \in \mathcal{M} \otimes M_2(\mathbb{C}).$$

We obtain the result that if $\Lambda_\theta^C((\mathcal{D}(\Lambda_\theta^C)^* \cap \mathcal{D}(\Lambda_\theta^C))^2)$ is total in $\mathcal{H}_\theta$ then π_φ and π_ψ are unitarily equivalent, and then the cocycle $[D\overline{\psi} : D\overline{\varphi}]$ associated with the quasi-weight $\overline{\psi}$ on $\mathcal{P}(\pi_\varphi(\mathcal{M})''_{\rm wc})$ with respect to the quasi-weight $\overline{\varphi}$ on $\mathcal{P}(\pi_\varphi(\mathcal{M})''_{\rm wc})$ is defined, but $\pi_\varphi(\mathcal{M})$ is not a generalized von Neumann algebra in general even if $\mathcal{M}$ is a generalized von Neumann algebra, and so the cocycle $[D\overline{\psi} : D\overline{\varphi}]$ for the generalized von Neumann algebra $\pi_\varphi(\mathcal{M})''_{\rm wc}$ does not necessarily induce the cocycle $[D\psi : D\varphi]$ associated with the (quasi-)weight ψ on $\mathcal{P}(\mathcal{M})$ with respect to the (quasi-) weight φ on $\mathcal{P}(\mathcal{M})$. We show that if $\mathcal{M}$ is a generalized von Neumann algebra with strongly dense bounded part and φ is strongly faithful, then $\pi_\varphi(\mathcal{M})$ is spatially isomorphic to $\mathcal{M}$, and so it is a generalized von Neumann algebra and the cocycle $[D\psi : D\varphi]$ for the generalized von Neumann algebra $\mathcal{M}$ is well-defined. In Section 3.5 we study the Radon-Nikodym theorem for (quasi-)weights on O*-algebras. Let φ be a (quasi-)weight on a closed O*-algebra $\mathcal{M}$ on $\mathcal{D}$ such that $\mathcal{M}'_{\rm w}\mathcal{D} \subset \mathcal{D}$. A (quasi-)weight ψ on $\mathcal{P}(\mathcal{M})$ is said to be φ-absolutely continuous if $\mathfrak{N}_\varphi \subset \mathfrak{N}_\psi$ and the map: $\lambda_\varphi(X) \to \lambda_\psi(X)$, $X \in \mathfrak{N}_\varphi$ is closable from the dense subspace $\lambda_\varphi(\mathfrak{N}_\varphi)$ in a Hilbert space $\mathcal{H}_\varphi$ to the Hilbert space $\mathcal{H}_\psi$, denoted by $K_{\varphi,\psi}$ the closure of this map. Using the map $K_{\varphi+\psi,\varphi}$, we characterize the φ-absolute continuity of ψ; in particular, in case that φ is standard. Suppose φ is standard. Then we show that ψ is a φ-absolutely continuous, $\{\sigma_t^\varphi\}$-KMS (quasi-)weight such that $\varphi+\psi$ is standard if and only if there exists a positive self-adjoint operator H in $\mathcal{H}_\varphi$ affiliated with $(\pi_\varphi(\mathcal{M})'_{\rm w})' \cap \pi_\varphi(\mathcal{M})'_{\rm w}$ such that $\mathcal{D}(H) \supset \lambda_\varphi(\mathfrak{N}_\varphi)$ and $\psi(X^\dagger X) = \|H\lambda_\varphi(X)\|^2$ for each $X \in \mathfrak{N}_\varphi$; and if there exists a standard, $\{\sigma_t^\varphi\}$-KMS (quasi-) weight τ on $\mathcal{P}(\mathcal{M})$ such that $\varphi+\psi \leq \tau$ and $\mathfrak{N}_\tau = \mathfrak{N}_\varphi$, then ψ is φ-absolutely continuous and $\{\sigma_t^\varphi\}$-invariant if and only if there exists a positive self-adjoint operator H' in $\mathcal{H}_\varphi$ affiliated with $\pi_\varphi(\mathcal{M})'^{\sigma^\varphi}_{\rm w} \equiv \{C \in \pi_\varphi(\mathcal{M})'_{\rm w}; \sigma_t^\varphi(C) = C, \ {}^\forall t \in \mathbb{R}\}$ such that $\lambda_\varphi(\mathfrak{N}_\varphi) \subset \mathcal{D}(H')$ and $\psi(X^\dagger X) = \|H'\lambda_\varphi(X)\|^2$ for each $X \in \mathfrak{N}_\varphi$. In Section 3.6 we give some concrete examples of regular (quasi-) weights, singular (quasi-)weights and standard (quasi-)weights. We first investigate the quasi-weights ω_ξ on $\mathcal{P}(\mathcal{M})$ defined by elements ξ in the Hilbert space. When is ω_ξ extended to a weight $\widetilde{\omega_\xi}$ on $\mathcal{P}(\mathcal{M})$ such that $\mathfrak{N}_{\widetilde{\omega_\xi}} = \mathfrak{N}_{\omega_\xi}$? We show that if $\mathcal{M}$ is commutative and integrable then the above question is affirmative. Further, we investigate the regularity, the singularity and the standardness of the quasi-weights ω_ξ. We next investigate the regualarity and the standardness of quasi-weights defined by density matrices which are important for the quantum physics. We shall apply these results to the physical models, namely the unbounded CCR algebra, a class of interacting boson model in the Fock space and the BCS-Bogolubov model of superconductivity in Chapter IV. And we shall give regular quasi-weighs and standard quasi-weights for the relative models.

3.1 Weights and quasi-weights on O*-algebras

In this section we define the notions of quasi-weights and weights on O*-algebras and give the fundamental examples. Throughout this section let $\mathcal{M}$ be a closed O^*-algebra on $\mathcal{D}$ in $\mathcal{H}$. For a subspace $\mathcal{N}$ of $\mathcal{M}$ we put

$$\mathcal{P}(\mathcal{N}) = \{\sum_{k=1}^{n} X_k^\dagger X_k;\ X_k \in \mathcal{N}\ \ (k = 1, 2, \cdots, n),\ \ n \in \mathbb{N}\}$$

and call it the *positive cone generated by* $\mathcal{N}$.

Definition 3.1.1. A map φ of $\mathcal{P}(\mathcal{M})$ into $\mathbb{R}_+ \cup \{+\infty\}$ is said to be a *weight on* $\mathcal{P}(\mathcal{M})$ if
(i) $\varphi(A + B) = \varphi(A) + \varphi(B), \quad A, B \in \mathcal{P}(\mathcal{M})$;
(ii) $\varphi(\alpha A) = \alpha\varphi(A), \quad A \in \mathcal{P}(\mathcal{M}),\ \alpha \geq 0$,
where $0 \cdot (+\infty) = 0$. A map φ of the positive cone $\mathcal{P}(\mathfrak{N}_\varphi)$ generated by a left ideal $\mathfrak{N}_\varphi$ of $\mathcal{M}$ into $\mathbb{R}_+$ is said to be a *quasi-weight* on $\mathcal{P}(\mathcal{M})$ if it satisfies the above conditions (i) and (ii) for $\mathcal{P}(\mathfrak{N}_\varphi)$.

Let φ be a quasi-weight on $\mathcal{P}(\mathcal{M})$. We denote by $\mathcal{D}(\varphi)$ the subspace of $\mathcal{M}$ generated by $\{X^\dagger X; X \in \mathfrak{N}_\varphi\}$. Since $\mathfrak{N}_\varphi$ is a left ideal of $\mathcal{M}$, we have

$$\mathcal{D}(\varphi) = \text{the linear span of } \{Y^\dagger X;\ X, Y \in \mathfrak{N}_\varphi\},$$

and so each $\sum_k \alpha_k Y_k^\dagger X_k$ ($\alpha_k \in \mathbb{C}$, $X_k, Y_k \in \mathfrak{N}_\varphi$) is represented as $\sum_j \beta_j Z_j^\dagger Z_j$ for some $\beta_j \in \mathbb{C}$ and $Z_j \in \mathfrak{N}_\varphi$. Then we can define a linear functional on $\mathcal{D}(\varphi)$ by

$$\sum_k \alpha_k Y_k^\dagger X_k \longrightarrow \sum_j \beta_j \varphi(Z_j^\dagger Z_j)$$

and write it by the same φ. It is easily shown that

$$|\varphi(Y^\dagger X)|^2 \leq \varphi(Y^\dagger Y)\varphi(X^\dagger X), \quad X, Y \in \mathfrak{N}_\varphi. \tag{3.1.1}$$

We put

$$N_\varphi = \{X \in \mathfrak{N}_\varphi; \varphi(X^\dagger X) = 0\}, \quad \lambda_\varphi(X) = X + N_\varphi \in \mathfrak{N}_\varphi/_{N_\varphi}, \ X \in \mathfrak{N}_\varphi.$$

Then it follows from (3.1.1) that N_φ is a left ideal of $\mathfrak{N}_\varphi$ and $\lambda_\varphi(\mathfrak{N}_\varphi) \equiv \mathfrak{N}_\varphi/_{N_\varphi}$ is a pre-Hilbert space with the inner product

$$(\lambda_\varphi(X) \mid \lambda_\varphi(Y)) = \varphi(Y^\dagger X), \quad X, Y \in \mathfrak{N}_\varphi.$$

We denote by $\mathcal{H}_\varphi$ the Hilbert space obtained by the completion of the pre-Hilbert space $\lambda_\varphi(\mathfrak{N}_\varphi)$. We define a $*$-representation π_φ^0 of $\mathcal{M}$ by

$$\pi_\varphi^0(A)\lambda_\varphi(X) = \lambda_\varphi(AX), \quad A \in \mathcal{M},\ X \in \mathfrak{N}_\varphi,$$

and denote by π_φ the closure of π_φ^0. We call the triple $(\pi_\varphi, \lambda_\varphi, \mathcal{H}_\varphi)$ the *GNS-construction* for φ. Let φ be a weight on $\mathcal{P}(\mathcal{M})$ and put

$$\mathfrak{N}_\varphi = \{X \in \mathcal{M}\ ;\ \varphi((AX)^\dagger(AX)) < \infty \ \text{ for all } \ A \in \mathcal{M}\}.$$

Then $\mathfrak{N}_\varphi$ is a left ideal of $\mathcal{M}$ and the restriction $\varphi\lceil \mathcal{P}(\mathfrak{N}_\varphi)$ of φ to the positive cone $\mathcal{P}(\mathfrak{N}_\varphi)$ is a quasi-weight on $\mathcal{P}(\mathcal{M})$ and it is called the *quasi-weight on* $\mathcal{P}(\mathcal{M})$ *generated by* φ and is denoted by φ_q. We denote by $(\pi_\varphi, \lambda_\varphi, \mathcal{H}_\varphi)$ the GNS-construction for the quasi-weight φ_q generated by φ. We remark that even if $\varphi \neq 0$ the case of $\varphi_q = 0$ arises (Example 3.6.2), and so the GNS-construction for such a weight is meaningless. We don't treat with such a weight. We next define a weight by another positive cone $\mathcal{M}_+ = \{X \in \mathcal{M};\ X \geq 0\}$.

Definition 3.1.2. A map φ of $\mathcal{M}_+$ into $\mathbb{R}_+ \cup \{+\infty\}$ is said to be a *weight on* $\mathcal{M}_+$ if

(i) $\varphi(X+Y) = \varphi(X) + \varphi(Y), \quad X, Y \in \mathcal{M}_+$

(ii) $\varphi(\alpha X) = \alpha\varphi(X), \quad X \in \mathcal{M}_+,\ \alpha \geq 0.$

A map φ of a hereditary positive subcone $\mathcal{D}(\varphi)_+$ of $\mathcal{M}_+$ into $\mathbb{R}_+$ is said to be a *quasi-weight* on $\mathcal{M}_+$ if it satisfies the above conditions (i) and (ii) for $\mathcal{D}(\varphi)_+$. A positive subcone $\mathcal{P}$ of $\mathcal{M}_+$ is said to be *hereditary* if any element X of $\mathcal{M}_+$ majorized by some element Y of $\mathcal{P}$ (that is, $X \leq Y$) belongs to $\mathcal{P}$.

It is clear that if φ is a weight on $\mathcal{M}_+$ then it is a weight on $\mathcal{P}(\mathcal{M})$. We denote by $\varphi\lceil \mathcal{P}(\mathcal{M})$ the restriction of φ to $\mathcal{P}(\mathcal{M})$. Suppose φ is a weight on $\mathcal{M}_+$. We define the finite part φ_q of φ by

$$\mathcal{D}(\varphi_q)_+ = \{X \in \mathcal{M}_+;\ \varphi(X) < \infty\},$$
$$\varphi_q(\sum_k \alpha_k X_k) = \sum_k \alpha_k \varphi(X_k), \qquad X_k \in \mathcal{D}(\varphi_q)_+,\ \alpha_k \geq 0.$$

Then $\mathcal{D}(\varphi_q)_+$ is a hereditary positive subcone of $\mathcal{M}_+$ and φ_q is a quasi-weight on $\mathcal{M}_+$. Suppose φ is a quasi-weight on $\mathcal{M}_+$. We put

$$\mathfrak{N}_\varphi = \{X \in \mathcal{M}\ ;\ (AX)^\dagger(AX) \in \mathcal{D}(\varphi)_+ \ \text{ for all } \ A \in \mathcal{M}\}.$$

Then $\mathfrak{N}_\varphi$ is a left ideal of $\mathcal{M}$ and the restriction of φ to $\mathcal{P}(\mathfrak{N}_\varphi)$ is a quasi-weight on $\mathcal{P}(\mathcal{M})$. In fact, for each $X_1, X_2 \in \mathfrak{N}_\varphi$ and $A \in \mathcal{M}$ we have

$$\begin{aligned}(X_1+X_2)^\dagger A^\dagger A(X_1+X_2) &+ (X_1-X_2)^\dagger A^\dagger A(X_1-X_2)\\ &= 2(X_1^\dagger A^\dagger A X_1 + X_2^\dagger A^\dagger A X_2) \in \mathcal{D}(\varphi)_+,\end{aligned}$$

and since $\mathcal{D}(\varphi)_+$ is a hereditary positive subcone of $\mathcal{M}_+$, it follows that $(X_1+X_2)^\dagger A^\dagger A(X_1+X_2) \in \mathcal{D}(\varphi)_+$, that is, $X_1+X_2 \in \mathfrak{N}_\varphi$. It is clear that $\alpha X, AX \in \mathfrak{N}_\varphi$ for all $\alpha \in \mathbb{C}$, $A \in \mathcal{M}$ and $X \in \mathfrak{N}_\varphi$. Thus, $\mathfrak{N}_\varphi$ is a left ideal of $\mathcal{M}$. Further, since $\mathcal{P}(\mathfrak{N}_\varphi) \subset \mathcal{D}(\varphi)_+$, the restriction of φ to $\mathcal{P}(\mathfrak{N}_\varphi)$ is a quasi-weight on $\mathcal{P}(\mathcal{M})$. We denote by $\varphi\lceil \mathcal{P}(\mathcal{M})$ the quasi-weight φ on $\mathcal{M}_+$ regarding it as the quasi-weight on $\mathcal{P}(\mathcal{M})$. The following diagram holds:

$$\begin{array}{ccc}
\underset{\text{weight on } \mathcal{M}_+}{\varphi} & \longrightarrow & \underset{\text{weight on } \mathcal{P}(\mathcal{M})}{\varphi\lceil \mathcal{P}(\mathcal{M})} \\
\swarrow & & \downarrow \\
\underset{\text{quasi-weight on } \mathcal{M}_+}{\varphi_q} & \longrightarrow & \underset{\text{quasi-weight on } \mathcal{P}(\mathcal{M})}{\varphi_q\lceil \mathcal{P}(\mathcal{M}) = (\varphi\lceil \mathcal{P}(\mathcal{M}))_q.}
\end{array}$$

The above equality $\varphi_q\lceil \mathcal{P}(\mathcal{M}) = (\varphi\lceil \mathcal{P}(\mathcal{M}))_q$ follows from

$$\mathfrak{N}_{\varphi_q\lceil \mathcal{P}(\mathcal{M})} = \mathfrak{N}_{\varphi_q} = \mathfrak{N}_\varphi = \mathfrak{N}_{\varphi\lceil \mathcal{P}(\mathcal{M})} = \mathfrak{N}_{(\varphi\lceil \mathcal{P}(\mathcal{M}))_q}.$$

This means that the GNS-constructions of all these (quasi-)weights coincide.

We give two kinds of important examples of weights and quasi-weights on $\mathcal{P}(\mathcal{M})$ or $\mathcal{M}_+$. We first give (quasi-)weights defined by vectors . Let $\xi \in \mathcal{H}\backslash\mathcal{D}$. We put

$$\mathfrak{N}_{\omega_\xi} = \{X \in \mathcal{M};\ \xi \in \mathcal{D}(X^{\dagger *}) \text{ and } X^{\dagger *}\xi \in \mathcal{D}\},$$

$$\omega_\xi(\sum_k X_k^\dagger X_k) = \sum_k \|X_k^{\dagger *}\xi\|^2, \quad X_k \in \mathfrak{N}_{\omega_\xi}.$$

Then ω_ξ is a quasi-weight on $\mathcal{P}(\mathcal{M})$. The following question arises: Is ω_ξ extended to a weight on $\mathcal{P}(\mathcal{M})$? In general, this question is inaffirmative, and so this is one of the reasons why we have to consider quasi-weights. In Section 3.6, we shall investigate such quasi-weights ω_ξ in more details.

We next give some (quasi-)weights defined by a net of positive linear functionals on $\mathcal{M}$. Let $\{f_\alpha\}$ be a net of positive linear functionals on $\mathcal{M}$. We put

$$\sup_\alpha f_\alpha : A \in \mathcal{P}(\mathcal{M}) \longrightarrow \sup_\alpha f_\alpha(A) \in [0, +\infty].$$

Then it is easily shown that

$$\max(\sup_\alpha f_\alpha(X^\dagger X), \sup_\alpha f_\alpha(Y^\dagger Y)) \tag{3.1.2}$$

$$\leq \sup_\alpha f_\alpha(X^\dagger X + Y^\dagger Y)$$

$$\leq \sup_\alpha f_\alpha(X^\dagger X) + \sup_\alpha f_\alpha(Y^\dagger Y) \tag{3.1.3}$$

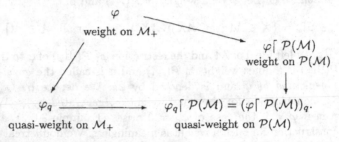

for all $X, Y \in \mathcal{M}$. We define the finite part of $\sup\limits_{\alpha} f_\alpha$ by

$$\mathfrak{N}^0_{\sup\limits_\alpha f_\alpha} \equiv \{X \in \mathcal{M};\ \sup_\alpha f_\alpha(X^\dagger X) < \infty\}.$$

Since

$$(X+Y)^\dagger(X+Y) + (X-Y)^\dagger(X-Y) = 2(X^\dagger X + Y^\dagger Y)$$

for each $X, Y \in \mathfrak{N}^0_{\sup\limits_\alpha f_\alpha}$, it follows that $\mathfrak{N}^0_{\sup\limits_\alpha f_\alpha}$ is a subspace of $\mathcal{M}$. But, $(\sup\limits_\alpha f_\alpha)(X^\dagger X + Y^\dagger Y) \neq \sup\limits_\alpha f_\alpha(X^\dagger X) + \sup\limits_\alpha f_\alpha(Y^\dagger Y)$ in general, and we have the following result:

Lemma 3.1.3. Let $\mathcal{N}$ be a subspace of $\mathfrak{N}^0_{\sup\limits_\alpha f_\alpha}$. The following statements are equivalent.

(1) $(\sup\limits_\alpha f_\alpha)(A+B) = (\sup\limits_\alpha f_\alpha)(A) + (\sup\limits_\alpha f_\alpha)(B)$ for all $A, B \in \mathcal{P}(\mathcal{N})$.

(2) For each finite subset $\{X_1, \cdots, X_m\}$ of $\mathcal{N}$ there exists a subsequence $\{\alpha_n\}$ of $\{\alpha\}$ such that

$$\lim_{n\to\infty} f_{\alpha_n}(X_k^\dagger X_k) = (\sup_\alpha f_\alpha)(X_k^\dagger X_k), \quad k = 1, 2, \cdots, m.$$

Proof. (1) $\Rightarrow$ (2) Take an arbitrary $\{X_1, \cdots, X_m\} \subset \mathcal{N}$. By (3.1.2), $(\sup\limits_\alpha f_\alpha)\ (\sum\limits_{k=1}^m X_k^\dagger X_k) < \infty$, and so there exists a subsequence $\{\alpha'_n\}$ of $\{\alpha\}$ such that

$$\lim_{n\to\infty} f_{\alpha'_n}(\sum_{k=1}^m X_k^\dagger X_k) = (\sup_\alpha f_\alpha)(\sum_{k=1}^m X_k^\dagger X_k).$$

Since $\sup\limits_n f_{\alpha'_n}(X_1^\dagger X_1) \leq (\sup\limits_\alpha f_\alpha)(\sum\limits_{k=1}^m X_k^\dagger X_k) < \infty$, there exists a subsequence $\{\alpha''_n\}$ of $\{\alpha'_n\}$ such that

$$\lim_{n\to\infty} f_{\alpha''_n}(X_1^\dagger X_1) = \sup_n f_{\alpha'_n}(X_1^\dagger X_1) \equiv \varphi(X_1^\dagger X_1).$$

Since $\{\alpha''_n\}$ is a subsequence of $\{\alpha'_n\}$, we have

$$\lim_{n\to\infty} f_{\alpha''_n}(\sum_{k=1}^m X_k^\dagger X_k) = (\sup_\alpha f_\alpha)(\sum_{k=1}^m X_k^\dagger X_k), \quad \lim_{n\to\infty} f_{\alpha''_n}(X_1^\dagger X_1) = \varphi(X_1^\dagger X_1).$$

Furthermore, since $\sup\limits_n f_{\alpha''_n}(X_2^\dagger X_2) < \infty$, there exists a subsequence $\{\alpha'''_n\}$ of $\{\alpha''_n\}$ such that

$$\lim_{n\mapsto\infty} f_{\alpha_n'''}(\sum_{k=1}^m X_k^\dagger X_k) = (\sup_\alpha f_\alpha)(\sum_{k=1}^m X_k^\dagger X_k),$$
$$\lim_{n\mapsto\infty} f_{\alpha_n'''}(X_1^\dagger X_1) = \varphi(X_1^\dagger X_1),$$
$$\lim_{n\mapsto\infty} f_{\alpha_n'''}(X_2^\dagger X_2) = (\sup_n f_{\alpha_n''})(X_2^\dagger X_2) \equiv \varphi(X_2^\dagger X_2).$$

Repeating this argument, there exists a subsequence $\{\alpha_n\}$ of $\{\alpha\}$ such that

$$\lim_{n\mapsto\infty} f_{\alpha_n}(\sum_{k=1}^m X_k^\dagger X_k) = (\sup_\alpha f_\alpha)(\sum_{k=1}^m X_k^\dagger X_k),$$
$$\lim_{n\mapsto\infty} f_{\alpha_n}(X_k^\dagger X_k) = \varphi(X_k^\dagger X_k), \quad k = 1,2,\cdots,m, \tag{3.1.4}$$

which implies by the assumption (1) that

$$\begin{aligned}\sum_{k=1}^m \varphi(X_k^\dagger X_k) &= \lim_{n\mapsto\infty} \sum_{k=1}^m f_{\alpha_n}(X_k^\dagger X_k) = \lim_{n\mapsto\infty} f_{\alpha_n}(\sum_{k=1}^m X_k^\dagger X_k)\\ &= (\sup_\alpha f_\alpha)(\sum_{k=1}^m (X_k^\dagger X_k))\\ &= \sum_{k=1}^m (\sup_\alpha f_\alpha)(X_k^\dagger X_k). \end{aligned} \tag{3.1.5}$$

Since $0 \leq \varphi(X_k^\dagger X_k) \leq (\sup_\alpha f_\alpha)(X_k^\dagger X_k)$, $k = 1,2,\cdots,m$, it follows from (3.1.4) that $\varphi(X_k^\dagger X_k) = (\sup_\alpha f_\alpha)(X_k^\dagger X_k)$, $k = 1,2,\cdots,m$. Therefore, we have by (3.1.3)

$$\lim_{n\mapsto\infty} f_{\alpha_n}(X_k^\dagger X_k) = (\sup_\alpha f_\alpha)(X_k^\dagger X_k), \quad k = 1,2,\cdots,m.$$

(2) ⇒ (1) Take an arbitrary subset $\{X_1, X_2, \cdots, X_m\}$ of $\mathcal{N}$. By the assumption (2) there exists a subsequence $\{\alpha_n\}$ of $\{\alpha\}$ such that

$$\lim_{n\mapsto\infty} f_{\alpha_n}(X_k^\dagger X_k) = (\sup_\alpha f_\alpha)(X_k^\dagger X_k), \quad k = 1,2,\cdots,m.$$

The statement (1) follows from

$$\begin{aligned}(\sup_\alpha f_\alpha)(\sum_{k=1}^m X_k^\dagger X_k) \leq \sum_{k=1}^m (\sup_\alpha f_\alpha)(X_k^\dagger X_k) &= \lim_{n\mapsto\infty} \sum_{k=1}^m f_{\alpha_n}(X_k^\dagger X_k)\\ &= \lim_{n\mapsto\infty} f_{\alpha_n}(\sum_{k=1}^m X_k^\dagger X_k)\\ &\leq (\sup_\alpha f_\alpha)(\sum_{k=1}^m X_k^\dagger X_k).\end{aligned}$$

When $\{f_\alpha\}$ satisfies the condition of Lemma 3.1.3, (2), we say that $\{f_\alpha\}$ *has the net property for* $\mathcal{P}(\mathcal{N})$ and then denote the restriction of the map $\sup_\alpha f_\alpha$ to $\mathcal{P}(\mathcal{N})$ by $\operatorname{Sup}_\alpha f_\alpha \lceil \mathcal{P}(\mathcal{N})$. In particular, when $\{f_\alpha\}$ has the net property for $\mathcal{P}(\mathfrak{N}^0_{\sup_\alpha f_\alpha})$, we simply say that $\{f_\alpha\}$ *has the net property* and then denote the map $\sup_\alpha f_\alpha$ by $\operatorname{Sup}_\alpha f_\alpha$. By Lemma 3.1.3 and (3.1.2) we have the following

Proposition 3.1.4. Let $\{f_\alpha\}$ be a net of positive linear functionals on $\mathcal{M}$. Suppose $\{f_\alpha\}$ has the net property for $\mathcal{P}(\mathcal{I})$, where $\mathcal{I}$ is a left ideal of $\mathcal{M}$ which is contained in $\mathfrak{N}^0_{\sup_\alpha f_\alpha}$. Then $\operatorname{Sup}_\alpha f_\alpha \lceil \mathcal{P}(\mathcal{I})$ is a quasi-weight on $\mathcal{P}(\mathcal{M})$. Suppose $\{f_\alpha\}$ has the net property. Then $\operatorname{Sup}_\alpha f_\alpha$ is a weight on $\mathcal{P}(\mathcal{M})$.

Let $\{f_\alpha\}$ be a net of strongly positive linear functionals on $\mathcal{M}$. A linear functional f on $\mathcal{M}$ is said to be *strongly positive* if $f(X) \geq 0$ for all $X \in \mathcal{M}_+$. We put

$$\sup_\alpha f_\alpha : X \in \mathcal{M}_+ \longrightarrow \sup_\alpha f_\alpha(X) \in [0, +\infty],$$
$$D(\sup_\alpha f_\alpha)_+ = \{X \in \mathcal{M}_+; \sup_\alpha f_\alpha(X) < \infty\}.$$

Then $D(\sup_\alpha f_\alpha)_+$ is a hereditary positive subcone of $\mathcal{M}_+$. Let $\mathcal{P}$ be a positive subcone of $D(\sup_\alpha f_\alpha)_+$. When $\{f_\alpha\}$ satisfies the condition of Lemma 3.1.3,(2) for $\mathcal{P}$, we say that $\{f_\alpha\}$ *has the net property for* $\mathcal{P}$ and then denote the restriction of the map $\sup_\alpha f_\alpha$ to $\mathcal{P}$ by $\operatorname{Sup}_\alpha f_\alpha \lceil \mathcal{P}$. In particular, when $\{f_\alpha\}$ has the net property for $D(\sup_\alpha f_\alpha)_+$, we simply say that $\{f_\alpha\}$ *has the net property* and then denote the map $\sup_\alpha f_\alpha$ by $\operatorname{Sup}_\alpha f_\alpha$. Similarly to the proofs of Lemma 3.1.3 and Proposition 3.1.4 we can show the following result:

Propositon 3.1.5. Let $\{f_\alpha\}$ be a net of strongly positive linear functionals on $\mathcal{M}$ and $\mathcal{P}$ a hereditary positive subcone of $D(\sup_\alpha f_\alpha)_+$. Then $\{f_\alpha\}$ has the net property for $\mathcal{P}$ if and only if $\operatorname{Sup}_\alpha f_\alpha \lceil \mathcal{P}$ is a quasi-weight on $\mathcal{M}_+$. Further, $\{f_\alpha\}$ has the net property if and only if $\operatorname{Sup}_\alpha f_\alpha$ is a weight on $\mathcal{M}_+$.

3.2 The regularity of quasi-weights and weights

In this section we define the notions of regularity and singularity of (quasi-)weights, and give the decomposition theorem of (quasi-)weights into the regular part and the singular part. Let $\mathcal{M}$ be a closed O*-algebra on $\mathcal{D}$ in $\mathcal{H}$.

Definition 3.2.1. A quasi-weight φ on $\mathcal{P}(\mathcal{M})$ is said to be *regular* if $\varphi = \operatorname{Sup}_{\alpha} f_\alpha \lceil \mathcal{P}(\mathfrak{N}_\varphi)$ ($= \sup_{\alpha} f_\alpha$ on $\mathcal{P}(\mathfrak{N}_\varphi)$ by Lemma 3.1.3) for some net $\{f_\alpha\}$ of positive linear functionals on $\mathcal{M}$, and it is said to be *singular* if there doesn't exist any positive linear functional f on $\mathcal{M}$ such that $f(X^\dagger X) \leq \varphi(X^\dagger X)$ for each $X \in \mathfrak{N}_\varphi$ and $f \neq 0$ on $\mathcal{P}(\mathfrak{N}_\varphi)$. A weight φ on $\mathcal{P}(\mathcal{M})$ is said to be *regular* if $\varphi = \operatorname{Sup}_{\alpha} f_\alpha$($= \sup_{\alpha} f_\alpha$ on $\mathcal{P}(\mathcal{M})$ by Lemma 3.1.3) for some net $\{f_\alpha\}$ of positive linear functionals on $\mathcal{M}$, and φ is said to be *quasi-regular* if the quasi-weight φ_q on $\mathcal{P}(\mathcal{M})$ defined by φ is regular. If there doesn't exist any positive linear functional f on $\mathcal{M}$ such that $f(X^\dagger X) \leq \varphi(X^\dagger X)$ for all $X \in \mathcal{M}$ and $f \neq 0$ on$\mathcal{P}(\mathcal{M})$, then φ is said to be *singular* .

We define trio-commutants $T(\varphi)'_\delta$ and $T(\varphi)'_c$ for a quasi-weight φ which play an important rule for the regularity of φ as follows :

$$\begin{aligned} T(\varphi)'_\delta = \{K = (C, \xi, \eta);\ \ & C \in \pi_\varphi(\mathcal{M})'_{\mathrm{w}}, \xi, \eta \in \mathcal{D}(\pi_\varphi^*) \\ & \text{s.t. } C\lambda_\varphi(X) = \pi_\varphi^*(X)\xi \text{ and} \\ & C^*\lambda_\varphi(X) = \pi_\varphi^*(X)\eta \text{ for all } X \in \mathfrak{N}_\varphi\}, \end{aligned}$$

$$T(\varphi)'_c = \{K = (C, \xi, \eta) \in T(\varphi)'_\delta;\ \ \xi, \eta \in \mathcal{D}(\pi_\varphi)\}.$$

For $K = (C, \xi, \eta) \in T(\varphi)'_\delta$ we put

$$\pi'(K) = C,\ \ \lambda'(K) = \xi,\ \ \lambda'_*(K) = \eta.$$

We have the following

Lemma 3.2.2. (1) $T(\varphi)'_\delta$ is a $*$-invariant vector space under the following operations and the involution :

$$\begin{aligned} K_1 + K_2 &= (C_1 + C_2, \xi_1 + \xi_2, \eta_1 + \eta_2),\quad \alpha K = (\alpha C, \alpha\xi, \bar{\alpha}\eta), \\ K^* &= (C^*, \eta, \xi) \end{aligned}$$

for $K_1 = (C_1, \xi_1, \eta_1), K_2 = (C_2, \xi_2, \eta_2)$ and $K = (C, \xi, \eta)$ in $T(\varphi)'_\delta$ and $\alpha \in \mathbb{C}$.

(2) $T(\varphi)'_c$ is a $*$-invariant subspace of $T(\varphi)'_\delta$. In particular, if $\pi_\varphi(\mathcal{M})'_{\mathrm{w}}\mathcal{D}(\pi_\varphi) \subset \mathcal{D}(\pi_\varphi)$, then $T(\varphi)'_c$ is a $*$-algebra under the following multiplication :

$$K_1K_2 = (C_1C_2, C_1\xi_2, C_2^*\eta_1)$$

for $K_1 = (C_1, \xi_1, \eta_1), K_2 = (C_2, \xi_2, \eta_2) \in T(\varphi)'_c$, and π' is a $*$-homomorphism of $T(\varphi)'_c$ into the von Neumann algebra $\pi_\varphi(\mathcal{M})'_{\mathrm{w}}$ and λ' is a linear map of $T(\varphi)'_c$ into $\mathcal{D}(\pi_\varphi)$ satisfying $\pi'(K_1)\lambda'(K_2) = \lambda'(K_1K_2)$ for all $K_1, K_2 \in T(\varphi)'_c$.

Lemma 3.2.3. Let φ be a quasi-weight on $\mathcal{P}(\mathcal{M})$. Suppose a linear functional f on $\mathfrak{N}_\varphi$ satisfies the following conditions (i) and (ii):

(i) $0 \leq f(X^\dagger X) \leq \varphi(X^\dagger X)$ for each $X \in \mathfrak{N}_\varphi$.

(ii) For any $A \in \mathcal{M}$ there exists $\gamma_A > 0$ such that $|f(A^\dagger X)|^2 \leq \gamma_A \varphi(X^\dagger X)$ for each $X \in \mathfrak{N}_\varphi$.

Then there exists an element $K \in T(\varphi)'_\delta$ such that $0 \leq \pi'(K) \leq I$ and $f(X) = (\lambda_\varphi(X)|\lambda'(K))$ for all $X \in \mathfrak{N}_\varphi$. Conversely, for each $K \in T(\varphi)'_\delta$ with $0 \leq \pi'(K) \leq I$ we put

$$f(X) = (\lambda_\varphi(X)|\lambda'(K)), \quad X \in \mathfrak{N}_\varphi.$$

Then f is a linear functional on $\mathfrak{N}_\varphi$ satisfying the above (i) and (ii).

Proof. Suppose f is a linear functional on $\mathfrak{N}_\varphi$ satisfying the conditions (i) and (ii). Similarly to the GNS-construction for quasi-weights, we can define the GNS-construction $(\pi_f, \lambda_f, \mathcal{H}_f)$ for f. By (i) there exists a bounded linear transform C from $\mathcal{H}_\varphi$ to $\mathcal{H}_f$ such that $C\lambda_\varphi(X) = \lambda_f(X)$ for all $X \in \mathfrak{N}_\varphi$. Further, we have

$$C^*C \in \pi_\varphi(\mathcal{M})'_{\rm w} \text{ and } f(Y^\dagger X) = (C^*C\lambda_\varphi(X) \mid \lambda_\varphi(Y)) \quad {}^\forall X, Y \in \mathfrak{N}_\varphi. \tag{3.2.1}$$

It follows from (ii) and the Riesz theorem that there exists an element ξ of $\mathcal{D}(\pi_\varphi^*)$ such that

$$f(X) = (\lambda_\varphi(X) \mid \xi), \quad {}^\forall X \in \mathfrak{N}_\varphi,$$

which implies by (3.2.1) that

$$(\lambda_\varphi(Y) \mid \pi_\varphi^*(X)\xi) = f(X^\dagger Y) = (\lambda_\varphi(Y) \mid C^*C\lambda_\varphi(X))$$

for all $X, Y \in \mathfrak{N}_\varphi$, and so $C^*C\lambda_\varphi(X) = \pi_\varphi^*(X)\xi$ for all $X \in \mathfrak{N}_\varphi$. Hence, $K = (C^*C, \xi, \xi) \in T(\varphi)'_\delta$, $0 \leq \pi'(K) \leq I$ and $f(X) = (\lambda_\varphi(X) \mid \lambda'(K))$ for all $X \in \mathfrak{N}_\varphi$.

We next show the converse. Take an arbitrary $K \in T(\varphi)'_\delta$ such that $0 \leq \pi'(K) \leq I$. Then it is clear that f is a linear functional on $\mathfrak{N}_\varphi$ and further, since

$$\begin{aligned} f(X^\dagger X) &= (\lambda_\varphi(X)|\pi_\varphi^*(X)\lambda'(K)) = (\lambda_\varphi(X)|\pi'(K)\lambda_\varphi(X)), \\ f(A^\dagger X) &= (\lambda_\varphi(X)|\pi_\varphi^*(A)\lambda'(K)) \end{aligned}$$

for all $X \in \mathfrak{N}_\varphi$ and $A \in \mathcal{M}$, it follows that f satisfies the conditions (i) and (ii).

Remark 3.2.4. For $K \in T(\varphi)'_\delta$ the linear functional $\omega_{\lambda'(K)} \circ \pi_\varphi^*$ on $\mathcal{M}$ defined by

$$(\omega_{\lambda'(K)} \circ \pi_\varphi^*)(X) = (\pi_\varphi^*(X)\lambda'(K)|\lambda'(K)), \quad X \in \mathcal{M}$$

is not necessarily positive in case π_φ^* is not a $*$-representation of $\mathcal{M}$. When $K \in T(\varphi)'_c$ and $0 \leq \pi'(K) \leq I$, $\pi_{\lambda'(K)} \circ \pi_\varphi$ is a positive linear functional on $\mathcal{M}$ satisfying

$$(\omega_{\lambda'(K)} \circ \pi_\varphi)(X^\dagger X) \leq \varphi(X^\dagger X), \quad {}^\forall X \in \mathfrak{N}_\varphi.$$

But, the above inequality does not hold for all $X \in \mathcal{M}$ because the equality $\pi_\varphi(X)\lambda'(K) = \pi'(K)\lambda_\varphi(X)$ holds for each $X \in \mathfrak{N}_\varphi$ but this doesn't hold for $X \in \mathcal{M} \setminus \mathfrak{N}_\varphi$ in general.

For the regularity and the singularity of quasi-weights we have the following

Theorem 3.2.5. Let φ be a quasi-weight on $\mathcal{P}(\mathcal{M})$.
(1) Consider the following statements:
(i) There exists a net $\{K_\alpha\}$ in $T(\varphi)'_c$ such that $0 \leq \pi'(K_\alpha) \leq I$ for each α and $\pi'(K_\alpha) \to I$ strongly.
(ii) $\varphi = \operatorname{Sup}_\alpha (\omega_{\xi_\alpha} \circ \pi_\varphi) \lceil \mathcal{P}(\mathfrak{N}_\varphi)$ for some net $\{\xi_\alpha\}$ in $D(\pi_\varphi)$.
(iii) φ is regular.
(iv) There exists a net $\{K_\alpha\}$ in $T(\varphi)'_\delta$ such that $0 \leq \pi'(K_\alpha) \leq I$ for each α and $\pi'(K_\alpha) \to I$ strongly.
Then the implications (i) $\Rightarrow$ (ii) $\Rightarrow$ (iii) $\Rightarrow$ (iv) hold. In particular, suppose π_φ is self-adjoint, then the statements (i) $\sim$ (iv) are equivalent.
(2) Suppose π_φ is self-adjoint. Then φ is singular if and only if there doesn't exist any element K of $T(\varphi)'_c$ such that $\pi'(K) \geq 0$ and $\pi'(K) \neq 0$.

Proof. (1) (i) $\Rightarrow$ (ii) We put $\xi_\alpha = \lambda'(K_\alpha)$. Since

$$(\omega_{\xi_\alpha} \circ \pi_\varphi)(X^\dagger X) = \| \pi'(K_\alpha)\lambda_\varphi(X) \|^2$$

for each $X \in \mathfrak{N}_\varphi$ and α, and $\pi'(K_\alpha) \to I$ strongly, it follows that the net $\{\omega_{\xi_\alpha} \circ \pi_\varphi\}$ of positive linear functionals on $\mathcal{M}$ has the net property for $\mathcal{P}(\mathfrak{N}_\varphi)$ and $\varphi = \operatorname{Sup}_\alpha (\omega_{\xi_\alpha} \circ \pi_\varphi) \lceil \mathcal{P}(\mathfrak{N}_\varphi)$.
(ii)$\Rightarrow$ (iii) This is trivial.
(iii) $\Rightarrow$(iv) This follows from Lemma 3.2.3.
Suppose π_φ is self-adjoint. Then, $T(\varphi)'_\delta = T(\varphi)'_c$, and so the implication (iv) $\Rightarrow$ (i) and the statement (2) follow from Lemma 3.2.3.

Similarly we have the following result for the regularity of weights :

Theorem 3.2.6. Let φ be a weight on $\mathcal{P}(\mathcal{M})$. Consider the following statements.
(i) $\varphi = \sup_\alpha (\omega_{\xi_\alpha} \circ \pi_\varphi)$ for some net $\{\xi_\alpha\}$ in $\mathcal{D}(\pi_\varphi)$.
(ii) φ is regular.
(iii) φ is quasi-regular.
(iv) There exists a net $\{K_\alpha\}$ in $T(\varphi)'_\delta$ such that $0 \leq \pi'(K_\alpha) \leq I$ for each α and $\pi'(K_\alpha) \longrightarrow I$ strongly.
Then the following implications (i) $\Rightarrow$ (ii) $\Rightarrow$ (iii) $\Rightarrow$ (iv) hold.

Let φ be a weight on $\mathcal{P}(\mathcal{M})$. It follows from the definition of $T(\varphi)'_c$ that the equality

$$\pi_\varphi(X)\lambda'(K) = \pi'(K)\lambda_\varphi(X), \ (X \in \mathfrak{N}_\varphi, K \in T(\varphi)'_c)$$

holds, but it doesn't hold for *all* $X \in \mathcal{M}$. For this reason, even if π_φ is self-adjoint, the quasi-regularity of φ doesn't necessarily imply the regularity of φ. So, we have defined the notions of normality and semifiniteness of φ and investigated the equivalence of the regularity and the quasi-regularity (refer to Inoue-Ogi [1])

As the decomposition theorem of quasi-weights we have the following

Theorem 3.2.7. Suppose φ is a quasi-weight on $\mathcal{P}(\mathcal{M})$ such that π_φ is self-adjoint. Then φ is decomposed into

$$\varphi = \varphi_r + \varphi_s,$$

where φ_r is a regular quasi-weight on $\mathcal{P}(\mathcal{M})$ and φ_s is a singular quasi-weight on $\mathcal{P}(\mathcal{M})$ such that π_{φ_r} and π_{φ_s} are self-adjoint.

Proof. We denote by P'_φ the projection from $\mathcal{H}_\varphi$ onto the closed subspace of $\mathcal{H}_\varphi$ generated by $\pi'(T(\varphi)'_c)\mathcal{H}_\varphi$. Then, $P'_\varphi \in \pi_\varphi(\mathcal{M})'_w$ and there exists a net $\{K_\alpha\}$ in $T(\varphi)'_c$ such that $0 \leq \pi'(K_\alpha) \leq P'_\varphi$ for each α and $\pi'(K_\alpha) \to P'_\varphi$ strongly. It is clear that the net $\{f_\alpha \equiv \omega_{\lambda'(K_\alpha)} \circ \pi_\varphi\}$ of positive linear functionals on $\mathcal{M}$ has the net property for $\mathcal{P}(\mathfrak{N}_\varphi)$, and so it follows from Lemma 3.1.3 that $\varphi_r \equiv \operatorname{Sup}_\alpha f_\alpha \lceil \mathcal{P}(\mathfrak{N}_\varphi)$ is a regular quasi-weight on $\mathcal{P}(\mathcal{M})$ such that $\mathfrak{N}_{\varphi_r} = \mathfrak{N}_\varphi$ and

$$\varphi_r(X^\dagger X) = \| P'_\varphi \lambda_\varphi(X) \|^2 \ \text{ for each } X \in \mathfrak{N}_\varphi. \qquad (3.2.2)$$

We put

$$\varphi_s = \varphi - \varphi_r.$$

Then φ_s is a quasi-weight on $\mathcal{P}(\mathcal{M})$ with $\mathfrak{N}_{\varphi_s} = \mathfrak{N}_\varphi$. It follows from (3.2.2) that π_{φ_r} (resp. π_{φ_s}) is unitarily equivalent to the induced representation $(\pi_\varphi)_{P'_\varphi}$ (resp. $(\pi_\varphi)_{I-P'_\varphi}$) of π_φ, so that π_{φ_r} and π_{φ_s} are self-adjoint. We show φ_s is singular. Suppose there exists a positive linear functional f on $\mathcal{M}$ such that $f(X^\dagger X) \leq \varphi_s(X^\dagger X)$ for all $X \in \mathfrak{N}_\varphi$ and $f(X_0^\dagger X_0) \neq 0$ for some $X_0 \in \mathfrak{N}_\varphi$. Since $\varphi_s \leq \varphi$, it follows from Lemma 3.2.3 that there exists an element K of $T(\varphi)'_c$ such that $0 \leq \pi'(K)$, $\pi'(K) \neq 0$ and $f(X) = (\lambda_\varphi(X) \mid \lambda'(K))$ for all $X \in \mathfrak{N}_\varphi$. Then we have

$$\begin{aligned} |(\pi'(K)\lambda_\varphi(X) \mid \lambda_\varphi(Y))|^2 &= |f(Y^\dagger X)|^2 \\ &\leq \gamma_Y \|(I - P'_\varphi)\lambda_\varphi(X)\|^2 \end{aligned}$$

for all $X, Y \in \mathfrak{N}_\varphi$, and so

$$\begin{aligned}|(\pi'(K)\lambda_\varphi(X) \mid \lambda_\varphi(Y))| &= |(P_\varphi'\lambda_\varphi(X) \mid \pi'(K)\lambda_\varphi(Y))| \\ &= \lim_{n\to\infty} |(\pi'(K)\lambda_\varphi(X_n) \mid \lambda_\varphi(Y))| \\ &\leq \gamma_Y \lim_{n\to\infty} \|(I - P_\varphi')\lambda_\varphi(X_n)\|^2 \\ &= 0\end{aligned}$$

for all $X, Y \in \mathfrak{N}_\varphi$, where $\{X_n\}$ is a sequence in $\mathfrak{N}_\varphi$ such that $\lim_{n\to\infty} \lambda_\varphi(X_n) = P_\varphi'\lambda_\varphi(X)$. Hence, $\pi'(K) = 0$, and so $f(X_0^\dagger X_0) = 0$. This is a contradiction. Hence, φ_s is singular.

We investigate the relation of quasi-weights and generalized vectors. Let φ be a quasi-weight on $\mathcal{P}(\mathcal{M})$. Suppose $\pi_\varphi(X) \to \lambda_\varphi(X)$, $X \in \mathfrak{N}_\varphi$ is a map and then put

$$\Lambda_\varphi(\pi_\varphi(X)) = \lambda_\varphi(X), \quad X \in \mathfrak{N}_\varphi.$$

Then Λ_φ is a generalized vector for the O*-algebra $\pi_\varphi(\mathcal{M})$, and it is called the *generalized vector induced by* φ . Then we have the following result for regularity of φ and Λ_φ:

Proposition 3.2.8. Let φ be a quasi-weight on $\mathcal{P}(\mathcal{M})$ such that π_φ is self-adjoint. The following statements are equivalent:

(i) φ is regular and $\lambda_\varphi(\mathfrak{N}_\varphi^\dagger \mathfrak{N}_\varphi)$ is total in $\mathcal{H}_\varphi$.

(ii) Λ_φ is well-defined and it is regular.

Proof. Suppose $\lambda_\varphi(\mathfrak{N}_\varphi^\dagger \mathfrak{N}_\varphi)$ is total in $\mathcal{H}_\varphi$. Then it is easily shown that $\mathcal{D}(\Lambda_\varphi^C) = \{\pi'(K); K \in T(\varphi)_C'\}$ and $\Lambda_\varphi^C(\pi'(K)) = \lambda'(K)$ for each $K \in T(\varphi)_C'$, which implies by Theorem 3.2.5 that the statements (i) and (ii) are equivalent.

Conversely, we consider when a generalized vector λ induces a quasi-weight on $\mathcal{P}(\mathcal{M})$. We put

$$\begin{cases}\mathcal{D}(\varphi_\lambda) = \mathcal{D}(\lambda) \\ \varphi_\lambda(\sum_k X_k^\dagger X_k) = \sum_k \|\lambda(X_k)\|^2, \quad \sum_k X_k^\dagger X_k \in \mathcal{D}(\varphi_\lambda).\end{cases}$$

When $\lambda = \lambda_\xi$ $(\xi \in \mathcal{H})$, φ_{λ_ξ} is a quasi-weight on $\mathcal{P}(\mathcal{M})$, and so the generalized vector λ_ξ always induces the quasi-weight φ_{λ_ξ} on $\mathcal{P}(\mathcal{M})$. But, for a general generalized vector λ φ_λ is not necessarily well-defined because $\sum_k X_k^\dagger X_k = 0$ $(\{X_k\} \subset \mathcal{D}(\lambda))$ doesn't imply $\sum_k \|\lambda(X_k)\|^2 = 0$. For this problem we have the following

Proposition 3.2.9. Let λ be a regular generalized vector for $\mathcal{M}$. Then φ_λ is a regular quasi-weight on $\mathcal{P}(\mathcal{M})$ such that $(\pi_{\varphi_\lambda}(\mathcal{M}), \Lambda_{\varphi_\lambda})$ is unitarily equivalent to $(\mathcal{M}_\lambda \equiv \mathcal{M}\lceil\mathcal{D}(\lambda), \lambda)$, that is, there exists a unitary transform U of $\mathcal{H}_\lambda$ onto $\mathcal{H}_{\varphi_\lambda}$ such that $U\mathcal{D}(\lambda) = \mathcal{D}(\Lambda_{\varphi_\lambda})$, $U\lambda(X) = \Lambda_{\varphi_\lambda}(\pi_\varphi(X))$, ${}^\forall X \in \mathcal{D}(\lambda)$ and $\pi_{\varphi_\lambda}(A) = U(A\lceil\mathcal{D}_\lambda)U^*, {}^\forall A \in \mathcal{M}$.

Proof. This follows from the equality:

$$\begin{aligned}\lim_\alpha(\sum_{k=1}^n \alpha_k X_k^\dagger X_k \lambda^c(K_\alpha)|\lambda^c(K_\alpha)) &= \lim_\alpha \sum_{k=1}^n \alpha_k \|K_\alpha\lambda(X_k)\|^2 \\ &= \sum_{k=1}^n \alpha_k \|\lambda(X_k)\|^2 \\ &= \varphi_\lambda(\sum_{k=1}^n \alpha_k X_k^\dagger X_k)\end{aligned}$$

for each $\{X_k\} \subset \mathcal{D}(\lambda)$ and $\{\alpha_k\} \subset \mathbb{R}_+$, where $\{K_\alpha\}$ is a net in $\mathcal{D}(\lambda^C)^* \cap \mathcal{D}(\lambda^C)$ such that $0 \leq K_\alpha \leq I$, ${}^\forall\alpha$ and $K_\alpha \to I$ strongly.

The φ_λ in Proposition 3.2.9 is said to be the *quasi-weight induced by* λ.

3.3 Standard weights

In this section we define and study an important class in regular (quasi-)weights which is possible to develop the Tomita-Takesaki theory in O*-algebras. Let $\mathcal{M}$ be a closed O*-algebra on $\mathcal{D}$ in $\mathcal{H}$. We need the notions of faithfulness and semifiniteness of (quasi-)weights :

Definition 3.3.1. Let φ be a (quasi-)weight on $\mathcal{P}(\mathcal{M})$. If $\varphi(A^\dagger A) = 0$, $A \in \mathcal{M}$ implies $A = 0$, then φ is said to be *faithful* . If there exists a net $\{U_\alpha\}$ in $\mathfrak{N}_\varphi^\dagger \cap \mathfrak{N}_\varphi$ such that $\| \overline{U_\alpha} \| \leq 1$ for each α and $\{U_\alpha\}$ converges strongly to I, then φ is said to be *semifinite*.

Let φ be a faithful semifinite (quasi-)weight on $\mathcal{P}(\mathcal{M})$. Then it is easily shown that π_φ is a $*$-isomorphism and the generalized vector Λ_φ for the O*-algebra $\pi_\varphi(\mathcal{M})$ is defined by

$$\Lambda_\varphi(\pi_\varphi(X)) = \lambda_\varphi(X), \quad X \in \mathfrak{N}_\varphi.$$

Suppose

(S)$_1$ $\pi_\varphi(\mathcal{M})'_{\mathrm{w}}\mathcal{D}(\pi_\varphi) \subset \mathcal{D}(\pi_\varphi)$,

(S)$_2$ $\lambda_\varphi((\mathfrak{N}_\varphi^\dagger \cap \mathfrak{N}_\varphi)^2)$ is total in $\mathcal{H}_\varphi$.

Then we can define a generalized vector $\Lambda_\varphi^{\mathrm{c}}$ for the von Neumann algebra $\pi_\varphi(\mathcal{M})'_{\mathrm{w}}$ by

$$\begin{cases} \mathcal{D}(\Lambda^{\mathrm{c}}_\varphi) = \{ K \in \pi_\varphi(\mathcal{M})'_{\mathrm{w}} \ ; \ ^\exists \xi_K \in \mathcal{D}(\pi_\varphi) \\ \qquad\qquad \text{s.t. } K\Lambda_\varphi(X) = \pi_\varphi(X)\xi_K, \ ^\forall X \in \mathfrak{N}_\varphi \} \\ \Lambda^{\mathrm{c}}_\varphi(K) = \xi_K, \ K \in \mathcal{D}(\Lambda^{\mathrm{c}}_\varphi). \end{cases}$$

Further, suppose

$(\mathrm{S})_3$ $\Lambda^{\mathrm{c}}_\varphi((\mathcal{D}(\Lambda^{\mathrm{c}}_\varphi)^* \cap \mathcal{D}(\Lambda^{\mathrm{c}}_\varphi))^2)$ is total in $\mathcal{H}_\varphi$.

Then, the generalized vector $\Lambda^{\mathrm{cc}}_\varphi$ for the von Neumann algebra $(\pi_\varphi(\mathcal{M})'_{\mathrm{w}})'$ is defined by

$$\begin{cases} \mathcal{D}(\Lambda^{\mathrm{cc}}_\varphi) = \{ A \in (\pi_\varphi(\mathcal{M})'_{\mathrm{w}})' \ ; \ ^\exists \xi_A \in \mathcal{H}_\varphi \\ \qquad\qquad \text{s.t. } A\Lambda^{\mathrm{c}}_\varphi(K) = K\xi_A, \ ^\forall K \in \mathcal{D}(\Lambda^{\mathrm{c}}_\varphi) \} \\ \Lambda^{\mathrm{cc}}_\varphi(A) = \xi_A, \ A \in \mathcal{D}(\Lambda^{\mathrm{cc}}_\varphi) \end{cases}$$

and $\Lambda^{\mathrm{cc}}_\varphi((\mathcal{D}(\Lambda^{\mathrm{cc}}_\varphi)^* \cap \mathcal{D}(\Lambda^{\mathrm{cc}}_\varphi))^2)$ is total in $\mathcal{H}_\varphi$. Hence, the maps $\lambda_\varphi(X) \longrightarrow \lambda_\varphi(X^\dagger)$, $X \in \mathfrak{N}^\dagger_\varphi \cap \mathfrak{N}_\varphi$ and $\Lambda^{\mathrm{cc}}_\varphi(A) \longrightarrow \Lambda^{\mathrm{cc}}_\varphi(A^*)$, $A \in \mathcal{D}(\Lambda^{\mathrm{cc}}_\varphi)^* \cap \mathcal{D}(\Lambda^{\mathrm{cc}}_\varphi)$ are closable in $\mathcal{H}_\varphi$ and their closures are denoted by S_φ and $S_{\Lambda^{\mathrm{cc}}_\varphi}$, respectively. Let $S_\varphi = J_\varphi \Delta^{\frac{1}{2}}_\varphi$ and $S_{\Lambda^{\mathrm{cc}}_\varphi} = J_{\Lambda^{\mathrm{cc}}_\varphi} \Delta^{\frac{1}{2}}_{\Lambda^{\mathrm{cc}}_\varphi}$ be the polar decompositions of S_φ and $S_{\Lambda^{\mathrm{cc}}_\varphi}$, respectively. Then we see that $S_\varphi \subset S_{\Lambda^{\mathrm{cc}}_\varphi}$, and by the Tomita fundamental theorem $J_{\Lambda^{\mathrm{cc}}_\varphi}(\pi_\varphi(\mathcal{M})'_{\mathrm{w}})' J_{\Lambda^{\mathrm{cc}}_\varphi} = \pi_\varphi(\mathcal{M})'_{\mathrm{w}}$ and $\Delta^{it}_{\Lambda^{\mathrm{cc}}_\varphi}(\pi_\varphi(\mathcal{M})'_{\mathrm{w}})' \Delta^{-it}_{\Lambda^{\mathrm{cc}}_\varphi} = (\pi_\varphi(\mathcal{M})'_{\mathrm{w}})'$ for all $t \in \mathbb{R}$. But, we don't know how the unitary group $\{\Delta^{it}_{\Lambda^{\mathrm{cc}}_\varphi}\}_{t\in\mathbb{R}}$ acts on the O*-algebra $\pi_\varphi(\mathcal{M})$, and so we define a system which has the best properties :

Definition 3.3.2. A faithful semifinite (quasi-)weight φ on $\mathcal{P}(\mathcal{M})$ is said to be *quasi-standard* if the above conditions $(\mathrm{S})_1$, $(\mathrm{S})_2$, $(\mathrm{S})_3$ and the following condition $(\mathrm{S})_4$ hold :

$(\mathrm{S})_4$ $\Delta^{it}_{\Lambda^{\mathrm{cc}}_\varphi}\mathcal{D}(\pi_\varphi) \subset \mathcal{D}(\pi_\varphi)$ for all $t \in \mathbb{R}$.

Further, if

$(\mathrm{S})_5$ $\Delta^{it}_{\Lambda^{\mathrm{cc}}_\varphi}\pi_\varphi(\mathcal{M})\Delta^{-it}_{\Lambda^{\mathrm{cc}}_\varphi} = \pi_\varphi(\mathcal{M})$ for all $t \in \mathbb{R}$,

then φ is said to be *essentially standard* , and in addition if

$(\mathrm{S})_6$ $\Delta^{it}_{\Lambda^{\mathrm{cc}}_\varphi}\pi_\varphi(\mathfrak{N}^\dagger_\varphi \cap \mathfrak{N}_\varphi)\Delta^{-it}_{\Lambda^{\mathrm{cc}}_\varphi} = \pi_\varphi(\mathfrak{N}^\dagger_\varphi \cap \mathfrak{N}_\varphi)$ for all $t \in \mathbb{R}$,

then φ is said to be *standard* .

We remark that a faithful semifinite (quasi-)weight φ is standard (resp. essentially standard, quasi-standard) if and only if the generalized vector Λ_φ for $\pi_\varphi(\mathcal{M})$ induced by φ is standard (resp. essentially standard, quasi-standard). Hence by Theorem 2.2.4 and Theorem 2.2.8 we have the following results for standard (quasi-)weights :

Theorem 3.3.3. Suppose φ is a faithful semifinite standard (quasi-)weight on $\mathcal{P}(\mathcal{M})$. Then the following statements hold :

(1) $S_\varphi = S_{\Lambda_\varphi^{cc}}$, and so $J_\varphi = J_{\Lambda_\varphi^{cc}}$ and $\Delta_\varphi = \Delta_{\Lambda_\varphi^{cc}}$.

(2) There exists a one-parameter group $\{\sigma_t^\varphi\}_{t\in\mathbb{R}}$ of $*$-automorphisms of $\mathcal{M}$ such that $\pi_\varphi(\sigma_t^\varphi(X)) = \Delta_\varphi^{it}\pi_\varphi(X)\Delta_\varphi^{-it}$ for all $X \in \mathcal{M}$ and $t \in \mathbb{R}$.

(3) φ is a $\{\sigma_t^\varphi\}$-KMS (quasi-)weight, that is, for any $X, Y \in \mathfrak{N}_\varphi^\dagger \cap \mathfrak{N}_\varphi$ there exists an element $f_{X,Y}$ of $A(0,1)$ such that $f_{X,Y}(t) = \varphi(Y\sigma_t^\varphi(X))$ and $f_{X,Y}(t+i) = \varphi(\sigma_t^\varphi(X)Y)$ for all $t \in \mathbb{R}$, where $A(0,1)$ is the set of all complex-valued functions, bounded and continuous on $0 \leq \text{Im}z \leq 1$ and analytic in the interior.

Using Theorem 2.2.8, we can show the following result for essentially standard quasi-weights:

Theorem 3.3.4. Suppose φ is an essentially standard quasi-weight on $\mathcal{P}(\mathcal{M})$. We put

$$\begin{cases} \mathfrak{N}_{\varphi_e} = \{\, X \in \mathcal{M}\,;\ {}^\exists\xi_X \in \mathcal{D}(\pi_\varphi) \\ \qquad\qquad \text{s.t. } \pi_\varphi(X)\Lambda_\varphi^c(K) = K\xi_X,\ \ {}^\forall K \in \mathcal{D}(\Lambda_\varphi^c)\,\}, \\ \varphi_e(\sum_k X_k^\dagger X_k) = \sum_k \|\,\xi_{X_k}\|^2,\ \sum_k X_k^\dagger X_k \in \mathcal{P}(\mathfrak{N}_{\varphi_e}). \end{cases}$$

Then φ_e is a standard quasi-weight on $\mathcal{P}(\mathcal{M})$ such that $\varphi \subset \varphi_e$ and $\pi_{\varphi_e} = \pi_\varphi$ and $S_{\varphi_e} = S_{\Lambda_\varphi^{cc}}$ as unitary equivalence. Here $\varphi \subset \varphi_e$ means that $\mathfrak{N}_\varphi \subset \mathfrak{N}_{\varphi_e}$ and $\varphi = \varphi_e \lceil \mathcal{P}(\mathfrak{N}_\varphi)$.

We next consider quasi-standard (quasi-)weights. Let φ be a quasi-standard (quasi-) weight on $\mathcal{P}(\mathcal{M})$. We put

$$\begin{cases} \mathcal{D}(\overline{\Lambda_\varphi}) = \{\, A \in \pi_\varphi(\mathcal{M})''_{\text{wc}}\,;\ {}^\exists\xi_A \in \mathcal{D}(\pi_\varphi) \\ \qquad\qquad \text{s.t. } A\Lambda_\varphi^c(K) = K\xi_A,\ \ {}^\forall K \in \mathcal{D}(\Lambda_\varphi^c)\,\}, \\ \overline{\Lambda_\varphi}(A) = \xi_A,\ \ A \in \mathcal{D}(\overline{\Lambda_\varphi}). \end{cases}$$

By Theorem 2.3.2, $\overline{\Lambda_\varphi}$ is a generalized vector for $\pi_\varphi(\mathcal{M})''_{\text{wc}}$ such that $\Lambda_\varphi \subset \overline{\Lambda_\varphi}$ and $\Lambda_\varphi^c = \overline{\Lambda_\varphi}^c$. We simply denote by $\overline{\varphi}$ the quasi-weight $\varphi_{\overline{\Lambda_\varphi}}$ on $\mathcal{P}(\pi_\varphi(\mathcal{M})''_{\text{wc}})$ induced by $\overline{\Lambda_\varphi}$. Then we have the following

Theorem 3.3.5. Suppose φ is a quasi-standard (quasi-)weight on $\mathcal{P}(\mathcal{M})$. Then the quasi-weight $\overline{\varphi}$ on $\mathcal{P}(\pi_\varphi(\mathcal{M})''_{\text{wc}})$ induced by φ is standard, and so it is a $\{\sigma_t^{\overline{\varphi}}\}_{t\in\mathbb{R}}$-KMS quasi-weight, where $\sigma_t^{\overline{\varphi}}(A) = \Delta_{\Lambda_\varphi^{cc}}^{it} A \Delta_{\Lambda_\varphi^{cc}}^{-it}$, $A \in \pi_\varphi(\mathcal{M})''_{\text{wc}}$, $t \in \mathbb{R}$.

Conversely we consider when a KMS (quasi-)weight is standard.

Theorem 3.3.6. Let $\{\alpha_t\}_{t\in\mathbb{R}}$ be a one-parameter group of $*$- automorphisms of $\mathcal{M}$. Suppose φ is a $\{\alpha_t\}$-KMS (quasi-)weight on $\mathcal{P}(\mathcal{M})$ such that $\lambda_\varphi((\mathfrak{N}_\varphi^\dagger \cap \mathfrak{N}_\varphi)^2)$ is total in $\mathcal{H}_\varphi$. Then the following statements hold:

(1) The map $\lambda_\varphi(X) \longrightarrow \lambda_\varphi(X^\dagger)$, $X \in \mathfrak{N}_\varphi^\dagger \cap \mathfrak{N}_\varphi$ is a closable conjugate-linear operator in $\mathcal{H}_\varphi$. Let S_φ be the closure of the above operator $\lambda_\varphi(X) \longrightarrow \lambda_\varphi(X^\dagger)$ and $S_\varphi = J_\varphi \Delta_\varphi^{\frac{1}{2}}$ the polar decomposition of S_φ.

(2) $\Delta_\varphi^{it}\lambda_\varphi(X) = \lambda_\varphi(\alpha_t(X))$, $^\forall X \in \mathfrak{N}_\varphi$, $^\forall t \in \mathbb{R}$.

(3) φ is standard if and only if the following statements hold:
(i) Λ_φ is well-defined.
(ii) $\Lambda_\varphi^c((\mathcal{D}(\Lambda_\varphi^c)^* \cap \mathcal{D}(\Lambda_\varphi^c))^2)$ is total in $\mathcal{H}_\varphi$.
(iii) $J_\varphi \Lambda_\varphi^c(\mathcal{D}(\Lambda_\varphi^c)^* \cap \mathcal{D}(\Lambda_\varphi^c)) \subset \Lambda_\varphi^{cc}(\mathcal{D}(\Lambda_\varphi^{cc})^* \cap \mathcal{D}(\Lambda_\varphi^{cc}))$.
(iv) $(J_\varphi \Lambda_\varphi^{cc}(A)|\Lambda_\varphi^{cc}(A^*)) \geq 0$, $^\forall A \in \mathcal{D}(\Lambda_\varphi^{cc})^* \cap \mathcal{D}(\Lambda_\varphi^{cc})$.

Proof. We put

$$U_t\lambda_\varphi(X) = \lambda_\varphi(\alpha_t(X)), \quad X \in \mathfrak{N}_\varphi.$$

Since φ is $\{\alpha_t\}$-KMS (quasi-)weight on $\mathcal{P}(\mathcal{M})$, for any $X, Y \in \mathfrak{N}_\varphi^\dagger \cap \mathfrak{N}_\varphi$ there exists an element $f_{X,Y}$ of $A(0,1)$ such that

$$f_{X,Y}(t) = \varphi(\alpha_t(X)Y) \text{ and } f_{X,Y}(t+i) = \varphi(Y\alpha_t(X)), \quad {}^\forall t \in \mathbb{R}.$$

We now have

$$\begin{aligned}
&\lim_{t\to 0} \| U_t\lambda_\varphi(X) - \lambda_\varphi(X) \|^2 \\
&= \lim_{t\to 0}\{ \varphi(\alpha_t(X)^\dagger\alpha_t(X)) - \varphi(\alpha_t(X)^\dagger Y) - \varphi(X^\dagger\alpha_t(X)) + \varphi(X^\dagger X) \} \\
&= \lim_{t\to 0}\{ 2\varphi(X^\dagger X) - f_{X^\dagger,X}(t) - f_{X,X^\dagger}(t+i)\} \\
&= 0
\end{aligned}$$

for each $X \in \mathfrak{N}_\varphi^\dagger \cap \mathfrak{N}_\varphi$, which implies that $\{\overline{U_t}\}_{t\in\mathbb{R}}$ is a strongly continuous one-parameter group of unitary operators on $\mathcal{H}_\varphi$. Let $\{X_n\}$ be any sequence in $\mathfrak{N}_\varphi^\dagger \cap \mathfrak{N}_\varphi$ such that $\lim_{n\to\infty} \lambda_\varphi(X_n) = 0$ and $\lim_{n\to\infty} \lambda_\varphi(X_n^\dagger) = \xi$. For any $Y \in \mathfrak{N}_\varphi^\dagger \cap \mathfrak{N}_\varphi$ we have

$$\begin{aligned}\lim_{n\to\infty}\sup_{t\in\mathbb{R}}|f_{X_n,Y}(t) - (\lambda_\varphi(Y)|\overline{U_t}\xi)| &= \lim_{n\to\infty}\sup_{t\in\mathbb{R}}|(\lambda_\varphi(Y)|\overline{U_t}(\lambda_\varphi(X_n^\dagger) - \xi))| \\ &\leq \lim_{n\to\infty} \| \lambda_\varphi(Y) \| \, \| \lambda_\varphi(X_n^\dagger) - \xi \| \\ &= 0,\end{aligned}$$

$$\lim_{n\to\infty}\sup_{t\in\mathbb{R}}|f_{X_n,Y}(t+i)| = 0,$$

and hence there exists an element f of $A(0,1)$ such that $f(t) = (\lambda_\varphi(Y)|\overline{U_t}\xi)$ and $f(t+i) = 0$ for all $t \in \mathbb{R}$. Hence we have $f = 0$, and so $\xi = 0$. Thus the statement (1) holds. The statement (2) is shown similarly to the proof of Theorem 2.2.4. We show the statement (3). It is clear that if φ is standard, then the statements (i) $\sim$ (iv) hold. Conversely suppose the statements (i) $\sim$ (iv) hold. We put

$$T\Lambda_\varphi^{cc}(A) = J_\varphi \Lambda_\varphi^{cc}(A^*), \quad A \in \mathcal{D}(\Lambda_\varphi^{cc})^* \cap \mathcal{D}(\Lambda_\varphi^{cc}).$$

Then T is a well-defined from (iii) that T is positive and $\overline{T} = J_\varphi S_{\Lambda_\varphi^{cc}} = J_\varphi J_{\Lambda_\varphi^{cc}} \Delta_{\Lambda_\varphi^{cc}}^{\frac{1}{2}}$. We put $U = J_\varphi J_{\Lambda_\varphi^{cc}}$. Then U is a unitary operator on $\mathcal{H}_\varphi$. Since $T^* = S^*_{\Lambda_\varphi^{cc}} J_\varphi$ and $J_\varphi \Lambda_\varphi^c(\mathcal{D}(\Lambda_\varphi^c)^* \cap \mathcal{D}(\Lambda_\varphi^c))$ is a core for $S^*_{\Lambda_\varphi^{cc}}$, it follows from (iv) that $\overline{T}$ is a positive self-adjoint operator in $\mathcal{H}_\varphi$, and so $U = I$ and $J_\varphi = J_{\Lambda_\varphi^{cc}}$. Hence we have $\Delta_\varphi = \Delta_{\Lambda_\varphi^{cc}}$, which implies by (2) that φ is standard. This completes the proof.

3.4 Generalized Connes cocycle theorem for weights

In this section we generalize the Connes cocycle theorem for weights on O*-algebras. In 2.5, 2.6 we studied to generalize the Connes cocycle theorem and the Pedersen-Takesaki Radon-Nikodym theorem to generalized von Neumann algebras in case of standard generalized vectors. As the notion of generalized vectors is spatial, such a generalization is possible to a certain extent, but the notion of (quasi-)weights is purely algebraic and not spatial and the algebraic properties don't reflect to the topological properties in general (for example, $\pi_\varphi(\mathcal{M})$ is not necessarily a generalized von Neumann algebra when $\mathcal{M}$ is a generalized von Neumann algebra), and so such generalizations for (quasi-)weights have some difficult problems. We first need the notions of semifiniteness and σ-weak continuity of (quasi-)weights. Let $\mathcal{M}$ be a closed O*-algebra on $\mathcal{D}$ in $\mathcal{H}$.

Definition 3.4.1. For any $X \in \mathfrak{N}_\varphi$ we put

$$\varphi_X(A) = \varphi(X^\dagger A X), \quad A \in \mathcal{M}.$$

Then φ_X is a positive linear functional on $\mathcal{M}$. If φ_X is σ-weakly continuous for each $X \in \mathfrak{N}_\varphi$, then φ is said to be *σ-weakly continuous* .

Lemma 3.4.2. Let φ be a (quasi-)weight on $\mathcal{P}(\mathcal{M})$. Then the following statements hold:

(1) φ is σ-weakly continuous if and only if $\varphi_{X,Y}$ is a σ-weakly continuous linear functional on $\mathcal{M}$ for each $X, Y \in \mathfrak{N}_\varphi$, where

$$\varphi_{X,Y}(A) = \varphi(Y^\dagger AX), \quad A \in \mathcal{M}.$$

(2) Suppose φ is σ-weakly continuous on $\mathcal{D}(\varphi)$, then φ is σ-weakly continuous.

(3) Suppose φ is faithful, σ-weakly continuous and semifinite. Then Λ_φ is a semifinite generalized vector for $\pi_\varphi(\mathcal{M})$ such that $\Lambda_\varphi((\mathcal{D}(\Lambda_\varphi)^\dagger \cap \mathcal{D}(\Lambda_\varphi))^2)$ is total in $\mathcal{H}_\varphi$.

Proof. (1) This follows since any $\varphi_{X,Y}$ is a linear combination of $\{\varphi_{X_k}; X_k \in \mathfrak{N}_\varphi\}$.

(2) This is almost trivial.

(3) Since φ is semifinite, there exists a net $\{U_\alpha\}$ in $\mathfrak{N}_\varphi^\dagger \cap \mathfrak{N}_\varphi$ such that $\| \overline{U_\alpha} \| \leq 1$, $^\forall\alpha$ and $\{U_\alpha\}$ converges strongly to I. Take an arbitrary $X \in \mathfrak{N}_\varphi$. Since φ_X is a σ-weakly continuous positive linear functional on the bounded part $\mathcal{M}_b$ of $\mathcal{M}$, it follows that φ_X can be extended to a σ-weakly continuous positive linear functional φ_X'' on the von Neumann algebra $\overline{\mathcal{M}_b}''$. Hence we have

$$\begin{aligned}
&\pi_\varphi(U_\alpha) \in \mathcal{D}(\Lambda_\varphi)^\dagger \cap \mathcal{D}(\Lambda_\varphi), \quad ^\forall\alpha, \\
&\| \pi_\varphi(U_\alpha)\lambda_\varphi(X) \|^2 = \varphi_X(U_\alpha^\dagger U_\alpha) \\
&\qquad = \varphi_X''(\overline{U_\alpha}^* \overline{U_\alpha}) \\
&\qquad \leq \| \overline{U_\alpha} \|^2 \varphi_X''(I) \\
&\qquad \leq \| \lambda_\varphi(X) \|^2, \quad ^\forall\alpha
\end{aligned}$$

and

$$\| \pi_\varphi(U_\alpha)\lambda_\varphi(X) - \lambda_\varphi(X) \|^2 = \varphi_X((U_\alpha - I)^\dagger (U_\alpha - I)) \xrightarrow{\alpha} 0,$$

which implies that Λ_φ is semifinite. Further, it follows that $\pi_\varphi(U_\beta U_\alpha X) \in (\mathcal{D}(\Lambda_\varphi)^\dagger \cap \mathcal{D}(\Lambda_\varphi))^2$ and

$$\begin{aligned}
\lim_{\alpha,\beta} \Lambda_\varphi(\pi_\varphi(U_\beta U_\alpha X)) &= \lim_{\alpha,\beta} \pi_\varphi(U_\beta)\pi_\varphi(U_\alpha)\lambda_\varphi(X) \\
&= \lambda_\varphi(X), \\
&= \Lambda_\varphi(\pi_\varphi(X)),
\end{aligned}$$

which implies that $\Lambda_\varphi((\mathcal{D}(\Lambda_\varphi)^\dagger \cap \mathcal{D}(\Lambda_\varphi))^2)$ is total in $\mathcal{H}_\varphi$.

Let φ and ψ be faithful, σ-weakly continuous semifinite (quasi-)weights on $\mathcal{P}(\mathcal{M})$ such that π_φ and π_ψ are self-adjoint. Let $M_2(\mathbb{C})$ be the 2×2-matrix algebra on $\mathbb{C}$ and put

$$E_{11} = \begin{pmatrix} 1 & 0 \\ 0 & 0 \end{pmatrix},\ E_{12} = \begin{pmatrix} 0 & 1 \\ 0 & 0 \end{pmatrix},\ E_{21} = \begin{pmatrix} 0 & 0 \\ 1 & 0 \end{pmatrix},\ E_{22} = \begin{pmatrix} 0 & 0 \\ 0 & 1 \end{pmatrix}.$$

Every element X of $\mathcal{M} \otimes M_2(\mathbb{C})$ is represented as

$$X = \begin{pmatrix} X_{11} & X_{12} \\ X_{21} & X_{22} \end{pmatrix} = X_{11} \otimes E_{11} + X_{12} \otimes E_{12} + X_{21} \otimes E_{21} + X_{22} \otimes E_{22}.$$

We put

$$\theta(X^\dagger X) = \varphi(X_{11}^\dagger X_{11} + X_{21}^\dagger X_{21}) + \psi(X_{12}^\dagger X_{12} + X_{22}^\dagger X_{22}),$$
$$X = (X_{ij}) \in \mathcal{M} \otimes M_2(\mathbb{C}).$$

Then we have the following

Lemma 3.4.3. (1) θ is a faithful, σ-weakly continuous, semifinite (quasi-)weight on $\mathcal{P}(\mathcal{M} \otimes M_2(\mathbb{C}))$ such that π_θ is self-adjoint and

$$\mathfrak{N}_\theta = \{X = (X)_{ij} \in \mathcal{M} \otimes M_2(\mathbb{C}); X_{11}, X_{21} \in \mathfrak{N}_\varphi \text{ and } X_{12}, X_{22} \in \mathfrak{N}_\psi\}.$$

(2) $\lambda_\varphi(\mathfrak{N}_\varphi \cap \mathfrak{N}_\psi^\dagger)$ is dense in $\mathcal{H}_\varphi$ and $\lambda_\psi(\mathfrak{N}_\psi \cap \mathfrak{N}_\varphi^\dagger)$ is dense in $\mathcal{H}_\psi$.

Proof. (1) It is easily shown that θ is a faithful, σ-weakly continuous (quasi-)weight on $\mathcal{P}(\mathcal{M} \otimes M_2(\mathbb{C}))$ such that π_θ is self-adjoint and

$$\mathfrak{N}_\theta = \{X = (X)_{ij} \in \mathcal{M} \otimes M_2(\mathbb{C}); X_{11}, X_{21} \in \mathfrak{N}_\varphi \text{ and } X_{12}, X_{22} \in \mathfrak{N}_\psi\}.$$

Let $\{U_\alpha\}$ and $\{V_\beta\}$ be nets in $\mathfrak{N}_\varphi^\dagger \cap \mathfrak{N}_\varphi$ and $\mathfrak{N}_\psi^\dagger \cap \mathfrak{N}_\psi$, respectively such that $\| \overline{U_\alpha} \| \leq 1$, ${}^\forall\alpha$ and $\| \overline{V_\beta} \| \leq 1$, ${}^\forall\beta$ and $\{U_\alpha\}$ and $\{V_\beta\}$ converge strongly to I. Considering

$$\begin{pmatrix} U_\alpha & 0 \\ 0 & V_\beta \end{pmatrix} \in \mathfrak{N}_\theta^\dagger \cap \mathfrak{N}_\theta, \quad {}^\forall\alpha, \beta,$$

we can show that θ is semifinite.
(2) Take an arbitrary $X \in \mathfrak{N}_\varphi$. We have $V_\beta X \in \mathfrak{N}_\varphi \cap \mathfrak{N}_\psi^\dagger$,

$$\| \lambda_\varphi(V_\beta X) - \lambda_\varphi(X) \|^2 = \varphi_X((V_\beta - I)^\dagger (V_\beta - I))$$

for ${}^\forall\beta$, and hence it follows from the σ-weak continuity of φ_X that $\lambda_\varphi(\mathfrak{N}_\varphi \cap \mathfrak{N}_\psi^\dagger)$ is dense in $\mathcal{H}_\varphi$. Similarly, $\lambda_\psi(\mathfrak{N}_\psi \cap \mathfrak{N}_\varphi^\dagger)$ is dense in $\mathcal{H}_\psi$.

By Lemma 3.4.2,(4) and Lemma 3.4.3,(1) we have

$$\lambda_\theta((\mathfrak{N}_\theta^\dagger \cap \mathfrak{N}_\theta)^2) \text{ is total in } \mathcal{H}_\theta. \tag{3.4.1}$$

Hence we can define the generalized vector Λ_θ^c for the von Neumann algebra $\pi_\theta(\mathcal{M} \otimes M_2(\mathbb{C}))'_{\mathrm{w}}$, and so to decide it we first define the following map $\Lambda_{\varphi,\psi}^c$:

$$\begin{cases} \mathcal{D}(\Lambda^c_{\varphi,\psi}) = \{ K \in \mathbb{I}(\pi_\varphi, \pi_\psi) \ ; \ ^\exists \eta \in \mathcal{D}(\pi_\psi) \\ \qquad\qquad\qquad\qquad \text{s.t. } K\lambda_\varphi(X) = \pi_\psi(X)\eta, \ ^\forall X \in \mathfrak{N}_\varphi \}, \\ \Lambda^c_{\varphi,\psi}(K) = \eta, \quad K \in \mathcal{D}(\Lambda^c_{\varphi,\psi}). \end{cases}$$

Then we have the following

Lemma 3.4.4. $\Lambda^c_{\varphi,\psi}$ is a linear map of $\mathcal{D}(\Lambda^c_{\varphi,\psi})$ into $\mathcal{D}(\pi_\psi)$ satisfying

(i) $CK \in \mathcal{D}(\Lambda^c_{\varphi,\psi})$ and $\Lambda^c_{\varphi,\psi}(CK) = C\Lambda^c_{\varphi,\psi}(K)$ for each $C \in \pi_\psi(\mathcal{M})'_{\mathrm{w}}$ and $K \in \mathcal{D}(\Lambda^c_{\varphi,\psi})$;

(ii) $CK \in \mathcal{D}(\Lambda^c_\varphi)$ and $\Lambda^c_\varphi(CK) = C\Lambda^c_{\varphi,\psi}(K)$ for each $C \in \mathbb{I}(\pi_\psi, \pi_\varphi)$ and $K \in \mathcal{D}(\Lambda^c_{\varphi,\psi})$.

We here put

$$\mathcal{H}_1 = \overline{\lambda_\theta(\mathfrak{N}_\varphi \otimes E_{11})},\ \mathcal{H}_2 = \overline{\lambda_\theta(\mathfrak{N}_\varphi \otimes E_{21})},$$
$$\mathcal{H}_3 = \overline{\lambda_\theta(\mathfrak{N}_\psi \otimes E_{12})},\ \mathcal{H}_4 = \overline{\lambda_\theta(\mathfrak{N}_\psi \otimes E_{22})}$$

and

$$\begin{aligned} U_1\lambda_\varphi(X) &= \lambda_\theta(X \otimes E_{11}), \quad X \in \mathfrak{N}_\varphi, \\ U_2\lambda_\varphi(X) &= \lambda_\theta(X \otimes E_{21}), \quad X \in \mathfrak{N}_\varphi, \\ U_3\lambda_\psi(X) &= \lambda_\theta(X \otimes E_{12}), \quad X \in \mathfrak{N}_\psi, \\ U_4\lambda_\psi(X) &= \lambda_\theta(X \otimes E_{22}), \quad X \in \mathfrak{N}_\psi. \end{aligned}$$

Then $\{\mathcal{H}_i\}_{i=1,\cdots,4}$ is a set of mutually orthogonal closed subspaces of $\mathcal{H}_\theta$ such that $\mathcal{H}_\theta = \mathcal{H}_1 \oplus \mathcal{H}_2 \oplus \mathcal{H}_3 \oplus \mathcal{H}_4$, and U_1 and U_2 (resp. U_3 and U_4) can be extended to the isometries from $\mathcal{H}_\varphi$ (resp. $\mathcal{H}_\psi$) to $\mathcal{H}_1$ and $\mathcal{H}_2$ (resp. $\mathcal{H}_3$ and $\mathcal{H}_4$), and they are also denoted by U_1 and U_2 (resp. U_3 and U_4). For $X = \begin{pmatrix} X_{11} & X_{12} \\ X_{21} & X_{22} \end{pmatrix} \in \mathcal{M} \otimes M_2(\mathbb{C})$, $\pi_\theta(X)$ is given by the matrix:

$$\begin{pmatrix} U_1\pi_\varphi(X_{11})U_1^* & U_1\pi_\varphi(X_{12})U_2^* & 0 & 0 \\ U_2\pi_\varphi(X_{21})U_1^* & U_2\pi_\varphi(X_{22})U_2^* & 0 & 0 \\ 0 & 0 & U_3\pi_\psi(X_{11})U_3^* & U_3\pi_\psi(X_{12})U_4^* \\ 0 & 0 & U_4\pi_\psi(X_{21})U_3^* & U_4\pi_\psi(X_{22})U_4^* \end{pmatrix}.$$

We now have the following results for the von Neumann algebras $\pi_\theta(\mathcal{M} \otimes M_2(\mathbb{C}))'_{\mathrm{w}}$ and $(\pi_\theta(\mathcal{M} \otimes M_2(\mathbb{C}))'_{\mathrm{w}})'$ and the generalized vector Λ^c_θ:

Lemma 3.4.5.

$\pi_\theta(\mathcal{M} \otimes M_2(\mathbb{C}))'_{\mathrm{w}}$

$$= \left\{ \begin{pmatrix} U_1C_1U_1^* & 0 & U_1C_2U_3^* & 0 \\ 0 & U_2C_1U_2^* & 0 & U_2C_2U_4^* \\ U_3C_3U_1^* & 0 & U_3C_4U_3^* & 0 \\ 0 & U_4C_3U_2^* & 0 & U_4C_4U_4^* \end{pmatrix} ; \ \begin{array}{l} C_1 \in \pi_\varphi(\mathcal{M})'_{\mathrm{w}}, \\ C_2 \in \mathbb{I}(\pi_\psi, \pi_\varphi), \\ C_3 \in \mathbb{I}(\pi_\varphi, \pi_\psi), \\ C_4 \in \pi_\psi(\mathcal{M})'_{\mathrm{w}} \end{array} \right\},$$

$(\pi_\theta(\mathcal{M} \otimes M_2(\mathbb{C}))'_{\rm w})'$

$$= \left\{ \begin{pmatrix} U_1A_{11}U_1^* & U_1A_{12}U_2^* & 0 & 0 \\ U_2A_{21}U_1^* & U_2A_{22}U_2^* & 0 & 0 \\ 0 & 0 & U_3B_{11}U_3^* & U_3B_{12}U_4^* \\ 0 & 0 & U_4B_{21}U_3^* & U_4B_{22}U_4^* \end{pmatrix} ; \begin{array}{c} (A_{ij}, B_{ij}) \in \mathfrak{A}_{\varphi,\psi}, \\ (i,j = 1,2) \end{array} \right\},$$

where

$$\mathfrak{A}_{\varphi,\psi} = \left\{ (A,B) \; ; \; \begin{array}{c} A \in (\pi_\varphi(\mathcal{M})'_{\rm w})',\; B \in (\pi_\psi(\mathcal{M})'_{\rm w})', \\ AC = CB \text{ and } A^*C = CB^* \\ \text{for all } C \in \mathbb{I}(\pi_\psi, \pi_\varphi) \end{array} \right\}$$

and

$$\mathcal{D}(\Lambda_\theta^c) = \left\{ C = \begin{pmatrix} U_1C_1U_1^* & 0 & U_1C_2U_3^* & 0 \\ 0 & U_2C_1U_2^* & 0 & U_2C_2U_4^* \\ U_3C_3U_1^* & 0 & U_3C_4U_3^* & 0 \\ 0 & U_4C_3U_2^* & 0 & U_4C_4U_4^* \end{pmatrix} ; \begin{array}{c} C_1 \in \mathcal{D}(\Lambda_\varphi^c), \\ C_2 \in \mathcal{D}(\Lambda_{\psi,\varphi}^c), \\ C_3 \in \mathcal{D}(\Lambda_{\varphi,\psi}^c), \\ C_4 \in \mathcal{D}(\Lambda_\psi^c) \end{array} \right\},$$

$$\Lambda_\theta^c = \begin{pmatrix} U_1\lambda_\varphi^c(C_1) \\ U_2\lambda_{\psi,\varphi}^c(C_2) \\ U_3\lambda_{\varphi,\psi}^c(C_3) \\ U_4\lambda_\psi^c(C_4) \end{pmatrix}, \quad C \in \mathcal{D}(\Lambda_\theta^c).$$

In bounded case $\Lambda_\theta^c((\mathcal{D}(\Lambda_\theta^c)^* \cap \mathcal{D}(\Lambda_\theta^c))^2)$ is total in $\mathcal{H}_\theta$, but in unbounded case this fact doesn't necessarily hold even if φ and ψ are standard. We have the following result for this problem:

Theorem 3.4.6. Let φ and ψ be faithful, σ-weakly continuous, semifinite (quasi-)weights on $\mathcal{P}(\mathcal{M})$ such that π_φ and π_ψ are self-adjoint. Suppose $\Lambda_\varphi^c((\mathcal{D}(\Lambda_\varphi^c)^* \cap \mathcal{D}(\Lambda_\varphi^c))^2)$ is total in $\mathcal{H}_\varphi$ and $\Lambda_\psi^c((\mathcal{D}(\Lambda_\psi^c)^* \cap \mathcal{D}(\Lambda_\psi^c))^2)$ is total in $\mathcal{H}_\psi$. The following statements are equivalent:

(i) π_φ and π_ψ are unitarily equivalent.

(ii) $\mathbb{I}(\pi_\varphi, \pi_\psi)^*\mathbb{I}(\pi_\varphi, \pi_\psi)$ and $\mathbb{I}(\pi_\psi, \pi_\varphi)^*\mathbb{I}(\pi_\psi, \pi_\varphi)$ are nondegenerate $*$-subalgebras of the von Neumann algebra $\pi_\varphi(\mathcal{M})'_{\rm w}$ and $\pi_\psi(\mathcal{M})'_{\rm w}$, respectively.

(iii) $\Lambda_{\psi,\varphi}^c(\mathcal{D}(\Lambda_{\psi,\varphi}^c))$ is dense in $\mathcal{H}_\varphi$ and $\Lambda_{\varphi,\psi}^c(\mathcal{D}(\Lambda_{\varphi,\psi}^c))$ is dense in $\mathcal{H}_\psi$.

(iv) $\Lambda_\theta^c((\mathcal{D}(\Lambda_\theta^c)^* \cap \mathcal{D}(\Lambda_\theta^c))^2)$ is total in $\mathcal{H}_\theta$.

Proof. (i) $\iff$ (ii) This follows from Theorem 1.8.9.

(i) $\Longrightarrow$ (iii) There exists a unitary transform W of $\mathcal{H}_\varphi$ onto $\mathcal{H}_\psi$ such that $W\mathcal{D}(\pi_\varphi) = \mathcal{D}(\pi_\psi)$ and $\pi_\varphi(X) = W^*\pi_\psi(X)W$ for all $X \in \mathcal{M}$. Then we have

$$\begin{cases} \mathcal{D}(\Lambda_{\varphi,\psi}^c) = \{ WC \; ; \; C \in \mathcal{D}(\Lambda_\varphi^c) \}, \\ \Lambda_{\varphi,\psi}^c(WC) = W\Lambda_\varphi^c(C), \; C \in \mathcal{D}(\Lambda_\varphi^c). \end{cases}$$

Hence, $\Lambda_{\varphi,\psi}^c(\mathcal{D}(\Lambda_{\varphi,\psi}^c))$ is dense in $\mathcal{H}_\psi$. Similarly, $\Lambda_{\psi,\varphi}^c(\mathcal{D}(\Lambda_{\psi,\varphi}^c))$ is dense in $\mathcal{H}_\varphi$.

(iii) $\Longrightarrow$ (iv) By Lemma 3.4.5 we have

$\Lambda_\theta^c((\mathcal{D}(\Lambda_\theta^c)^* \cap \mathcal{D}(\Lambda_\theta^c))^2)$

$$= \left\{ \begin{pmatrix} U_1 C_1 \Lambda^c_\varphi(D_1) + U_1 C_2 \Lambda^c_{\varphi,\psi}(D_3) \\ U_2 C_1 \Lambda^c_{\psi,\varphi}(D_2) + U_2 C_2 \Lambda^c_\psi(D_4) \\ U_3 C_3 \Lambda^c_\varphi(D_1) + U_3 C_4 \Lambda^c_{\varphi,\psi}(D_3) \\ U_4 C_3 \Lambda^c_{\psi,\varphi}(D_2) + U_4 C_4 \Lambda^c_\psi(D_4) \end{pmatrix} ; \begin{array}{l} C_1, D_1 \in \mathcal{D}(\Lambda^c_\varphi)^* \cap \mathcal{D}(\Lambda^c_\varphi), \\ C_2, D_2, C_3^*, D_3^* \in \mathcal{D}(\Lambda^c_{\psi,\varphi}), \\ C_3, D_3, C_2^*, D_2^* \in \mathcal{D}(\Lambda^c_{\varphi,\psi}), \\ C_4, D_4 \in \mathcal{D}(\Lambda^c_\psi)^* \cap \mathcal{D}(\Lambda^c_\psi) \end{array} \right\}, \tag{3.4.2}$$

which implies since $\mathcal{D}(\Lambda^c_\varphi)^* \cap \mathcal{D}(\Lambda^c_\varphi)$ and $\mathcal{D}(\Lambda^c_\psi)^* \cap \mathcal{D}(\Lambda^c_\psi)$ are nondegenerate that $\lambda^c_\theta((\mathcal{D}(\Lambda^c_\theta)^* \cap \mathcal{D}(\Lambda^c_\theta))^2)$ is total in $\mathcal{H}_\theta$.

(iv) $\Longrightarrow$ (i) Since $\lambda^c_\theta((\mathcal{D}(\Lambda^c_\theta)^* \cap \mathcal{D}(\Lambda^c_\theta))^2)$ is total in $\mathcal{H}_\theta$, it follows that Λ^{cc}_θ is well-defined and

$$\mathcal{D}(\Lambda^{cc}_\theta)$$
$$= \left\{ A = \begin{pmatrix} U_1 A_{11} U_1^* & U_1 A_{12} U_2^* & 0 & 0 \\ U_2 A_{21} U_1^* & U_2 A_{22} U_2^* & 0 & 0 \\ 0 & 0 & U_3 B_{11} U_3^* & U_3 B_{12} U_4^* \\ 0 & 0 & U_4 B_{21} U_3^* & U_4 B_{22} U_4^* \end{pmatrix} ; \begin{array}{l} A_{ij} \in \mathcal{D}(\Lambda^{cc}_\varphi), \\ B_{ij} \in \mathcal{D}(\Lambda^{cc}_\psi) \ (i,j = 1,2) \\ \qquad \text{s.t.} \\ B_{11} \Lambda^c_{\varphi,\psi}(C_3) = C_3 \Lambda^{cc}_\varphi(A_{11}), \\ B_{21} \Lambda^c_{\varphi,\psi}(C_3) = C_3 \Lambda^{cc}_\varphi(A_{21}), \\ A_{12} \Lambda^c_{\psi,\varphi}(C_2) = C_2 \Lambda^{cc}_\psi(B_{12}), \\ A_{12} \Lambda^c_{\psi,\varphi}(C_2) = C_2 \Lambda^{cc}_\psi(B_{22}) \\ \text{for } {}^\forall C_2 \in \mathcal{D}(\Lambda^c_{\psi,\varphi}) \\ \qquad \text{and } {}^\forall C_3 \in \mathcal{D}(\Lambda^c_{\varphi,\psi}) \end{array} \right\},$$

$$\Lambda^{cc}_\theta(A) = \begin{pmatrix} U_1 \Lambda^{cc}_\varphi(A_{11}) \\ U_2 \Lambda^{cc}_\varphi(A_{21}) \\ U_3 \Lambda^{cc}_\psi(B_{12}) \\ U_4 \Lambda^{cc}_\psi(B_{22}) \end{pmatrix}, \quad A \in \mathcal{D}(\Lambda^{cc}_\theta).$$

Then we have

$$S_{\Lambda^{cc}_\theta} = \begin{pmatrix} U_1 S_{11} U_1^* & 0 & 0 & 0 \\ 0 & 0 & U_2 S_{21} U_3^* & 0 \\ 0 & U_3 S_{12} U_2^* & 0 & 0 \\ 0 & 0 & 0 & U_4 S_{22} U_4^* \end{pmatrix},$$

where S_{ij} $(i,j = 1,2)$ is a closed operator defined by

$$S_{11} : \Lambda^{cc}_\varphi(A_{11}) \longrightarrow \Lambda^{cc}_\varphi(A_{11}^*),$$
$$S_{22} : \Lambda^{cc}_\psi(B_{22}) \longrightarrow \Lambda^{cc}_\psi(B_{22}^*),$$
$$S_{12} : \Lambda^{cc}_\psi(B_{12}) \longrightarrow \Lambda^{cc}_\varphi(A_{12}^*),$$
$$S_{21} : \Lambda^{cc}_\varphi(A_{21}) \longrightarrow \Lambda^{cc}_\psi(B_{21}^*).$$

Let $S_{ij} = J_{ij} \Delta_{ij}^{\frac{1}{2}}$ be the polar decomposition of S_{ij} $(i,j = 1,2)$. Then we have

$$\Delta_{\Lambda_\theta^{cc}} = \begin{pmatrix} U_1\Delta_{11}U_1^* & 0 & 0 & 0 \\ 0 & U_2\Delta_{21}U_2^* & 0 & 0 \\ 0 & 0 & U_3\Delta_{12}U_3^* & 0 \\ 0 & 0 & 0 & U_4\Delta_{22}U_4^* \end{pmatrix}, \tag{3.4.3}$$

$$J_{\Lambda_\theta^{cc}} = \begin{pmatrix} U_1J_{11}U_1^* & 0 & 0 & 0 \\ 0 & 0 & U_2J_{12}U_3^* & 0 \\ 0 & U_3J_{21}U_2^* & 0 & 0 \\ 0 & 0 & 0 & U_4J_{22}U_4^* \end{pmatrix}.$$

Then it follows from Lemma 3.4.5 that

$$\begin{aligned} C \equiv J_\theta^{cc} & \begin{pmatrix} 0 & U_1U_2^* & 0 & 0 \\ U_2U_1^* & 0 & 0 & 0 \\ 0 & 0 & 0 & U_3U_4^* \\ 0 & 0 & U_4U_3^* & 0 \end{pmatrix} J_\theta^{cc} \\ &= \begin{pmatrix} 0 & 0 & U_1J_{11}J_{12}U_3^* & 0 \\ 0 & 0 & 0 & U_1J_{12}J_{22}U_4^* \\ U_3J_{21}J_{11}U_1^* & 0 & 0 & 0 \\ 0 & U_4J_{22}J_{21}U_2^* & 0 & 0 \end{pmatrix} \\ &\in \pi_\theta(\mathcal{M}\otimes M_2(\mathbb{C}))'_{\mathrm{w}}. \end{aligned}$$

Hence we have

$$C\pi_\theta(X)C = \pi_\theta(X), \quad {}^\forall X \in \mathcal{M}\otimes M_2(\mathbb{C}),$$

which implies that $W \equiv J_{22}J_{21}$ is a unitary tranform of $\mathcal{H}_\varphi$ onto $\mathcal{H}_\psi$ such that

$$\pi_\varphi(X) = W^*\pi_\psi(X)W, \quad {}^\forall X \in \mathcal{M}.$$

This completes the proof.

Proposition 3.4.7. Let φ and ψ be faithful, σ-weakly continuous, semifinite (quasi-)weights on $\mathcal{P}(\mathcal{M})$ such taht π_φ and π_ψ are self-adjoint. The following statements are equivalent:

(i) φ and ψ are quasi-standard (quasi-)weights which satisfy one of (i) $\sim$ (iv) in Theorem 3.4.6.

(ii) θ is quasi-standard.

Proof. (i) $\Longrightarrow$ (ii) By Theorem 3.4.6 there exists a unitary transform W of $\mathcal{H}_\varphi$ onto $\mathcal{H}_\psi$ such that $W\mathcal{D}(\pi_\varphi) = \mathcal{D}(\pi_\psi)$ and $\pi_\varphi(X) = W^*\pi_\psi(X)W$ for all $X \in \mathcal{M}$. Since $W^* \in \mathbb{I}(\pi_\psi, \pi_\varphi)$, it follows that

$$\mathfrak{A}_{\varphi,\psi} = \{\ (A, WAW^*)\ ;\ A \in (\pi_\varphi(\mathcal{M})'_{\mathrm{w}})'\ \},$$

which implies by Lemma 3.4.5 and (3.4.3) that for each $A_{ij} \in (\pi_\varphi(\mathcal{M})'_{\mathrm{w}})'$ $(i,j = 1,2)$

$$\Delta_{\Lambda_\theta^{cc}}^{it} \begin{pmatrix} U_1A_{11}U_1^* & U_1A_{12}U_2^* & 0 & 0 \\ U_2A_{21}U_1^* & U_2A_{22}U_2^* & 0 & 0 \\ 0 & 0 & U_3WA_{11}W^*U_3^* & U_3WA_{12}W^*U_4^* \\ 0 & 0 & U_4WA_{21}W^*U_3^* & U_4WA_{22}W^*U_4^* \end{pmatrix} \Delta_{\Lambda_\theta^{cc}}^{-it} =$$

$$\begin{pmatrix} U_1\Delta_{11}^{it}A_{11}\Delta_{11}^{-it}U_1^* & U_1\Delta_{11}^{it}A_{12}\Delta_{21}^{-it}U_2^* & 0 & 0 \\ U_2\Delta_{21}^{it}A_{21}\Delta_{11}^{-it}U_1^* & U_2\Delta_{21}^{it}A_{22}\Delta_{21}^{-it}U_2^* & 0 & 0 \\ 0 & 0 & U_3\Delta_{12}^{it}WA_{11}W^*\Delta_{12}^{-it}U_3^* & U_3\Delta_{12}^{it}WA_{12}W^*\Delta_{22}^{-it}U_4^* \\ 0 & 0 & U_4\Delta_{22}^{it}WA_{21}W^*\Delta_{12}^{-it}U_3^* & U_4\Delta_{22}^{it}WA_{22}W^*\Delta_{22}^{-it}U_4^* \end{pmatrix}$$

$\in (\pi_\theta(\mathcal{M}\otimes M_2(\mathbb{C}))'_{\mathrm{w}})'$.

Hence we have by Lemma 3.4.5

$$W\Delta_{11}^{it}A_{11}\Delta_{11}^{-it}W^* = \Delta_{12}^{it}WA_{11}W^*\Delta_{12}^{-it}, \tag{3.4.4}$$

$$W\Delta_{21}^{it}A_{22}\Delta_{21}^{-it}W^* = \Delta_{22}^{it}WA_{22}W^*\Delta_{22}^{-it}, \tag{3.4.5}$$

$$W\Delta_{11}^{it}A_{12}\Delta_{21}^{-it}W^* = \Delta_{12}^{it}WA_{12}W^*\Delta_{22}^{-it}, \tag{3.4.6}$$

$$W\Delta_{21}^{it}A_{21}\Delta_{11}^{-it}W^* = \Delta_{22}^{it}WA_{21}W^*\Delta_{12}^{-it}. \tag{3.4.7}$$

It follows from (3.4.4) and (3.4.5) that $\Delta_{11}^{-it}W^*\Delta_{12}^{it}W, \Delta_{21}^{-it}W^*\Delta_{22}^{it}W \in \pi_\varphi(\mathcal{M})'_{\mathrm{w}}$ for all $t\in\mathbb{R}$, and hence

$$\begin{aligned}\Delta_{12}^{it}\mathcal{D}(\pi_\psi) &= W\Delta_{11}^{it}(\Delta_{11}^{-it}W^*\Delta_{12}^{it}W)W^*\mathcal{D}(\pi_\psi)\\ &\subset \mathcal{D}(\pi_\psi)\end{aligned}$$

for all $t\in\mathbb{R}$. Similary,

$$\Delta_{21}^{it}\mathcal{D}(\pi_\varphi)\subset\mathcal{D}(\pi_\varphi),\quad {}^\forall t\in\mathbb{R}.$$

Hence we have

$$\Delta_{\Lambda_\theta^{cc}}^{it}\mathcal{D}(\pi_\theta)\subset\mathcal{D}(\pi_\theta),\quad {}^\forall t\in\mathbb{R}.$$

Therefore, θ is quasi-standard.

(ii) $\Longrightarrow$ (i) Since $\Lambda_\theta^c((\mathcal{D}(\Lambda_\theta^c)^*\cap\mathcal{D}(\Lambda_\theta^c))^2)$ is total in $\mathcal{H}_\theta$, it follows from (3.4.2) that $\Lambda_\varphi^c\,((\mathcal{D}(\Lambda_\varphi^c)^*\cap\mathcal{D}(\Lambda_\varphi^c))^2)$ and $\Lambda_\psi^c((\mathcal{D}(\Lambda_\psi^c)^*\cap\mathcal{D}(\Lambda_\psi^c))^2)$ are total in $\mathcal{H}_\varphi$ and $\mathcal{H}_\psi$, respectively. Furthermore, since $\Delta_{\Lambda_\theta^{cc}}^{it}\mathcal{D}(\pi_\theta)\subset\mathcal{D}(\pi_\theta)$ for all $t\in\mathbb{R}$, it follows from (3.4.3) that $\Delta_{\Lambda_\varphi^{cc}}^{it}\mathcal{D}(\pi_\varphi)\subset\mathcal{D}(\pi_\varphi)$ and $\Delta_{\Lambda_\psi^{cc}}^{it}\mathcal{D}(\pi_\psi)\subset\mathcal{D}(\pi_\psi)$ for all $t\in\mathbb{R}$. Therefore, φ and ψ are quasi-standard. This completes the proof.

Theorem 3.4.8. (Generalized Connes cocycle theorem) Suppose φ and ψ are faithful, σ-weakly continuous, semifinite, quasi-standard (quasi-)weights on $\mathcal{P}(\mathcal{M})$ which satisfy one of (i) $\sim$ (iv) in Theorem 3.4.6. Then there exists a strongly continuous map $t\in\mathbb{R}\longrightarrow U_t\in\pi_\varphi(\mathcal{M})''_{\mathrm{wc}}$, uniquely determined, such that

(i) $\overline{U_t}$ is unitary, ${}^\forall t\in\mathbb{R}$;

(ii) $U_{s+t}=U_t\sigma_t^{\Lambda_\varphi^{cc}}(U_s)$, ${}^\forall s,t\in\mathbb{R}$;

(iii) $\sigma_t^{\Lambda_\psi^{cc}}(WAW^\dagger)=WU_t\sigma_t^{\Lambda_\varphi^{cc}}(A)U_t^*W^*$, ${}^\forall A\in\pi_\varphi(\mathcal{M})''_{\mathrm{wc}}$, ${}^\forall t\in\mathbb{R}$,

where W is a unitary transform of $\mathcal{H}_\varphi$ onto $\mathcal{H}_\psi$ such that $W\mathcal{D}(\pi_\varphi)=\mathcal{D}(\pi_\psi)$ and $\pi_\psi(X)=W\pi_\varphi(X)W^\dagger$ for all $X\in\mathcal{M}$;

(iv) for any $A \in (W^\dagger \mathfrak{N}_{\overline{\psi}} W) \cap \mathfrak{N}_{\overline{\varphi}}^\dagger$ and $B \in \mathfrak{N}_{\overline{\varphi}} \cap (W^\dagger \mathfrak{N}_{\overline{\psi}}^\dagger W)$ there exists an element $F_{A,B}$ of $A(0,1)$ such that

$$F_{A,B}(t) = \overline{\varphi}(AU_t \sigma_t^{\Lambda_\varphi^{cc}}(B)),$$
$$F_{A,B}(t+i) = \overline{\psi}(\sigma_t^{\Lambda_\psi^{cc}}(WBW^\dagger)WU_tAW^\dagger),$$

for all $t \in \mathbb{R}$, where $\overline{\varphi}$ and $\overline{\psi}$ are the quasi-weights induced by φ and ψ, respectively.

Proof. We put

$$\begin{cases} \mathcal{D}(\Lambda_\varphi^\psi) = \{ \pi_\varphi(X) \; ; \; X \in \mathfrak{N}_\psi \}, \\ \Lambda_\varphi^\psi(\pi_\varphi(X)) = W^* \lambda_\psi(X), \quad X \in \mathfrak{N}_\psi. \end{cases}$$

Then it is easily shown that Λ_φ^ψ is a generalized vector for $\pi_\varphi(\mathcal{M})$ such that

$$\begin{cases} \mathcal{D}((\Lambda_\varphi^\psi)^c) = \{ W^*KW \; ; \; K \in \mathcal{D}(\Lambda_\psi^c) \}, \\ (\Lambda_\varphi^\psi)^c(W^*KW) = W^*\Lambda_\psi^c(K), \quad K \in \mathcal{D}(\Lambda_\psi^c) \; ; \end{cases}$$

$$\begin{cases} \mathcal{D}((\Lambda_\varphi^\psi)^{cc}) = \{ W^*AW \; ; \; A \in \mathcal{D}(\Lambda_\psi^{cc}) \}, \\ (\Lambda_\varphi^\psi)^{cc}(W^*AW) = W^*\Lambda_\psi^{cc}(A), \quad A \in \mathcal{D}(\Lambda_\psi^{cc}) \; ; \end{cases}$$

$$S_{(\Lambda_\varphi^\psi)^{cc}} = W^* S_{\Lambda_\psi^{cc}} W.$$

Hence we have

$$\Delta_{(\Lambda_\varphi^\psi)^{cc}}^{it} \mathcal{D}(\pi_\varphi) = W^* \Delta_{\Lambda_\psi^{cc}}^{it} W \mathcal{D}(\pi_\varphi) \subset \mathcal{D}(\pi_\varphi)$$

for all $t \in \mathbb{R}$, and so Λ_φ^ψ is quasi-standard. By Theorem 3.3.4 $\overline{\Lambda_\varphi}$ and $\overline{\Lambda_\varphi^\psi}$ are standard generalized vectors for the generalized von Neumann algebra $\pi_\varphi(\mathcal{M})''_{\mathrm{wc}}$, and so it follows from Theorem 2.5.6 that there exists a strongly continuous map $t \in \mathbb{R} \longrightarrow U_t \in \pi_\varphi(\mathcal{M})''_{\mathrm{wc}}$ satisfying the conditions (i) ~ (iv) and it is identical with the Connes cocycle $[D\overline{\Lambda_\varphi^\psi} : D\overline{\Lambda_\varphi}]_t$ $(= \Delta_{21}^{it}\Delta_{11}^{-it})$ associated with $\overline{\Lambda_\varphi^\psi}$ with respect to $\overline{\Lambda_\varphi}$. This completes the proof.

The map $t \in \mathbb{R} \longrightarrow U_t \in \pi_\varphi(\mathcal{M})''_{\mathrm{wc}}$, uniquely determined by the above theorem, is called *the cocycle associated with the quasi-weight* $\overline{\psi}$ *with respect to the quasi-weight* $\overline{\varphi}$, and denoted by $[D\overline{\psi} : D\overline{\varphi}]$. It follows from (3.4.6) that the cocycle $[D\overline{\varphi} : D\overline{\psi}]_t$ associated with $\overline{\varphi}$ with respect to $\overline{\psi}$ equals $W[D\overline{\psi} : D\overline{\varphi}]_t^* W^*$. By (iii) and (iv) in Theorem 3.4.8 we have

(iii)′ $\sigma_t^{\Lambda_\psi^{cc}}(\pi_\psi(X)) = W[D\overline{\psi} : D\overline{\varphi}]_t \sigma_t^{\Lambda_\varphi^{cc}}(\pi_\varphi(X))[D\overline{\psi} : D\overline{\varphi}]_t^* W^*$, $\forall X \in \mathcal{M}$, $\forall t \in \mathbb{R}$;

(iv)′ for any $X \in \mathfrak{N}_\varphi^\dagger \cap \mathfrak{N}_\psi$ and $Y \in \mathfrak{N}_\varphi \cap \mathfrak{N}_\psi^\dagger$ there exists an element $F_{X,Y}$ of $A(0,1)$ such that

$$F_{X,Y}(t) = \overline{\varphi}(\pi_\varphi(X)[D\overline{\psi} : D\overline{\varphi}]_t \sigma_t^{\Lambda_\varphi^{cc}}(\pi_\varphi(Y))),$$

$$F_{X,Y}(t+i) = \overline{\psi}(\sigma_t^{\Lambda_\psi^{cc}}(\pi_\psi(Y))[D\overline{\varphi} : D\overline{\psi}]_t^*(\pi_\psi(X)),$$

for all $t \in \mathbb{R}$.

Corollary 3.4.9. Suppose φ and ψ are faithful, σ-weakly continuous, semifinite, standard (quasi-)weights which satisfy one of (i) ~ (iv) in Theorem 3.4.6 and $\pi_\varphi(\mathcal{M})$ is a generalized von Neumann algebra. Then there exists a strongly continuous map $t \in \mathbb{R} \longrightarrow [D\psi : D\varphi]_t \in \mathcal{M}$, uniquely determined, such that

(i) $[D\psi : D\varphi]_t$ is unitary, $^\forall t \in \mathbb{R}$;

(ii) $[D\psi : D\varphi]_{s+t} = [D\psi : D\varphi]_t \sigma_t^\varphi([D\psi : D\varphi]_s)$;

(iii) $\sigma_t^\psi(X) = [D\psi : D\varphi]_t \sigma_t^\varphi(X)[D\psi : D\varphi]_t^*$, $^\forall X \in \mathcal{M}$, $^\forall t \in \mathbb{R}$;

(iv) for any $X \in \mathfrak{N}_\varphi^\dagger \cap \mathfrak{N}_\psi$ and $Y \in \mathfrak{N}_\varphi \cap \mathfrak{N}_\psi^\dagger$ there exists an element $F_{X,Y}$ of $A(0,1)$ such that

$$F_{X,Y}(t) = \varphi(X[D\psi : D\varphi]_t \sigma_t^\varphi(Y)),$$

$$F_{X,Y}(t+i) = \psi(\sigma_t^\psi(Y)[D\psi : D\varphi]_t X)$$

for all $t \in \mathbb{R}$.

This $[D\psi : D\varphi]$ is called *the cocycle associated with the (quasi-) weight* ψ *with respect to the (quasi-)weight* φ.

As seen in Corollary 3.4.9, if $\pi_\varphi(\mathcal{M})$ is a generalized von Neumann algebra, then the generalized Connes cocycle theorem for weights on O*-algebras becomes the best form. Here we show that if $\mathcal{M}$ is a generalized von Neumann algebra with strongly dense bounded part and φ is a strongly faithful, σ-weakly continuous (quasi-)weight on $\mathcal{P}(\mathcal{M})$, then $\pi_\varphi(\mathcal{M})$ is spatially isomorphic to $\mathcal{M}$, and so it is a generalized von Neumann algebra.

Lemma 3.4.10. Let $\mathcal{M}$ be a self-adjoint O*-algebra on $\mathcal{D}$ in $\mathcal{H}$ such that $\mathcal{M}_b'' = (\mathcal{M}_{\mathrm{w}}')'$ and φ a σ-weakly continuous (quasi-)weight on $\mathcal{P}(\mathcal{M})$. Then there exists a normal $*$-homomorphism $\overline{\pi_\varphi}$ of $(\mathcal{M}_{\mathrm{w}}')'$ onto $(\pi_\varphi(\mathcal{M})_{\mathrm{w}}')'$ such that $\overline{\pi_\varphi}(A) = \overline{\pi_\varphi(A)}$ for all $A \in \mathcal{M}_b$.

Proof. Since φ_X can be extended to a σ-weakly continuous positive linear functional on $\mathcal{M}_b''$ for each $X \in \mathfrak{N}_\varphi$, it follows that

$$\varphi_X(A^\dagger A) \leq \| \overline{A} \|^2 \varphi_X(I), \quad {}^\forall A \in \mathcal{M}_b,$$

which implies

$$\pi_\varphi(\mathcal{M}_b) \subset \pi_\varphi(\mathcal{M})_b,$$
$$\| \overline{\pi_\varphi(A)} \| \leq \| \overline{A} \|, \quad {}^\forall A \in \mathcal{M}_b. \tag{3.4.8}$$

We now have the following:

$$\begin{aligned}&\text{If } \{A_\alpha\} \text{ is any uniformly bounded net in } \mathcal{M}_b \text{ such that}\\ &\overline{A_\alpha} \longrightarrow A \in \mathcal{B}(\mathcal{H}) \text{ weakly (resp. strongly, strongly}^*\text{)},\\ &\text{then } \{\overline{\pi_\varphi(A_\alpha)}\} \text{ converges weakly (resp.strongly, strongly}^*\text{)}\\ &\text{to an element of } \mathcal{B}(\mathcal{H}_\varphi).\end{aligned} \tag{3.4.9}$$

In fact, for each $X, Y \in \mathfrak{N}_\varphi$ we have

$$\begin{aligned}\lim_{\alpha,\beta} ((\pi_\varphi(A_\alpha) - \pi_\varphi(A_\beta))\lambda_\varphi(X) \mid \lambda_\varphi(Y)) &= \lim_{\alpha,\beta} \varphi_{X,Y}(A_\alpha - A_\beta)\\ &= 0,\end{aligned}$$

and so we put

$$B(\lambda_\varphi(X), \lambda_\varphi(Y)) = \lim_\alpha (\pi_\varphi(A_\alpha)\lambda_\varphi(X) \mid \lambda_\varphi(Y)) \quad X, Y \in \mathfrak{N}_\varphi.$$

By (3.4.8) B is a bounded sesquilinear form on $\lambda_\varphi(\mathfrak{N}_\varphi) \times \lambda_\varphi(\mathfrak{N}_\varphi)$, and so it can be extended to a bounded sesquilinear form on $\mathcal{H}_\varphi \times \mathcal{H}_\varphi$. It hence follows from the Riesz theorem that $\{\overline{\pi_\varphi(A_\alpha)}\}$ converges weakly (resp. strongly, strongly*) to an element of $\mathcal{B}(\mathcal{H}_\varphi)$. Since $\mathcal{M}_b'' = (\mathcal{M}_{\mathrm{w}}')'$, it follows from the Kaplansky density theorem that for each $A \in (\mathcal{M}_{\mathrm{w}}')'$ there exists a net $\{A_\alpha\}$ in $\mathcal{M}_b$ such that $\| \overline{A_\alpha} \| \leq \| A \|$ for all α and $\overline{A_\alpha} \longrightarrow A$ strongly*, and so we put

$$\overline{\pi_\varphi}(A) = s^* - \lim_\alpha \overline{\pi_\varphi(A_\alpha)}, \quad A \in (\mathcal{M}_{\mathrm{w}}')'.$$

By (3.4.9) $\overline{\pi_\varphi}(A)$ is well-defined, i.e., it is independent for taking a net $\{A_\alpha\}$ in $\mathcal{M}_b$, and $\overline{\pi_\varphi}$ is a normal $*$-homomorphism of $(\mathcal{M}_{\mathrm{w}}')'$ to $\mathcal{B}(\mathcal{H}_\varphi)$. Hence, it follows that

$$\overline{\pi_\varphi}((\mathcal{M}_{\mathrm{w}}')') \text{ is a von Neumann algebra .} \tag{3.4.10}$$

We finally show that

$$\pi_\varphi(\mathcal{M}_b)'' = \overline{\pi_\varphi}((\mathcal{M}_{\mathrm{w}}')') = (\pi_\varphi(\mathcal{M})_{\mathrm{w}}')'. \tag{3.4.11}$$

In fact, take an arbitrary $C \in \pi_\varphi(\mathcal{M}_b)'$. Since $\overline{X}$ is affiliated with $(\mathcal{M}_{\mathrm{w}}')' = \mathcal{M}_b''$ for each $X \in \mathcal{M}$, there exists a net $\{A_\alpha\}$ in $\mathcal{M}_b$ which converges σ-strongly* to X. Hence we have

$$\begin{aligned}&\lim_\alpha \| \pi_\varphi(A_\alpha)\lambda_\varphi(Y) - \pi_\varphi(X)\lambda_\varphi(Y) \|^2\\ &\quad = \lim_\alpha \varphi_Y((A_\alpha - X)^\dagger(A_\alpha - X))\\ &\quad = 0\end{aligned}$$

$$\lim_{\alpha} \| \pi_\varphi(A_\alpha^\dagger)\lambda_\varphi(Y) - \pi_\varphi(X^\dagger)\lambda_\varphi(Y) \| = 0$$

for each $Y \in \mathfrak{N}_\varphi$, and so

$$\begin{aligned}(C\pi_\varphi(X)\lambda_\varphi(Y) \mid \lambda_\varphi(Z)) &= \lim_{\alpha} (C\pi_\varphi(A_\alpha)\lambda_\varphi(Y) \mid \lambda_\varphi(Z)) \\ &= \lim_{\alpha} (C\lambda_\varphi(Y) \mid \pi_\varphi(A_\alpha^\dagger)\lambda_\varphi(Z)) \\ &= (C\lambda_\varphi(Y) \mid \pi_\varphi(X^\dagger)\lambda_\varphi(Z))\end{aligned}$$

for all $Y, Z \in \mathfrak{N}_\varphi$. Hence, $C \in \pi_\varphi(\mathcal{M})'_{\mathrm{w}}$. Thus we have $\pi_\varphi(\mathcal{M}_b)' \subset \pi_\varphi(\mathcal{M})'_{\mathrm{w}}$, which implies by (3.4.10) that

$$\pi_\varphi(\mathcal{M}_b)'' \subset \overline{\pi_\varphi}((\mathcal{M}'_{\mathrm{w}})') \subset (\pi_\varphi(\mathcal{M})'_{\mathrm{w}})' \subset \pi_\varphi(\mathcal{M}_b)''.$$

Therefore, the statement (3.4.11) holds. This completes the proof.

As shown in Lemma 3.4.2, if φ is a faithful, semifinite (quasi-)weight on $\mathcal{P}(\mathcal{M})$, then π_φ is a $*$-isomorphism, but we don't know whether $\overline{\pi_\varphi}$ is a $*$-isomorphism in general. For this we have the following

Lemma 3.4.11. Suppose φ is a faithful, σ-weakly continuous (quasi-)weight on $\mathcal{P}(\mathcal{M})$. Then the following statements are equivalent:
(i) $\overline{\pi_\varphi}$ is a $*$-isomorphism.
(ii) The map π_φ^{-1} from $\pi_\varphi(\mathcal{M}_b)[\tau_{\sigma s}]$ to $(\mathcal{M}'_{\mathrm{w}})'\lceil_{\mathcal{D}}[\tau_{\sigma s}]$ is closable.

Proof. (i) $\Longrightarrow$ (ii) Let $\{A_\alpha\}$ be any net in $\mathcal{M}_b$ such that $\tau_{\sigma s}-\lim_{\alpha} \pi_\varphi(A_\alpha) = 0$ and $\tau_{\sigma s} - \lim_{\alpha} A_\alpha = A \in (\mathcal{M}'_{\mathrm{w}})'\lceil_{\mathcal{D}}$. By Lemma 3.4.10 we have

$$\overline{\pi_\varphi}(\overline{A}) = \tau_s - \lim_{\beta} \overline{\pi_\varphi(B_\beta)},$$

where $\{B_\beta\}$ is a uniformly bounded net in $\mathcal{M}_b$ which converges σ-strongly* to A. And we have

$$\begin{aligned}&\lim_{\alpha,\beta} \| \pi_\varphi(A_\alpha)\lambda_\varphi(X) - \pi_\varphi(B_\beta)\lambda_\varphi(X) \|^2 \\ &= \lim_{\alpha,\beta} \varphi_X((A_\alpha - B_\beta)^\dagger(A_\alpha - B_\beta)) \\ &= 0\end{aligned}$$

for all $X \in \mathfrak{N}_\varphi$. Hence we have

$$\begin{aligned}\pi_\varphi(\overline{A})\lambda_\varphi(X) &= \lim_{\alpha} \pi_\varphi(B_\beta)\lambda_\varphi(X) \\ &= \lim_{\alpha} \pi_\varphi(A_\alpha)\lambda_\varphi(X) \\ &= 0\end{aligned}$$

for all $X \in \mathfrak{N}_\varphi$, and so $\overline{\pi_\varphi}(\overline{A}) = 0$. Since $\overline{\pi_\varphi}$ is a $*$-isomorphism, we have $\overline{A} = 0$.

(ii) $\Longrightarrow$ (i) Suppose $\overline{\pi_\varphi}(A) = 0$, $A \in (\mathcal{M}'_{\rm w})'$. Then there exists a net $\{A_\alpha\}$ in $\mathcal{M}_b$ such that $\| \overline{A_\alpha} \| \leq r$ for all α and $\tau_{\sigma s} - \lim_\alpha \overline{A_\alpha} = A$. By Lemma 3.4.10 we have

$$\tau_{\sigma s} - \lim_\alpha \overline{\pi_\varphi}(A_\alpha) = \overline{\pi_\varphi}(A) = 0.$$

Hence, $A = 0$. This completes the proof.

Definition 3.4.12. A σ-weakly continuous (quasi-)weight φ on $\mathcal{P}(\mathcal{M})$ is said to be *strongly faithful* if φ is faithful and one of the conditions (i) and (ii) of Lemma 3.4.11 holds.

Proposition 3.4.13. Let $\mathcal{M}$ be a self-adjoint O*-algebra on $\mathcal{D}$ in $\mathcal{H}$ such that $\mathcal{M}''_b = (\mathcal{M}'_{\rm w})'$, and φ a strongly faithful, σ-weakly continuous (quasi-)weight on $\mathcal{P}(\mathcal{M})$ such that π_φ is self-adjoint. Suppose $(\mathcal{M}'_{\rm w})'$ and $(\pi_\varphi(\mathcal{M})'_{\rm w})'$ satisfy one of the following statements:

(i) they are standard von Neumann algebras.

(ii) $\mathcal{M}'_{\rm w}$ and $\pi_\varphi(\mathcal{M})'_{\rm w}$ are properly infinite and of countable type.

(iii) $\mathcal{H}$ and $\mathcal{H}_\varphi$ are separable and $(\mathcal{M}'_{\rm w})'$ and $(\pi_\varphi(\mathcal{M})'_{\rm w})'$ are von Neumann algebras of type III.

Then the O*-algebras $\mathcal{M}$ and $\pi_\varphi(\mathcal{M})$ are spatially isomorphic.

Proof. It follows from Lemma 3.4.10, 3.4.11 and (Stratila-Zsido [1] §8) that $\overline{\pi_\varphi}$ is spatial, that is, there exists a unitary transform U of $\mathcal{H}$ onto $\mathcal{H}_\varphi$ such that $\overline{\pi_\varphi}(A) = UAU^*$ for all $A \in (\mathcal{M}'_{\rm w})'$. This implies that

$$U\mathcal{D} = \mathcal{D}(\pi_\varphi) \text{ and } \pi_\varphi(X) = UXU^* \tag{3.4.12}$$

for all $X \in \mathcal{M}$. Take an arbitrary $X \in \mathcal{M}$. For each $\xi \in \mathcal{D}$ and $Y \in \mathfrak{N}_\varphi$ we have

$$\begin{aligned}(\pi_\varphi(X^\dagger)\lambda_\varphi(Y) \mid U\xi) &= \lim_\alpha (\pi_\varphi(A_\alpha^\dagger)\lambda_\varphi(Y) \mid U\xi) \\ &= \lim_\alpha (UA_\alpha^\dagger U^*\lambda_\varphi(Y) \mid U\xi) \\ &= (\lambda_\varphi(Y) \mid UX\xi),\end{aligned}$$

where $\{A_\alpha\}$ is a net in $\mathcal{M}_b$ which converges σ-strongly* to X. By the self-adjointness of π_φ we have

$$U\xi \in \mathcal{D}(\pi_\varphi) \text{ and } \pi_\varphi(X)U\xi = UX\xi, \tag{3.4.13}$$

and further

$$\begin{aligned}(X^\dagger\xi \mid U^*\eta) &= (UX^\dagger\xi \mid \eta) \\ &= (\pi_\varphi(X^\dagger)U\xi \mid \eta) \\ &= (\xi \mid U^*\pi_\varphi(X)\eta)\end{aligned}$$

for all $\xi \in \mathcal{D}$ and $\eta \in \mathcal{D}(\pi_\varphi)$. Hence it follows from the self-adjointness of

$\mathcal{M}$ that $U^*\mathcal{D}(\pi_\varphi) \subset \mathcal{D}$, which implies that the statement 3.4.13 holds. This completes the proof.

Throughout the rest of this section let $\mathcal{M}$ be a self-adjoint generalized von Neumann algebra on $\mathcal{D}$ in $\mathcal{H}$ such that $\mathcal{M}_b'' = (\mathcal{M}_w')'$ and $(\mathcal{M}_w')'$ is a standard form. We denote by $W_s(\mathcal{M})$ the set of all strongly faithful, σ-weakly continuous, semifinite, quasi-standard quasi-weights φ on $\mathcal{P}(\mathcal{M})$ such that π_φ are self-adjoint. Suppose $\varphi \in W_s(\mathcal{M})$. By Proposition 3.4.13 $\pi_\varphi(\mathcal{M})$ is a generalized von Neumann algebra on $\mathcal{D}(\pi_\varphi)$ in $\mathcal{H}_\varphi$, and so φ is standard. By Theorem 3.3.2 we have the following

Corollary 3.4.14. For every $\varphi \in W_s(\mathcal{M})$ there exists a one-parameter group $\{\sigma_t^\varphi\}_{t\in\mathbb{R}}$ of $*$-automorphisms of $\mathcal{M}$ such that
(i) $\pi_\varphi(\sigma_t^\varphi(X)) = \Delta_\varphi^{it}\pi_\varphi(X)\Delta_\varphi^{-it}, \quad X \in \mathcal{M},\ t \in \mathbb{R}$;
(ii) φ is a $\{\sigma_t^\varphi\}$-KMS quasi-weight on $\mathcal{P}(\mathcal{M})$.

Suppose $\varphi, \psi \in W_s(\mathcal{M})$. By Proposition 3.4.13 $\pi_\varphi(\mathcal{M})$ and $\pi_\psi(\mathcal{M})$ are generalized von Neumann algebras, and φ and ψ are standard. Hence, by Corollary 3.4.9 we have the following

Corollary 3.4.15. Suppose $\varphi, \psi \in W_s(\mathcal{M})$. Then, the cocycle $[D\psi : D\varphi]$ associated with the quasi-weight ψ with respect to the quasi-weight φ is well-defined in $\mathcal{M}$, that is, $t \longrightarrow [D\psi : D\varphi]_t$ is a strongly continuous map of $\mathbb{R}$ into $\mathcal{M}$ satisfying the conditions (i) $\sim$ (iv) in Corollary 3.4.9.

We generalize the Pedersen-Takesaki theorem (Pedersen-Takesaki [1], Stratila [1]) for standard weights on von Neumann algebras to those on O*-algebras. Let $\varphi \in W_s(\mathcal{M})$. Since $\mathcal{M}$ is a generalized von Neumann algebra, the quasi-weights φ_e and $\overline{\varphi}$ on $\mathcal{P}(\mathcal{M})$ defined by Theorem 3.3.3 and Theorem 3.3.4 coincide, that is,

$$\begin{cases} \mathfrak{N}_{\overline{\varphi}} = \{\ X \in \mathcal{M}\ ;\ {}^\exists\xi_X \in \mathcal{D}(\pi_\varphi) \\ \qquad\qquad\qquad \text{s.t. } \pi_\varphi(X)\Lambda_\varphi^c(K) = K\xi_X,\ {}^\forall K \in \mathcal{D}(\Lambda_\varphi^c)\ \}, \\ \overline{\varphi}(\sum_k X_k^\dagger X_k) = \sum_k \|\ \xi_{X_k}\ \|^2,\ \sum_k X_k^\dagger X_k \in \mathcal{P}(\mathfrak{N}_{\overline{\varphi}}). \end{cases}$$

Using Theorem 2.6.2 and Theorem 2.6.6, we can show the following results:

Corollary 3.4.16. Suppose $\varphi, \psi \in W_s(\mathcal{M})$. The following statements are equivalent:
(i) $\overline{\psi} \circ \sigma_t^\varphi = \overline{\psi}, \quad {}^\forall t \in \mathbb{R}$.
(ii) $\overline{\varphi} \circ \sigma_t^\psi = \overline{\varphi}, \quad {}^\forall t \in \mathbb{R}$.
(iii) $[D\psi : D\varphi]_t \in \mathcal{M}^{\sigma^\psi}, \quad {}^\forall t \in \mathbb{R}$.
(iv) $[D\psi : D\varphi]_t \in \mathcal{M}^{\sigma^\varphi}, \quad {}^\forall t \in \mathbb{R}$.

(v) $\{ [D\psi : D\varphi]_t \}_{t\in\mathbb{R}}$ is a strongly continuous group of unitary elements of $\mathcal{M}$.

Corollary 3.4.17. Suppose $\varphi, \psi \in W_s(\mathcal{M})$. The following statements are equivalent:

(i) ψ satisfies the KMS-condition with respect to $\{\sigma_t^{\varphi}\}_{t\in\mathbb{R}}$.

(ii) $\sigma_t^{\psi} = \sigma_t^{\varphi}, \quad {}^{\forall}t \in \mathbb{R}$.

(iii) There exists a positive self-adjoint operator A in $\mathcal{H}$ affiliated with $(\pi_{\varphi}(\mathcal{M})'_{\mathrm{w}})' \cap \pi_{\varphi}(\mathcal{M})'_{\mathrm{w}}$ such taht $\overline{\overline{\psi}} = \overline{\overline{\varphi_A}}$, where $\overline{\varphi_A}$ is the quasi-weight on $\mathcal{P}(\pi_{\varphi}(\mathcal{M})''_{\mathrm{wc}})$ induced by the quasi-weight on $\mathcal{P}(\mathcal{M})$ defined by

$$\begin{cases} \mathfrak{N}_{\overline{\varphi_A}} = \{ X \in \mathcal{M} \; ; \; \overline{\Lambda_{\varphi}}(X) \in \mathcal{D}(A) \}, \\ \overline{\varphi_A}(X^{\dagger}X) = \| A\overline{\Lambda_{\varphi}}(X) \|^2, \quad X \in \mathfrak{N}_{\overline{\varphi_A}}. \end{cases}$$

3.5 Radon-Nikodym theorem for weights

In this section we extend the Radon-Nikodym theorem for positive linear functionals (Gudder [1], Inoue [8, 10]) to for weights on O*-algebras.

Throughout this section, let $\mathcal{M}$ be a closed O*-algebra on $\mathcal{D}$ in $\mathcal{H}$.

Definition 3.5.1. Let φ and ψ be (quasi-)weights on $\mathcal{P}(\mathcal{M})$. If $\mathfrak{N}_{\varphi} \subset \mathfrak{N}_{\psi}$ and $\psi(X^{\dagger}X) \leq \gamma\varphi(X^{\dagger}X), \quad {}^{\forall}X \in \mathfrak{N}_{\varphi}$ for some constant $\gamma > 0$, then ψ is said to be *φ-dominated* and denoted by $\psi \leq \gamma\varphi$. If $\mathfrak{N}_{\varphi} \subset \mathfrak{N}_{\psi}$ and the map $K_{\varphi,\psi}$: $\lambda_{\varphi}(X) \to \lambda_{\psi}(X), \; X \in \mathfrak{N}_{\varphi}$ is closable from the dense subspace $\lambda_{\varphi}(\mathfrak{N}_{\varphi})$ in a Hilbert space $\mathcal{H}_{\varphi}$ to the Hilbert space $\mathcal{H}_{\psi}$, then ψ is said to be *φ-absolutely continuous* . If $\mathfrak{N}_{\varphi} \subset \mathfrak{N}_{\psi}$ and for any $X \in \mathfrak{N}_{\varphi}$ there exists a sequence $\{X_n\}$ in $\mathfrak{N}_{\varphi}$ such that $\lim_{n\to\infty} \varphi(X_n^{\dagger}X_n) = 0$ and $\lim_{n\to\infty} \psi((X_n - X)^{\dagger}(X_n - X)) = 0$, then ψ is said to be *φ-singular* . If $\mathfrak{N}_{\varphi} \subset \mathfrak{N}_{\psi}$ and $\varphi(X^{\dagger}X) = \psi(X^{\dagger}X)$ for each $X \in \mathfrak{N}_{\varphi}$, then ψ is said to be an *extension* of φ and denoted by $\varphi \subset \psi$.

We have the Radon-Nikodym theorem and the Lebesque decomposition theorem for weights similarly proofs to Theorem 3.2, Theorem 3.3 in Inoue [8]:

Theorem 3.5.2. (Radon-Nikodym theorem) Let φ and ψ be (quasi-) weights on $\mathcal{P}(\mathcal{M})$. Then the following statements hold:

(1) ψ is φ-dominated if and only if there exists a positive operator H' in $\pi_{\varphi}(\mathcal{M})'_{\mathrm{w}}$ such that $\psi(X^{\dagger}X) = (H'\lambda_{\varphi}(X)|\lambda_{\varphi}(X))$ for all $X \in \mathfrak{N}_{\varphi}$.

(2) Suppose $\mathfrak{N}_{\varphi} \subset \mathfrak{N}_{\psi}$ and $\pi_{\varphi+\psi}(\mathcal{M})'_{\mathrm{w}}$ is a von Neumann algebra. Then the following statements are equivalent:

(i) ψ is φ-absolutely continuous.

(ii) There exists an increasing sequence $\{H'_n\}$ of positive operators in

$\pi_\varphi(\mathcal{M})'_{\rm w}$ such that $\lim_{n\to\infty} H'_n\lambda_\varphi(X)$ exists in $\mathcal{H}_\varphi$ for each $X \in \mathfrak{N}_\varphi$ and $\psi(X^\dagger X) = \lim_{n\to\infty} \|H'_n\lambda_\varphi(X)\|^2$ for each $X \in \mathfrak{N}_\varphi$.

(iii) There exists a positive self-adjoint operator H' in $\mathcal{H}_\varphi$ affiliated with $(\pi_\varphi(\mathcal{M})'_{\rm w})''$ such that $\mathcal{D}(H') \supset \lambda_\varphi(\mathfrak{N}_\varphi)$ and $\psi(X^\dagger X) = \|H'\lambda_\varphi(X)\|^2$ for each $X \in \mathfrak{N}_\varphi$.

Proof. Here we simply state these proofs.

(1) Suppose ψ is φ-dominated and put $H' = K^*_{\varphi,\psi}K_{\varphi,\psi}$. Then $H' \in \pi_\varphi(\mathcal{M})'_{\rm w}$ and $\psi(X^\dagger X) = (H'\lambda_\varphi(X)|\lambda_\varphi(X))$ for each $X \in \mathfrak{N}_\varphi$. The converse is trivial.

(2) (i) $\Rightarrow$ (ii) We put

$$K' = (K^*_{\varphi+\psi,\varphi}K_{\varphi+\psi,\varphi})^{\frac{1}{2}},$$
$$K'_n = \int_{\frac{1}{n}}^{1} t^{-1}(1-t)dE(t),\ n \in \mathbb{N},$$

where $K' = \displaystyle\int_0^1 tdE(t)$ is the spectral resolution of K'. Let U denote the isometry of $\mathcal{H}_\varphi$ into $\mathcal{H}_{\varphi+\psi}$ defined by $U\lambda_\varphi(X) = K'\lambda_{\varphi+\psi}(X)$, $X \in \mathfrak{N}_\varphi$, and put

$$H'_n = U^*K'_nU,\ n \in \mathbb{N}.$$

Then we can show that $\{H'_n\}$ satisfies our assertion in (ii).

(ii) $\Rightarrow$ (iii) We put

$$\begin{cases} \mathcal{D}(H'_0) = \{\xi \in \mathcal{H}_\varphi;\ \lim_{n\to\infty} H'_n\xi \text{ exists in } \mathcal{H}_\varphi\} \\ H'_0\xi = \lim_{n\to\infty} H'_n\xi,\ \xi \in \mathcal{D}(H'_0). \end{cases}$$

Then H'_0 is a positive operator in $\mathcal{H}_\varphi$ such that $\mathcal{D}(H'_0) \supset \lambda_\varphi(\mathfrak{N}_\varphi)$ and $\overline{H'_0}$ is affiliated with $(\pi_\varphi(\mathcal{M})'_{\rm w})''$, so that the Friedrichs self-adjoint extension H' of H'_0 satisfies our assertion in (iii).

(iii) $\Rightarrow$ (ii) This is trivial.

Theorem 3.5.3. (Lebesgue decomposition theorem) Let φ and ψ be (quasi-) weights on $\mathcal{P}(\mathcal{M})$ such that $\mathfrak{N}_\varphi \subset \mathfrak{N}_\psi$ and $\pi_{\varphi+\psi}(\mathcal{M})'_{\rm w}$ is a von Neumann algebra. Then ψ is decomposed into the sum: $\psi \supset \psi^\varphi_c + \psi^\varphi_s$, where ψ^φ_c is a φ-absolutely continuous quasi-weight on $\mathcal{P}(\mathcal{M})$ with $\mathfrak{N}_{\psi^\varphi_c} = \mathfrak{N}_\varphi$ and ψ^φ_s is a φ-singular quasi-weight on $\mathcal{P}(\mathcal{M})$ with $\mathfrak{N}_{\psi^\varphi_s} = \mathfrak{N}_\varphi$.

Proof. Let $\{H'_n\}$ be in the proof of (i) $\Rightarrow$ (ii) in Theorem 3.5.2, and let $P_{\varphi+\psi,\varphi}$ be the projection of $\mathcal{H}_{\varphi+\psi}$ onto Ker $K^*_{\varphi+\psi,\varphi}K_{\varphi+\psi,\varphi}$. We here put

$$\psi^\varphi_c(X^\dagger X) = \lim_{n\to\infty} \|H'_n\lambda_\varphi(X)\|^2,$$
$$\psi^\varphi_s(X^\dagger X) = \|P_{\varphi+\psi,\varphi}\lambda_\varphi(X)\|^2,\ X \in \mathfrak{N}_\varphi.$$

Then ψ_c^φ and ψ_s^φ imply our assertions. This completes the proof.

As seen in Example 3.5.15, the Lebesque decomposition of a (quasi-) weight is not unique in general.

Proposition 3.5.4. Let φ and ψ be (quasi-)weights on $\mathcal{P}(\mathcal{M})$ such that $\mathfrak{N}_\varphi \subset \mathfrak{N}_\psi$ and $\pi_{\varphi+\psi}(\mathcal{M})'_{\mathrm{w}}$ is a von Neumann algebra. The following statements are equivalent:

(i) ψ is φ-singular.

(ii) $\psi_c^\varphi = 0$

(iii) If χ is a (quasi-) weight on $\mathcal{P}(\mathcal{M})$ such that $\chi \leq \varphi$ and $\chi \leq \psi$, then $\chi = 0$ on $\mathcal{P}(\mathcal{M})$.

Proof. (ii) $\Rightarrow$ (i) This is trivial.

(i) $\Rightarrow$ (iii) Since ψ is φ-singular and $\chi \leq \psi$, it follows that χ is φ-singular. On the other hand, since $\chi \leq \varphi$, it follows that χ is φ-absolutely continuous, which implies $\chi = 0$ on $\mathcal{P}(\mathfrak{N}_\varphi)$.

(iii) $\Rightarrow$ (ii) By Theorem 3.5.2, 3.5.3 ψ_c^φ is represented as

$$\psi_c^\varphi(X^\dagger X) = \lim_{n\to\infty} \|H'_n \lambda_\varphi(X)\|^2, \ X \in \mathfrak{N}_\varphi$$

for some increasing sequence $\{H'_n\}$ of positive operators in $\pi_\varphi(\mathcal{M})'_{\mathrm{w}}$. For any $n \in \mathbb{N}$ we put

$$\varphi_n(X^\dagger X) = \|H'_n \lambda_\varphi(X)\|^2, \ X \in \mathfrak{N}_\varphi.$$

Then φ_n is a quasi-weight on $\mathcal{P}(\mathcal{M})$ such that $\varphi_n \leq \psi_c^\varphi \leq \psi$ and $\varphi_n \leq \|H'_n\|^2 \varphi$. By the assumption (iii) we have $\varphi_n = 0$ for each $n \in \mathbb{N}$. Hence it follows from $\mathfrak{N}_{\psi_c^\varphi} = \mathfrak{N}_\varphi$ that $\psi_c^\varphi = 0$. This completes the proof.

In Section 3.2 we have obtained the generalized Pedersen-Takesaki Radon-Nikodym theorem for strongly faithful, normal, semifinite weights on generalized von Neumann algebras with strongly dense bounded part by using the generalized Connes cocyle theorem. Here we consider the generalized Pedersen-Takesaki Radon-Nikodym theorem in more general cases.

Theorem 3.5.5. Let φ be a standard (quasi-) weight on $\mathcal{P}(\mathcal{M})$ and ψ a (quasi-) weight on $\mathcal{P}(\mathcal{M})$. The following statements hold:

(1) ψ is a φ-dominated, $\{\sigma_t^\varphi\}$-KMS (quasi-) weight on $\mathcal{P}(\mathcal{M})$ if and only if there exists a positive operator H in $(\pi_\varphi(\mathcal{M})'_{\mathrm{w}})' \cap \pi_\varphi(\mathcal{M})'_{\mathrm{w}}$ such that $\psi(X^\dagger X) = \|H\lambda_\varphi(X)\|^2$ for each $X \in \mathfrak{N}_\varphi$.

(2) The following statements are equivalent:

(i) ψ is a φ-absolutely continuous, $\{\sigma_t^\varphi\}$-KMS (quasi-) weight on $\mathcal{P}(\mathcal{M})$ such that $\varphi + \psi$ is standard.

(ii) There exists an increasing sequence $\{H'_n\}$ of positive operators in

$(\pi_\varphi(\mathcal{M})'_{\rm w})' \cap \pi_\varphi(\mathcal{M})'_{\rm w}$ such that $\lim_{n\to\infty} H'_n\lambda_\varphi(X)$ exists in $\mathcal{H}_\varphi$ for each $X \in \mathfrak{N}_\varphi$ and $\psi(X^\dagger X) = \lim_{n\to\infty} \|H'_n\lambda_\varphi(X)\|^2$ for each $X \in \mathfrak{N}_\varphi$.

(iii) There exists a positive self-adjoint operator H in $\mathcal{H}_\varphi$ affiliated with $(\pi_\varphi(\mathcal{M})'_{\rm w})' \cap \pi_\varphi(\mathcal{M})'_{\rm w}$ such that $\mathcal{D}(H) \supset \lambda_\varphi(\mathfrak{N}_\varphi)$ and $\psi(X^\dagger X) = \|H\lambda_\varphi(X)\|^2$ for each $X \in \mathfrak{N}_\varphi$.

(3) Suppose ψ is a $\{\sigma_t^\varphi\}$-KMS (quasi-) weight on $\mathcal{P}(\mathcal{M})$ such that $\mathfrak{N}_\varphi \subset \mathfrak{N}_\psi$ and $\varphi + \psi$ is standard. Then ψ is decomposed into the sum: $\psi \supset \psi_c^\varphi + \psi_s^\varphi$, where ψ_c^φ is a φ-absolutely continuous, $\{\sigma_t^\varphi\}$-KMS quasi-weight on $\mathcal{P}(\mathcal{M})$ with $\mathfrak{N}_{\psi_c^\varphi} = \mathfrak{N}_\varphi$ and ψ_s^φ is a φ-singular, $\{\sigma_t^\varphi\}$-KMS quasi-weight on $\mathcal{P}(\mathcal{M})$ with $\mathfrak{N}_{\psi_s^\varphi} = \mathfrak{N}_\varphi$.

Proof. (1) Suppose ψ is a φ-deminated, $\{\sigma_t^\varphi\}$-KMS (quasi-) weight on $\mathcal{P}(\mathcal{M})$. By Theorem 3.5.2 there exists a positive operator H in $\pi_\varphi(\mathcal{M})'_{\rm w}$ such that $\psi(X^\dagger X) = \|H\lambda_\varphi(X)\|^2$ for each $X \in \mathfrak{N}_\varphi$. We put

$$\psi''(A^*A) = \begin{cases} \|H\Lambda_\varphi^{cc}(A)\|^2, & \text{if } A \in \mathcal{D}(\Lambda_\varphi^{cc}) \\ \infty & \text{if otherwise.} \end{cases}$$

Then ψ'' is a normal, semifinite weight on $(\pi_\varphi(\mathcal{M})'_{\rm w})'_+$. Take arbitrary $A, B \in \mathcal{D}(\Lambda_\varphi^{cc})^* \cap \mathcal{D}(\Lambda_\varphi^{cc})$. Since $S_\varphi = S_{\Lambda_\varphi^{cc}}$, there exist sequences $\{X_n\}$ and $\{Y_n\}$ in $\mathfrak{N}_\varphi^\dagger \cap \mathfrak{N}_\varphi$ such that

$$\lim_{n\to\infty} \lambda_\varphi(X_n) = \Lambda_\varphi^{cc}(A), \ \lim_{n\to\infty} \lambda_\varphi(X_n^\dagger) = \Lambda_\varphi^{cc}(A^*),$$
$$\lim_{n\to\infty} \lambda_\varphi(Y_n) = \Lambda_\varphi^{cc}(B), \ \lim_{n\to\infty} \lambda_\varphi(Y_n^\dagger) = \Lambda_\varphi^{cc}(B^*).$$

Since ψ is a $\{\sigma_t^\varphi\}$-KMS (quasi-) weight on $\mathcal{P}(\mathcal{M})$, there exists a sequence $\{f_{X_n,Y_n}\}$ in $A(0,1)$ such that

$$f_{X_n,Y_n}(t) = \psi(\sigma_t^\varphi(X_n)Y_n) = (H^2\lambda_\varphi(Y_n)|\Delta_\varphi^{it}\lambda_\varphi(X_n^\dagger)),$$
$$f_{X_n,Y_n}(t+i) = \psi(Y_n\sigma_t^\varphi(X_n)) = (H^2\Delta_\varphi^{it}\lambda_\varphi(X_n)|\lambda_\varphi(Y_n^\dagger))$$

for all $t \in \mathbb{R}$ and $n \in \mathbb{N}$, which implies that

$$\lim_{n\to\infty}\sup_{t\in\mathbb{R}}|f_{X_n,Y_n}(t) - (H^2\Lambda_\varphi^{cc}(B)|\Delta_\varphi^{it}\Lambda_\varphi^{cc}(A^*))| = 0,$$
$$\lim_{n\to\infty}\sup_{t\in\mathbb{R}}|f_{X_n,Y_n}(t+i) - (H^2\Delta_\varphi^{it}\Lambda_\varphi^{cc}(A)|\Lambda_\varphi^{cc}(B^*))| = 0.$$

Hence there exists a function $f_{A,B}$ in $A(0,1)$ such that

$$f_{A,B}(t) = (H^2\Lambda_\varphi^{cc}(B)|\Delta_\varphi^{it}\Lambda_\varphi^{cc}(A^*)) = \psi''(\sigma_t^\varphi(A)B),$$
$$f_{A,B}(t+i) = (H^2\Delta_\varphi^{it}\Lambda_\varphi^{cc}(A)|\Lambda_\varphi^{cc}(B^*)) = \psi''(B\sigma_t^\varphi(A))$$

for all $t \in \mathbb{R}$, which means that ψ'' satisfies the KMS-condition with respect to $\{\sigma_t^\varphi\}$. By Theorem 15.4 in Takesaki [1] we have $H \in (\pi_\varphi(\mathcal{M})'_{\rm w})' \cap \pi_\varphi(\mathcal{M})'_{\rm w}$.

Conversely suppose $H \in (\pi_\varphi(\mathcal{M})'_{\rm w})' \cap \pi_\varphi(\mathcal{M})'_{\rm w}$. Then ψ'' is a normal, semifinite weight on $(\pi_\varphi(\mathcal{M})'_{\rm w})'_+$ which satisfies the KMS-condition with respect to $\{\sigma_t^\varphi\}$. Since $S_\varphi = S_{\Lambda_\varphi^{cc}}$, we can show similarly to the above proof that ψ satisfies the KMS-condition with respect to $\{\sigma_t^\varphi\}$.

(2) (i) $\Rightarrow$ (ii) Let K', U and $H'_n, n \in \mathbb{N}$ be in Theorem 3.5.2. By (1) we have $K' \in (\pi_\varphi(\mathcal{M})'_{\rm w})' \cap \pi_\varphi(\mathcal{M})'_{\rm w}$. We show $H'_n \in (\pi_\varphi(\mathcal{M})'_{\rm w})'$ for each $n \in \mathbb{N}$. Take an arbitrary $C \in \pi_\varphi(\mathcal{M})'_{\rm w}$. Since

$$\begin{aligned}
&(UCU^*K'\pi_{\varphi+\psi}(A)\lambda_{\varphi+\psi}(X) \mid \lambda_{\varphi+\psi}(Y)) \\
&= (C\pi_\varphi(A)\lambda_\varphi(X) \mid U^*\lambda_{\varphi+\psi}(Y)) \\
&= (C\lambda_\varphi(X) \mid U^*\pi_{\varphi+\psi}(A^\dagger)\lambda_{\varphi+\psi}(Y)) \\
&= (UCU^*K'\lambda_{\varphi+\psi}(X) \mid \pi_{\varphi+\psi}(A^\dagger)\lambda_{\varphi+\psi}(Y))
\end{aligned}$$

for each $A \in \mathcal{M}$ and $X, Y \in \mathfrak{N}_\varphi$, it follows that $UCU^*K' \in \pi_{\varphi+\psi}(\mathcal{M})'_{\rm w}$, which implies

$$\begin{aligned}
CH'_n\lambda_\varphi(X) &= CU^*\Big(\int_{\frac{1}{n}}^1 t^{-1}(1-t)dE(t)\Big)U\lambda_\varphi(X) \\
&= U^*(UCU^*)K'\Big(\int_{\frac{1}{n}}^1 t^{-1}(1-t)dE(t)\Big)\lambda_{\varphi+\psi}(X) \\
&= U^*\Big(\int_{\frac{1}{n}}^1 t^{-1}(1-t)dE(t)\Big)(UCU^*K')\lambda_{\varphi+\psi}(X) \\
&= H'_nC\lambda_\varphi(X)
\end{aligned}$$

for each $C \in \pi_\varphi(\mathcal{M})'_{\rm w}$, $X \in \mathfrak{N}_\varphi$ and $n \in \mathbb{N}$. Hence we have $H'_n \in (\pi_\varphi(\mathcal{M})'_{\rm w})'$ for all $n \in \mathbb{N}$.

(ii) $\Rightarrow$ (iii) This is shown similarly to the proof of (ii) $\Rightarrow$ (iii) in Theorem 3.5.2.

(iii) $\Rightarrow$ (i) It is clear that ψ is a φ-absolutely continuous, $\{\sigma_t^\varphi\}$-KMS (quasi-) weight on $\mathcal{P}(\mathcal{M})$ and

$$(\varphi+\psi)(X^\dagger X) = \|(I+H^2)^{\frac{1}{2}}\lambda_\varphi(X)\|^2, \; X \in \mathfrak{N}_\varphi. \tag{3.5.1}$$

We put

$$\begin{cases}
\mathcal{D}((I+H^2)^{\frac{1}{2}}\Lambda_\varphi^{cc}) = \{A \in \mathcal{D}(\Lambda_\varphi^{cc});\ \Lambda_\varphi^{cc}(A) \in \mathcal{D}((I+H^2)^{\frac{1}{2}})\} \\
((I+H^2)^{\frac{1}{2}}\Lambda_\varphi^{cc})(A) = (I+H^2)^{\frac{1}{2}}\Lambda_\varphi^{cc}(A),\ A \in \mathcal{D}((I+H^2)^{\frac{1}{2}}\Lambda_\varphi^{cc}).
\end{cases}$$

Then $\Lambda \equiv (I+H^2)^{\frac{1}{2}}\Lambda_\varphi^{cc}$ is a generalized vector for the von Neumann algebra $(\pi_\varphi(\mathcal{M})'_{\rm w})'$. Since $(I+H^2)^{-\frac{1}{2}}A \in \mathcal{D}(\Lambda)^* \cap \mathcal{D}(\Lambda)$ for each $A \in \mathcal{D}(\Lambda_\varphi^{cc})^* \cap \mathcal{D}(\Lambda_\varphi^{cc})$, it follows that

$$\Lambda((\mathcal{D}(\Lambda)^* \cap \mathcal{D}(\Lambda))^2) \text{ is total in } \mathcal{H}_\varphi, \tag{3.5.2}$$

and further it is easily shown that

$$\left.\begin{array}{l} \{(I+H^2)^{-\frac{1}{2}}K;\ K \in \mathcal{D}(\Lambda_\varphi^{\mathrm{c}})\} \subset \mathcal{D}(\Lambda') \subset \mathcal{D}(\Lambda_\varphi^{\mathrm{c}}), \\ \Lambda'((I+H^2)^{-\frac{1}{2}}K) = \Lambda_\varphi^{\mathrm{c}}(K), \quad {}^\forall K \in \mathcal{D}(\Lambda_\varphi^{\mathrm{c}}), \\ (I+H^2)^{-\frac{1}{2}}\Lambda'(K) = \Lambda_\varphi^{\mathrm{c}}(K), \quad {}^\forall K \in \mathcal{D}(\Lambda'), \end{array}\right\} \tag{3.5.3}$$

so that

$$\Lambda'((\mathcal{D}(\Lambda')^* \cap \mathcal{D}(\Lambda'))^2) \text{ is total in } \mathcal{H}_\varphi. \tag{3.5.4}$$

By (3.5.2) and (3.5.3), Λ is a standard generalized vector for the von Neumann algebra $(\pi_\varphi(\mathcal{M})'_{\mathrm{w}})'$. By (3.5.3) we have

$$\begin{aligned} S^*_{\Lambda_\varphi^{\mathrm{cc}}}\Lambda_\varphi^{\mathrm{c}}(K) = \Lambda_\varphi^{\mathrm{c}}(K^*) &= \Lambda'((I+H^2)^{-\frac{1}{2}}K^*) \\ &= S^*_\Lambda \Lambda'((I+H^2)^{-\frac{1}{2}}K) \\ &= S^*_\Lambda \Lambda_\varphi^{\mathrm{c}}(K) \end{aligned}$$

for each $K \in \mathcal{D}(\Lambda_\varphi^{\mathrm{c}})^* \cap \mathcal{D}(\Lambda_\varphi^{\mathrm{c}})$, and hence $S^*_{\Lambda_\varphi^{\mathrm{cc}}} \subset S^*_\Lambda$. Further, since $\mathcal{D}(\Lambda') \subset \mathcal{D}(\Lambda_\varphi^{\mathrm{c}})$ by (3.5.3), it follows that $S^*_{\Lambda_\varphi^{\mathrm{cc}}} = S^*_\Lambda$, and so $S_{\Lambda_\varphi^{\mathrm{cc}}} = S_\Lambda$, which implies by (3.5.1) and the standardness of $\Lambda_\varphi^{\mathrm{cc}}$ that $\varphi + \psi$ is a standard (quasi-) weight on $\mathcal{P}(\mathcal{M})$.

(3) This follows from Theorem 3.5.2 and the statement (2). This completes the proof.

We next study the Radon-Nikodym theorem for $\{\sigma_t^\varphi\}$-invariant (quasi-) weights on $\mathcal{P}(\mathcal{M})$:

Theorem 3.5.6. Let φ be a standard (quasi-)weight on $\mathcal{P}(\mathcal{M})$ and ψ a (quasi-)weight on $\mathcal{P}(\mathcal{M})$. The following statements are equivalent:

(i) ψ is φ-dominated and $\{\sigma_t^\varphi\}$-invariant.

(ii) There exists a positive operator H' in $\pi_\varphi(\mathcal{M})'^{\sigma^\varphi}_{\mathrm{w}}$ such that $\psi(X^\dagger X) = \|H'\lambda_\varphi(X)\|^2$ for each $X \in \mathfrak{N}_\varphi$.

Further, suppose φ and ψ are positive linear functionals. Then the above equivalent statements are equivalent to the following (iii):

(iii) There exists a positive operator H in $(\pi_\varphi(\mathcal{M})'_{\mathrm{w}})'^{\sigma^\varphi}$ such that $H\lambda_\varphi(I) \in \mathcal{D}(\pi_\varphi)$ and $\psi(X) = (\pi_\varphi(X)H\lambda_\varphi(I)|H\lambda_\varphi(I))$ for each $X \in \mathfrak{N}_\varphi$.

Proof. (i) $\Leftrightarrow$ (ii) This follows from Theorem 3.5.2. Suppose φ and ψ are positive linear functionals on $\mathcal{M}$.

(ii) $\Rightarrow$ (iii) We put $H = J_\varphi H' J_\varphi$. Then it is easily shown that H is a positive operator in $(\pi_\varphi(\mathcal{M})'_{\mathrm{w}})'^{\sigma^\varphi}$ such that $H\lambda_\varphi(I) = H'\lambda_\varphi(I)$, and hence $H\lambda_\varphi(I) \in \mathcal{D}(\pi_\varphi)$ and $\psi(X) = (\pi_\varphi(X)H\lambda_\varphi(I)|H\lambda_\varphi(I))$ for each $X \in \mathfrak{N}_\varphi$.

(iii) $\Rightarrow$ (ii) This is proved similarly to the proof of (ii) $\Rightarrow$ (iii).

Theorem 3.5.7. Let φ be a standard (quasi-) weight on $\mathcal{P}(\mathcal{M})$ and ψ a (quasi-) weight on $\mathcal{P}(\mathcal{M})$. Suppose there exists a standard, $\{\sigma_t^\varphi\}$-KMS (quasi-) weight τ on $\mathcal{P}(\mathcal{M})$ such that $\varphi+\psi \leq \tau$ and $\mathfrak{N}_\tau = \mathfrak{N}_\varphi$. Then the following statements hold:

(1) Suppose ψ is $\{\sigma_t^\varphi\}$-invariant. Then ψ is decomposed into the sum: $\psi \supset \psi_c^\varphi + \psi_s^\varphi$, where ψ_c^φ is a φ-absolutely continuous, $\{\sigma_t^\varphi\}$-invariant quasi-weight on $\mathcal{P}(\mathcal{M})$ with $\mathfrak{N}_{\psi_c^\varphi} = \mathfrak{N}_\varphi$ and ψ_s^φ is a φ-singular, $\{\sigma_t^\varphi\}$-invariant quasi-weight on $\mathcal{P}(\mathcal{M})$ with $\mathfrak{N}_{\psi_s^\varphi} = \mathfrak{N}_\varphi$. If ψ is φ-absolutely continuous, then $\psi \supset \psi_c^\varphi$, and if ψ is φ-singular, then $\psi \supset \psi_s^\varphi$.

(2) The following statements are equivalent:

(i) ψ is φ-absolutely continuous and $\{\sigma_t^\varphi\}$-invariant.

(ii) There exists a positive self-adjoint operator H' in $\mathcal{H}_\varphi$ affiliated with $\pi_\varphi(\mathcal{M})_{\mathrm{w}}^{\prime\sigma^\varphi}$ such that $\lambda_\varphi(\mathfrak{N}_\varphi) \subset \mathcal{D}(H')$ and $\psi(X^\dagger X) = \|H'\lambda_\varphi(X)\|^2$ for each $X \in \mathfrak{N}_\varphi$.

Further, suppose φ and ψ are positive linear functionals on $\mathcal{M}$. Then the above equivalent statements (i) and (ii) are equivalent to the following (iii):

(iii) There exists a positive self-adjoint operator H in $\mathcal{H}_\varphi$ affiliated with $(\pi_\varphi(\mathcal{M})_{\mathrm{w}}')^{\prime\sigma^\varphi}$ such that $\lambda_\varphi(I) \in \mathcal{D}(H), H\lambda_\varphi(I) \in \mathcal{D}(\pi_\varphi)$ and $\psi(X) = (\pi_\varphi(X)H\lambda_\varphi(I)|H\lambda_\varphi(I))$ for each $X \in \mathcal{M}$.

Proof. (1) By Theorem 3.3.6 we have

$$\Delta_\tau^{it}\lambda_\tau(X) = \lambda_\tau(\sigma_t^\varphi(X)), \quad {}^\forall X \in \mathfrak{N}_\tau^\dagger \cap \mathfrak{N}_\tau = \mathfrak{N}_\varphi^\dagger \cap \mathfrak{N}_\varphi, \quad {}^\forall t \in \mathbb{R}. \tag{3.5.5}$$

Since $\varphi \leq \tau$ and $\psi \leq \tau$, there exist R' and K' in $\pi_\tau(\mathcal{M})_{\mathrm{w}}'$ such that $0 \leq R', K' \leq I$ and

$$\varphi(X^\dagger X) = \|R'\lambda_\tau(X)\|^2 \text{ and } \psi(X^\dagger X) = \|K'\lambda_\tau(X)\|^2$$

for each $X \in \mathfrak{N}_\tau = \mathfrak{N}_\varphi$. Using (3.5.5) and the standardness of τ, we can prove in the same way as in Theorem 3.5.5 that the generalized vector $R'\Lambda_\tau^{cc}$ for the von Neumann algebra $(\pi_\tau(\mathcal{M})_{\mathrm{w}}')'$ satisfies the KMS-condition with respect to $\{\sigma_t^\tau\}$, so that $R' \in (\pi_\tau(\mathcal{M})_{\mathrm{w}}')' \cap \pi_\tau(\mathcal{M})_{\mathrm{w}}'$. Further, since ψ is $\{\sigma_t^\varphi\}$-invariant, it follows from (3.5.5) that $K' \in \pi_\tau(\mathcal{M})_{\mathrm{w}}^{\prime\sigma^\tau}$. We denote by U the isometry of $\mathcal{H}_\varphi$ into $\mathcal{H}_\tau$ defined by $U\Lambda_\varphi(X) = R'\lambda_\tau(X)$, $X \in \mathfrak{N}_\tau = \mathfrak{N}_\varphi$. We now put

$$H_n' = U^*\left(\int_{\frac{1}{n}}^1 \frac{1}{t} dE'(t)\right)K'U, \; n \in \mathbb{N},$$

where $R' = \displaystyle\int_0^1 tdE'(t)$ is the spectral resolution of R'. Since R' and K' commute, it follows that $\{H_n'\}$ is an increasing sequence of positive operators in $\pi_\varphi(\mathcal{M})_{\mathrm{w}}^{\prime\sigma^\varphi}$ and $\lim_{n\to\infty} H_n'\lambda_\varphi(X)$ exists in $\mathcal{H}_\varphi$ for each $X \in \mathfrak{N}_\varphi$. We put

$$\psi_c^\varphi(X^\dagger X) = \lim_{n\to\infty} \|H_n'\lambda_\varphi(X)\|^2,$$
$$\psi_s^\varphi(X^\dagger X) = \|K'E'(0)\lambda_\tau(X)\|^2, \; X \in \mathfrak{N}_\tau = \mathfrak{N}_\varphi.$$

Then it is easily shown that ψ_c^φ is a φ-absolutely continuous, $\{\sigma_t^\varphi\}$-invariant quasi-weight on $\mathcal{P}(\mathcal{M})$ with $\mathfrak{N}_{\psi_c^\varphi} = \mathfrak{N}_\varphi$, ψ_s^φ is a φ-singular, $\{\sigma_t^\varphi\}$-invariant quasi-weight on $\mathcal{P}(\mathcal{M})$ with $\mathfrak{N}_{\psi_s^\varphi} = \mathfrak{N}_\varphi$ and $\psi \supset \psi_c^\varphi + \psi_s^\varphi$. Suppose ψ is φ-absolutely continuous. For any $X \in \mathfrak{N}_\tau$ there is a sequence $\{X_n\}$ in $\mathfrak{N}_\tau$ such that $\lim_{n\to\infty} \lambda_\tau(X_n) = E'(0)\lambda_\tau(X)$. Since $\mathfrak{N}_\tau = \mathfrak{N}_\varphi \subset \mathfrak{N}_\psi$, we have

$$\lim_{n\to\infty} \lambda_\varphi(X_n) = \lim_{n\to\infty} U^* R' \lambda_\tau(X_n) = U^* R' E'(0)\lambda_\tau(X) = 0,$$

$$\begin{aligned}\lim_{n\to\infty} (\lambda_\psi(X_n)|\lambda_\psi(Y)) &= \lim_{n\to\infty} (K'\lambda_\tau(X_n)|K'\lambda_\tau(Y)) \\ &= (K'E'(0)\lambda_\tau(X)|K'\lambda_\tau(Y))\end{aligned}$$

for each $Y \in \mathfrak{N}_\tau$. Further, since $\psi \le \tau$, we have $\lim_{n\to\infty} \lambda_\psi(X_n) = 0$, and so it follows from the φ-absolute continuity of ψ that $K'^2 E'(0)\lambda_\tau(X) = 0$. Hence we have $\psi_s^\varphi(X^\dagger X) = 0$, and so $\psi \supset \psi_c^\varphi$. Similarly, if ψ is φ-singular, then $\psi \supset \psi_s^\varphi$.

(2) (i) $\Leftrightarrow$ (ii) Using the statement (1), this is proved similarly to the proof of Theorem 3.5.2. Suppose φ and ψ are positive linear functionals on $\mathcal{M}$.

(ii) $\Rightarrow$ (iii) We put $H = J_\varphi H' J_\varphi$. Then H is a positive self-adjoint operator in $\mathcal{H}_\varphi$ affiliated with $(\pi_\varphi(\mathcal{M})'_{\mathrm{w}})'^{\sigma^\varphi}$ such that $\lambda_\varphi(I) \in \mathcal{D}(H)$, $H\lambda_\varphi(I) = H'\lambda_\varphi(I) \in \mathcal{D}(\pi_\varphi)$ and $\psi(X) = (\pi_\varphi(X)H\lambda_\varphi(I)|H\lambda_\varphi(I))$ for each $X \in \mathcal{M}$.

(iii) $\Rightarrow$ (ii) This is proved similarly the proof of (ii) $\Rightarrow$ (iii). This completes the proof.

It seems to be difficult to show directly the standardness of $\varphi + \psi$ in Theorem 3.5.5 and the existence of $\{\sigma_t^\varphi\}$-KMS (quasi-) weight τ with $\varphi + \psi \le \tau$ in Theorem 3.5.7, and so we consider when Theorem 3.5.5 and Theorem 3.5.7 hold without these assumptions.

Proposition 3.5.8. Let φ be a standard (quasi-)weight on $\mathcal{P}(\mathcal{M})$ and ψ a (quasi-)weight on $\mathcal{P}(\mathcal{M})$. Suppose ψ is φ-absolutely continuous, $\pi_{\varphi+\psi}(\mathcal{M})'_{\mathrm{w}}$ is a von Neumann algebra and

$$\psi(X^\dagger X) \le \gamma\{\varphi(X^\dagger X) + \varphi(XX^\dagger)\}, \quad {}^\forall X \in \mathfrak{N}_\varphi^\dagger \cap \mathfrak{N}_\varphi \tag{3.5.6}$$

for some constant $\gamma > 0$. Then the following statements hold:

(1) If ψ satisfies the KMS-condition with respet to $\{\sigma_t^\varphi\}$, then there exist a positive self-adjoint operator H in $\mathcal{H}_\varphi$ affiliated with $(\pi_\varphi(\mathcal{M})'_{\mathrm{w}})' \cap \pi_\varphi(\mathcal{M})'_{\mathrm{w}}$ such that $\lambda_\varphi(\mathfrak{N}_\varphi) \subset \mathcal{D}(H)$ and $\psi(X^\dagger X) = \|H\lambda_\varphi(X)\|^2$ for each $X \in \mathfrak{N}_\varphi$.

(2) If ψ is $\{\sigma_t^\varphi\}$-invariant, then there exists a positive self-adjoint operator H' in $\mathcal{H}_\varphi$ affiliated with $\pi_\varphi(\mathcal{M})'^{\sigma^\varphi}_{\mathrm{w}}$ such that $\lambda_\varphi(\mathfrak{N}_\varphi) \subset \mathcal{D}(H')$ and $\psi(X^\dagger X) = \|H'\lambda_\varphi(X)\|^2$ for each $X \in \mathfrak{N}_\varphi$.

Proof. (1) By Theorem 3.5.2 ψ is represented as

$$\psi(X^\dagger X) = \|H'\lambda_\varphi(X)\|^2, \ X \in \mathfrak{N}_\varphi \tag{3.5.7}$$

for some positive self-adjoint operator H' in $\mathcal{H}_\varphi$ affiliated with $\pi_\varphi(\mathcal{M})'_{\mathrm{w}}$ such that $\lambda_\varphi(\mathfrak{N}_\varphi) \subset \mathcal{D}(H')$. Since ψ is $\{\sigma_t^\varphi\}$-invariant, we have

$$\|H'\Delta_\varphi^{it}\lambda_\varphi(X)\| = \|H'\lambda_\varphi(X)\|, \quad {}^\forall X \in \mathfrak{N}_\varphi. \tag{3.5.8}$$

Take an arbitrary $A \in \mathcal{D}(\Lambda_\varphi^{cc})^* \cap \mathcal{D}(\Lambda_\varphi^{cc})$. Since $S_{\Lambda_\varphi^{cc}} = S_\varphi$, there exists a sequence $\{X_n\}$ in $\mathfrak{N}_\varphi^\dagger \cap \mathfrak{N}_\varphi$ such that $\lim_{n\to\infty} \lambda_\varphi(X_n) = \Lambda_\varphi^{cc}(A)$ and $\lim_{n\to\infty} \lambda_\varphi(X_n^\dagger) = \Lambda_\varphi^{cc}(A^*)$. By (3.5.6), (3.5.7) and (3.5.8) we have

$$\begin{aligned} &\Lambda_\varphi^{cc}(A) \in \mathcal{D}(H'), \ \lim_{n\to\infty} H'\lambda_\varphi(X_n) = H'\Lambda_\varphi^{cc}(A), \\ &\lim_{n\to\infty} H'\Delta_\varphi^{it}\lambda_\varphi(X_n) = H'\Delta_\varphi^{it}\Lambda_\varphi^{cc}(A), \quad {}^\forall t \in \mathbb{R}, \end{aligned} \tag{3.5.9}$$

which implies by (3.5.7) and (3.5.8) that

$$\begin{aligned} &\|H'\Delta_\varphi^{it}\Lambda_\varphi^{cc}(A)\| = \|H'\Lambda_\varphi^{cc}(A)\|, \\ &(H'\Delta_\varphi^{it}\lambda_\varphi(X)|H'\Delta_\varphi^{it}\Lambda_\varphi^{cc}(A)) = (H'\lambda_\varphi(X)|H'\Lambda_\varphi^{cc}(A)) \end{aligned}$$

for each $X \in \mathfrak{N}_\varphi^\dagger \cap \mathfrak{N}_\varphi$ and $t \in \mathbb{R}$, so that

$$\|H'\Delta_\varphi^{it}\lambda_\varphi(X) - H'\Delta_\varphi^{it}\Lambda_\varphi^{cc}(A)\| = \|H'\lambda_\varphi(X) - H'\Lambda_\varphi^{cc}(A)\| \tag{3.5.10}$$

for each $A \in \mathcal{D}(\Lambda_\varphi^{cc})^* \cap \mathcal{D}(\Lambda_\varphi^{cc})$ and $X \in \mathfrak{N}_\varphi^\dagger \cap \mathfrak{N}_\varphi$. Since ψ is a $\{\sigma_t^\varphi\}$-KMS (quasi-) weight, it follows from (3.5.9) and (3.5.10) that the generalized vector $H'\Lambda_\varphi^{cc}$ for the von Neumann algebra $(\pi_\varphi(\mathcal{M})'_{\mathrm{w}})'$ satisfies the KMS-condition with respect to $\{\sigma_t^\varphi\}$. Hence there exists a positive self-adjoint operator H in $\mathcal{H}_\varphi$ affiliated with $(\pi_\varphi(\mathcal{M})'_{\mathrm{w}})' \cap \pi_\varphi(\mathcal{M})'_{\mathrm{w}}$ such that

$$\Lambda_\varphi^{cc}(A) \in \mathcal{D}(H) \text{ and } \|H'\Lambda_\varphi^{cc}(A)\| = \|H\Lambda_\varphi^{cc}(A)\|$$

for each $A \in \mathcal{D}(\Lambda_\varphi^{cc})^* \cap \mathcal{D}(\Lambda_\varphi^{cc})$, which implies by the polar decomposition theorem that

$$\Lambda_\varphi^{cc}(A) \in \mathcal{D}(H) \text{ and } \|H'\Lambda_\varphi^{cc}(A)\| = \|H\Lambda_\varphi^{cc}(A)\| \tag{3.5.11}$$

for each $A \in \mathcal{D}(\Lambda_\varphi^{cc})$. Take an arbitrary $X \in \mathfrak{N}_\varphi$. Let $\overline{X} = U|\overline{X}|$ be the polar decomposition of $\overline{X}$ and $|\overline{X}| = \int_0^\infty t dE(t)$ the spectral resolution of $|\overline{X}|$. We put $X_n = \overline{X}E(n)$, $n \in \mathbb{N}$. Similarly to (2.1.1) we can show $X_n \in \mathcal{D}(\Lambda_\varphi^{cc})$, $n \in \mathbb{N}$ and $\lim_{n,m\to\infty} \Lambda_\varphi^{cc}(X_n) = \lambda_\varphi(X)$, and further by (3.5.11)

$$\begin{aligned}\lim_{n,m\to\infty}\|H\Lambda_\varphi^{cc}(X_n)-H\Lambda_\varphi^{cc}(X_m)\| &= \lim_{n,m\to\infty}\|H'\Lambda_\varphi^{cc}(X_n)-H'\Lambda_\varphi^{cc}(X_m)\|\\ &= \lim_{n,m\to\infty}\|(E(n)-E(m))H'\lambda_\varphi(X)\|\\ &=0,\end{aligned}$$

which implies $\lambda_\varphi(X)\in\mathcal{D}(H)$ and $\lim_{n\to\infty}H\Lambda_\varphi^{cc}(X_n)=H\lambda_\varphi(X)$. Hence we have

$$\begin{aligned}\psi(X^\dagger X)=\|H'\lambda_\varphi(X)\|^2 &= \lim_{n\to\infty}\|H'\Lambda_\varphi^{cc}(X_n)\|^2\\ &= \lim_{n\to\infty}\|H\Lambda_\varphi^{cc}(X_n)\|^2\\ &= \|H\lambda_\varphi(X)\|^2\end{aligned}$$

for each $X\in\mathfrak{N}_\varphi$.

(2) This is proved similarly to the proof of (1).

Proposition 3.5.9. Let φ be a standard positive linear functional on $\mathcal{M}$. Suppose ψ is a φ-absolutely continuous positive linear functional on $\mathcal{M}$ such that $\pi_{\varphi+\psi}(\mathcal{M})'_{\mathrm{w}}$ is a von Neumann algebra and

$$\psi(X^\dagger X)\le\sum_{k=1}^{n}\varphi(X^\dagger Y_k^\dagger Y_kX),\quad {}^\forall X\in\mathcal{M} \tag{3.5.12}$$

for some finite subset $\{Y_1,Y_2,\cdots,Y_n\}$ of $\mathcal{M}$. Then the following statements hold:

(1) If ψ satisfies the KMS-condition w.r.t. $\{\sigma_t^\varphi\}$, then there exists a positive self-adjoint operator H in $\mathcal{H}_\varphi$ affiliated with $(\pi_\varphi(\mathcal{M})'_{\mathrm{w}})'\cap\pi_\varphi(\mathcal{M})'_{\mathrm{w}}$ such that $\lambda_\varphi(I)\in\mathcal{D}(H)$, $H\lambda_\varphi(I)\in\mathcal{D}(\pi_\varphi)$ and $\psi(X)=(\pi_\varphi(X)H\lambda_\varphi(I)|H\lambda_\varphi(I))$ for each $X\in\mathcal{M}$.

(2) If ψ is $\{\sigma_t^\varphi\}$-invariant, then there exists a positive self-adjoint operator H in $\mathcal{H}_\varphi$ affiliated with $(\pi_\varphi(\mathcal{M})'_{\mathrm{w}})'^{\sigma^\varphi}$ such that $\lambda_\varphi(I)\in\mathcal{D}(H)$, $H\lambda_\varphi(I)\in\mathcal{D}(\pi_\varphi)$ and $\psi(X)=(\pi_\varphi(X)H\lambda_\varphi(I)|H\lambda_\varphi(I))$ for each $X\in\mathcal{M}$.

Proof. (1) By Theorem 3.5.2 ψ is represented as

$$\psi(X^\dagger X)=\|H'\lambda_\varphi(X)\|^2,\ X\in\mathcal{M}$$

for some positive self-adjoint operator H' in $\mathcal{H}_\varphi$ affiliated with $\pi_\varphi(\mathcal{M})'_{\mathrm{w}}$ such that $\lambda_\varphi(\mathcal{M})\subset\mathcal{D}(H')$, and further by (3.5.12) we have $\mathcal{D}(\pi_\varphi)\subset\mathcal{D}(H')$, and so $\lambda_\varphi(\mathcal{M})\subset\mathcal{D}(H'^2)$. Hence, since ψ is $\{\sigma_t^\varphi\}$-invariant, it follows that

$$H'^2\Delta_\varphi^{it}A\lambda_\varphi(I)=\Delta_\varphi^{it}H'^2A\lambda_\varphi(I),\quad {}^\forall A\in\pi_\varphi(\mathcal{M})'_{\mathrm{w}}\cap(\pi_\varphi(\mathcal{M})'_{\mathrm{w}})',\quad {}^\forall t\in\mathbb{R},$$

which implies since ψ is a $\{\sigma_t^\varphi\}$-KMS positive linear functional on $\mathcal{M}$ that the normal positive linear fuctional $\omega''_{H'^2\lambda_\varphi(I)}$ on the von Neumann algebra $(\pi_\varphi(\mathcal{M})'_{\mathrm{w}})'$ satisfies the KMS-condition with respect to $\{\sigma_t^\varphi\}$, so that the statement (1) is shown similarly to the proof of Proposition 3.5.7,(1).

(2) We can show similarly to the proof of the above (1) and Proposition 3.5.8, (2) that there exists a positive self-adjoint operator H' affiliated with $\pi_\varphi(\mathcal{M})_{\mathrm{w}}^{'\sigma^\varphi}$ such that $\lambda_\varphi(\mathcal{M}) \subset \mathcal{D}(H')$ and $\psi(X) = (\pi_\varphi(X)H'\lambda_\varphi(I)|H'\lambda_\varphi(I))$ for each $X \in \mathcal{M}$. We put $H = J_\varphi H' J_\varphi$. Then H satisfies our assertion in (2). This completes the proof.

We next consider the Radon-Nikodym theorem for positive linear functionals on O*-algebras with the von Neumann density type property .

Let $\mathcal{M}$ be a closed O*-algebra on $\mathcal{D}$ in $\mathcal{H}$ such that $\mathcal{M}'_{\mathrm{w}}\mathcal{D} \subset \mathcal{D}$. Suppose that the O*-algebra $\mathcal{M}$ has the von Neumann density type property, that is, $\overline{\mathcal{M}}^{\tau^*_{\sigma s}} = \mathcal{M}''_{\mathrm{w}\sigma}$. Let ξ_0 be a standard vector for $\mathcal{M}$ and put $\varphi_0 = \omega_{\xi_0}$. We denote by $\mathcal{P}^\natural_{(\mathcal{M}'_{\mathrm{w}})'\xi_0}$ the natural positive cone associated with the left Hilbert algebra $(\mathcal{M}'_{\mathrm{w}})'\xi_0$, that is, $\mathcal{P}^\natural_{(\mathcal{M}'_{\mathrm{w}})'\xi_0}$ is the closure of $\{AJ''_{\xi_0}AJ''_{\xi_0}\xi_0; A \in (\mathcal{M}'_{\mathrm{w}})'\}$, and denote $\mathcal{P}^\natural_{(\mathcal{M}'_{\mathrm{w}})'\xi_0} \cap \mathcal{D}$ by $\mathcal{P}^\natural_{\xi_0}$. Let $\mathcal{M}^+_*$ be the set of all σ-weakly continuous positive linear functionals on $\mathcal{M}$.

Lemma 3.5.10. (1) Any element ψ of $\mathcal{M}^+_*$ is represented as $\psi = \omega_{\xi_\psi}$ for the unique vector ξ_ψ in $\mathcal{P}^\natural_{\xi_0}$.

(2) Let ψ be any element of $\mathcal{M}^+_*$. Then $(\pi_{\varphi_0+\psi}(\mathcal{M})'_{\mathrm{w}})'$ is unitarily equivalent to $(\mathcal{M}'_{\mathrm{w}})'$, and so $\pi_{\varphi_0+\psi}(\mathcal{M})'_{\mathrm{w}}$ is a von Neumann algebra.

Proof. (1) By Lemma 5.2 in Inoue-Ueda-Yamauchi [1], ψ is represented as $\psi = \omega_\xi$ for some ξ in $\mathcal{D}$. Hence it follows from Theorem 10.25 in Stratila-Zsido [1] that

$$(A\xi|\xi) = (A\xi_\psi|\xi_\psi), \quad {}^\forall A \in (\mathcal{M}'_{\mathrm{w}})' \tag{3.5.13}$$

for the unique vector ξ_ψ in $\mathcal{P}^\natural_{(\mathcal{M}'_{\mathrm{w}})'\xi_0}$. Take an arbitray $X \in \mathcal{M}$. Let $|\overline{X}| = \int_0^\infty t dE(t)$ be the spectral resolution of $|\overline{X}|$ and $E_n = \int_0^n dE(t)$, $n \in \mathbb{N}$. Since $\mathcal{M}'_{\mathrm{w}}\mathcal{D} \subset \mathcal{D}$, it follows that $E_n, \overline{X}E_n \in (\mathcal{M}'_{\mathrm{w}})'$ for $n \in \mathbb{N}$, so that by (3.5.13)

$$\lim_{n\to\infty} E_n\xi_\psi = \xi_\psi,$$
$$\lim_{m,n\to\infty} \|\overline{X}E_m\xi_\psi - \overline{X}E_n\xi_\psi\| = \lim_{m,n\to\infty} \|\overline{X}E_m\xi - \overline{X}E_n\xi\| = 0.$$

Hence, $\xi_\psi \in \mathcal{D}$ and $\psi = \omega_{\xi_\psi}$. Suppose $\psi = \omega_{\xi_1} = \omega_{\xi_2}$ for $\xi_1, \xi_2 \in \mathcal{P}^\natural_{\xi_0}$. By $\overline{\mathcal{M}}^{\tau^*_{\sigma s}} = \mathcal{M}''_{\mathrm{w}\sigma}$ and (3.5.13), we have $\xi_1 = \xi_2$.

(2) By (1), $\varphi_0 + \psi$ is represented as

$$\varphi_0 + \psi = \omega_{\xi_{\varphi_0+\psi}} \tag{3.5.14}$$

for the unique vector $\xi_{\varphi_0+\psi} \in \mathcal{P}^\sharp_{\xi_0}$, which implies by $\overline{\mathcal{M}}^{\tau^*_{\sigma s}} = \mathcal{M}''_{w\sigma}$ that $(\pi_{\varphi_0+\psi}(\mathcal{M})'_w)'$ is unitarily equivalent to $(\mathcal{M}'_w)'$. This completes the proof.

Proposition 3.5.11. Let $\mathcal{M}$ be a closed O*-algebra on $\mathcal{D}$ in $\mathcal{H}$ such that $\mathcal{M}'_w\mathcal{D} \subset \mathcal{D}$ and $\overline{\mathcal{M}}^{\tau^*_{\sigma s}} = \mathcal{M}''_{w\sigma}, \xi_0$ a standard vector for $\mathcal{M}$ and $\psi \in \mathcal{M}^+_*$. Then ψ is φ_0-absolutely continuous if and only if there exists a positive self-adjoint operator H' affiliated with $\mathcal{M}'_w$ such that $\mathcal{M}\xi_0$ is a core for H' and $\psi = \omega_{H'\xi_0}$. If this is true, such an operator H' for ψ is unique and it is denoted by H'_ψ.

Proof. Suppose ψ is φ_0-absolutely continuous. By Lemma 3.5.10, $\psi = \omega_{\xi_\psi}$ for some $\xi_\psi \in \mathcal{P}^\sharp_{\xi_0}$. We denote by $K_{\varphi_0,\psi}$ the closure of a closable map: $X\xi_0 \to X\xi_\psi$, $X \in \mathcal{M}$. Then it follows from $\overline{\mathcal{M}}^{\tau^*_{\sigma s}} = \mathcal{M}''_{w\sigma}$ that $(\mathcal{M}'_w)'\xi_0 \subset \mathcal{D}(K_{\varphi_0,\psi})$ and $K_{\varphi_0,\psi}A\xi_0 = A\xi_\psi$ for all $A \in (\mathcal{M}'_w)'$, which implies that $(\mathcal{M}'_w)'\xi_0$ is a core for $K_{\varphi_0,\psi}$ and $K_{\varphi_0,\psi}$ is affiliated with $\mathcal{M}'_w$. We put $H' = (K^*_{\varphi_0,\psi}K_{\varphi_0,\psi})^{\frac{1}{2}}$. Then, H' is a positive self-adjoint operator affiliated with $\mathcal{M}'_w$ such that $\mathcal{M}\xi_0$ is a core for H' and $\psi = \omega_{H'\xi_0}$. The uniqueness of H' follows from that of the polar decomposition. The converse follows from Theorem 3.5.2. This completes the proof.

Proposition 3.5.12. Let $\mathcal{M}$ be a closed O*-algebra on $\mathcal{D}$ in $\mathcal{H}$ such that $\mathcal{M}'_w\mathcal{D} \subset \mathcal{D}$ and $\overline{\mathcal{M}}^{\tau^*_{\sigma s}} = \mathcal{M}''_{w\sigma}, \xi_0$ a standard vector for $\mathcal{M}$ and $\psi \in \mathcal{M}^+_*$. The following statements hold:

(1) ψ satisfies the KMS-condition with respect to $\{\sigma^{\xi_0}_t\}$ if and only if $\psi = \omega_{H\xi_0}$ for some positive self-adjoint operator H affiliated with $(\mathcal{M}'_w)' \cap \mathcal{M}'_w$ such that $\xi_0 \in \mathcal{D}(H)$ and $H\xi_0 \in \mathcal{D}$.

(2) ψ is $\{\sigma^{\xi_0}_t\}$-invariant if and only if $\psi = \omega_{H\xi_0}$ for some positive self-adjoint operator H affiliated with $(\mathcal{M}'_w)'^{\sigma^{\xi_0}}$ such that $\xi_0 \in \mathcal{D}(H)$ and $H\xi_0 \in \mathcal{D}$.

Proof. By Lemma 3.5.10, $\psi = \omega_{\xi_\psi}$ for some $\xi_\psi \in \mathcal{P}^\sharp_{\xi_0}$. Suppose ψ is a $\{\sigma^{\xi_0}_t\}$-KMS (resp. $\{\sigma^{\xi_0}_t\}$-invariant) positive linear functional on $\mathcal{M}$. Then, since $\overline{\mathcal{M}}^{\tau^*_{\sigma s}} = \mathcal{M}''_{w\sigma}$, it follows that the normal positive linear functional ω''_{ξ_ψ} on the von Neumann algebra $(\mathcal{M}'_w)'$ satifsies the KMS-condition with respect to $\{\sigma^{\xi_0}_t\}$ (resp. $\{\sigma^{\xi_0}_t\}$-invariant), so that by Theorem 15.4 in Takesaki [1] (resp. Theorem 15.2 in Takesaki [1]) there exists a positive self-adjoint operator H affiliated with $(\mathcal{M}'_w)' \cap \mathcal{M}'_w$ (resp. $(\mathcal{M}'_w)'^{\sigma^{\xi_0}}$) such that

$$\xi_0 \in \mathcal{D}(H) \text{ and } (A\xi_\psi|\xi_\psi) = (AH\xi_0|H\xi_0), \quad {}^\forall A \in (\mathcal{M}'_w)'. \tag{3.5.15}$$

We denote by U' the partial isometry of $\mathcal{H}$ defined by $A\xi_\psi \to AH\xi_0$, $A \in (\mathcal{M}'_w)'$. Then $U' \in \mathcal{M}'_w$, and so $H\xi_0 = U'\xi_\psi \in \mathcal{D}$. Further, since $\overline{X}$ is affiliated with $(\mathcal{M}'_w)'$ for each $X \in \mathcal{M}$, it follows from (3.5.15) that $\psi = \omega_{H\xi_0}$. This completes the proof.

We finally investigate the absolute continuity and the singularity of positive linear functionals on the O*-algebra generated by the differential operator, on the O*-algebra defined by the Schrödinger representation and on the maximal O*-algebra $\mathcal{L}^\dagger(\mathcal{S}(\mathbb{R}))$ on the Schwatz space $\mathcal{S}(\mathbb{R})$.

Example 3.5.13. We put

$$\mathcal{D} = \{\xi \in C^\infty[0,1];\ \xi^{(n)}(0) = \xi^{(n)}(1), n = 0,1,2,\cdots\},$$
$$X_0 = -i\frac{d}{dt}\lceil\mathcal{D},$$
$$\xi_0(t) = [\exp\{-\exp(-\frac{d^2}{dt^2})\}](5 - 4\cos 2\pi t)^{-1},\ t \in [0,1].$$

Then the polynomial algebra $\mathcal{P}(X_0)$ generated by X_0 is an integrable, commutative O*-algebra on $\mathcal{D}$ and ξ_0 is a standard, strongly cyclic vector for $\mathcal{P}(X_0)$ (Takesue [1]). We consider positive linear functionals on $\mathcal{P}(X_0)$ defined by

$$\varphi_a^b(p(X_0)) = (p(aX_0 + b)\xi_0|\xi_0),\ a \neq 0,\ b \in \mathbb{R}.$$

Then the following statements hold:

(1) For any $n \neq 0$, $m \in \mathbb{Z}$, $\varphi_n^{2\pi m}$ is ω_{ξ_0}-absolutely continuous.

(2) For any bounded subset B of $\mathbb{R}$ and $a \neq 0, b \in \mathbb{R}$ we define positive linear functionals on $\mathcal{P}(X_0)$ by

$$(\omega_{\xi_0} \circ \chi_B)(p(X_0)) = (\chi_B(X_0)p(X_0)\xi_0|\xi_0),$$
$$(\varphi_a^b \circ \chi_B)(p(X_0)) = (\chi_B(X_0)p(aX_0 + b)\xi_0|\xi_0).$$

Then $\varphi_a^b \circ \chi_B$ is $(\omega_{\xi_0} \circ \chi_B)$-singular for each $a \in \mathbb{Z}$ or $b \notin 2\pi\mathbb{Z}$.

We show the statement (1). By $\varphi_n^{2\pi m}$ is represented as

$$\varphi_n^{2\pi m}(p(X_0)) = (p(X_0)U\xi_0|U\xi_0)$$

for some $U \in \mathcal{L}^\dagger(\mathcal{D})_i \equiv \{U \in \mathcal{L}^\dagger(\mathcal{D});\ \overline{U}$ is an isometry $\}$. We put

$$(\varphi_n^{2\pi m})''(A) = (AU\xi_0|U\xi_0),\ A \in (\mathcal{P}(X_0)'_{\mathrm{w}})'.$$

Since $(\mathcal{P}(X_0)'_{\mathrm{w}})'$ is a commutative von Neumann algebra and ξ_0 is a cyclic tracial vector for $(\mathcal{P}(X_0)'_{\mathrm{w}})'$, it is easily shown that $(\varphi_n^{2\pi m})''$ is ω_{ξ_0}-absolutely continuous. Hence we have

$$(AU\xi_0|U\xi_0) = (AH'\xi_0|H'\xi_0),\ A \in (\mathcal{P}(X_0)'_{\mathrm{w}})'$$

for some positive self-adjoint operator H' in $L^2[0,1]$ affiliated with $\mathcal{P}(X_0)'_{\mathrm{w}}$, which implies

$$H'\xi_0 \in \mathcal{D} \text{ and } \varphi_n^{2\pi m}(p(X_0)) = (p(X_0)H'\xi_0|H'\xi_0).$$

Hence $\varphi_n^{2\pi m}$ is ω_{ξ_0}-absolutely continuous.

We next show the statement (2). For any polynomial p and $n \in \mathbb{N}$ we define a polynomial p_n by

$$p_n(t) = \sum_{k=1}^{2n+1} \alpha_k \{(t+2n\pi)(t+2(n-1))\pi)\cdots(t+2\pi)t(t-2\pi)\cdots(t-2n\pi)\}^k,$$

where $\{\alpha_1, \alpha_2, \cdots, \alpha_{2n+1}\}$ is the unique solution of the equation:

$$p_n(2m\pi a + b) = p(2m\pi a + b), \ m = -n, \cdots, -1, 0, 1, \cdots, n$$

(the existence of the unique solution dues to $a \neq 0$). Since B is a bounded subset of $\mathbb{R}$, it follows that

$$\begin{aligned}(\omega_{\xi_0} \circ \chi_B)(p_n(X_0)^\dagger p_n(X_0)) = 0,\\ (\varphi_a^b \circ \chi_B)((p_n(X_0) - p(X_0))^\dagger (p_n(X_0) - p(X_0))) = 0\end{aligned}$$

for sufficient large all $n \in \mathbb{N}$. Hence, $\varphi_a^b \circ \chi_B$ is $\omega_{\xi_0} \circ \chi_B$-singular.

Let $L^2 \otimes \overline{L^2}$, $\mathcal{S} \otimes \overline{L^2}$, $(\mathcal{S} \otimes \overline{L^2})_+$, s_+, $\Omega_{\{\alpha_n\}}$ ($\{\alpha_n\} \in s_+$) and $\{f_n\}$ be in Example 2.4.24. Let π be the self-adjoint representation of $\mathcal{L}^\dagger(\mathcal{S}(\mathbb{R}))$ in the Hilbert space $L^2 \otimes \overline{L^2}$ defined by

$$\pi(X)T = XT, \ X \in \mathcal{L}^\dagger(\mathcal{S}(\mathbb{R})), \ T \in \mathcal{S} \otimes \overline{L^2},$$

and π'' and π' *-representations of $\mathcal{B}(L^2(\mathbb{R}))$ on $L^2 \otimes \overline{L^2}$ defined by

$$\pi''(A)T = AT \text{ and } \pi'(A)T = TA, \ A \in \mathcal{B}(L^2(\mathbb{R})), \ T \in L^2 \otimes \overline{L^2}.$$

We here consider strongly positive linear functionals on $\mathcal{L}^\dagger(\mathcal{S}(\mathbb{R}))$ defined by

$$\begin{aligned}\varphi_\rho(X) &= tr\rho^2 X = <\pi(X)\rho|\rho>, \qquad (3.5.16)\\ \varphi_{\{\alpha_n\}} &\equiv \varphi_{\Omega_{\{\alpha_n\}}}, \ \{\alpha_n\} \in s_+.\end{aligned}$$

Since $\Omega_{\{\alpha_n\}}$ is a standard, strongly cyclic vector for $\pi(\mathcal{L}^\dagger(\mathcal{S}(\mathbb{R})))$ with $\Delta^{it}_{\Omega_{\{\alpha_n\}}} = \pi'(\Omega^{-2it}_{\{\alpha_n\}})\pi''(\Omega^{2it}_{\{\alpha_n\}})$ ($^\forall t \in \mathbb{R}$) for each $\{\alpha_n\} \in s_+$ (Example 2.4.24,(3)), it follows from (3.5.16) that $\varphi_{\{\alpha_n\}}$ is a standard, strongly positive linear functional on $\mathcal{L}^\dagger(\mathcal{S}(\mathbb{R}))$. Let $\mathcal{M}$ be a self-adjoint O*-algebra on $\mathcal{S}(\mathbb{R})$ defined by the Schrödinger representation of the canonical algebra for one degree of freedom. Since $\Omega_{\{e^{-n\beta}\}}$ is a standard, strongly cyclic vector for $\pi(\mathcal{M})$ for each $\beta > 0$ (Example 2.4.24,(5)), it follows that $\varphi_{\{e^{-n\beta}\}}$ is a standard, strongly positive linear functional on $\mathcal{M}$. In next Example 3.5.14 we consider the $\varphi_{\{e^{-n\beta}\}}$-absolute continuity, the $\varphi_{\{e^{-n\beta}\}}$-singularity and the $\{\sigma_t^{\varphi_{\{e^{-n\beta}\}}}\}$-invariance of strongly positive linear functionals on $\mathcal{M}$.

Example 3.5.14. Let $\rho \in (\mathcal{S} \otimes \overline{L^2})_+$, $\{\alpha_n\} \in s_+$ and $\beta > 0$.

(1) Suppose $\Omega^{-1}_{\{\alpha_n\}}\rho$ is densely defined. Then φ_ρ is a $\varphi_{\{\alpha_n\}}$-absolutely continuous, strongly positive linear functional on $\mathcal{M}$.

(2) $\varphi_{\{\alpha_n\}}$ is a $\{\sigma_t^{\varphi_{\{e^{-n\beta}\}}}\}$-invariant, strongly positive linear functional on $\mathcal{M}$. Conversely, suppose $\Omega^{-1}_{\{e^{-n\beta}\}}\rho$ is densely defined and $\overline{\rho^2 \Omega^{-1}_{\{e^{-n\beta}\}}} \in \mathcal{S} \otimes \overline{L^2}$ (in particular, φ_ρ is $\varphi_{\{e^{-n\beta}\}}$-dominated) and φ_ρ is a $\{\sigma_t^{\varphi_{\{e^{-n\beta}\}}}\}$-invariant positive linear functional on $\mathcal{M}$. Then $\varphi_\rho = \varphi_{\{\alpha_n\}}$ for some $\{\alpha_n\} \in s_+$.

(3) Every $\{\sigma_t^{\varphi_{\{e^{-n\beta}\}}}\}$-KMS positive linear functional φ on $\mathcal{M}$ is represented as $\varphi = \gamma\varphi_{\{e^{-n\beta}\}}$ for some constant $\gamma > 0$.

The statement (1) follows since φ_ρ is represented as

$$\varphi_\rho(X) = < \pi(X)|\pi'(\Omega^{-1}_{\{\alpha_n\}}\rho)|\Omega_{\{\alpha_n\}} \mid |\pi'(\Omega^{-1}_{\{\alpha_n\}}\rho)|\Omega_{\{\alpha_n\}} >, \; X \in \mathcal{M}$$

for a posiitve self-adjoint operator $|\pi'(\Omega^{-1}_{\{\alpha_n\}}\rho)|$ affiliated with $\pi'(\mathcal{B}(L^2(\mathbb{R})))$ such that $|\pi'(\Omega^{-1}_{\{\alpha_n\}}\rho)|\Omega_{\{\alpha_n\}} \in \mathcal{S} \otimes \overline{L^2}$.

We show the statement (2). It is clear that $\varphi_{\{\alpha_n\}}$ is a $\{\sigma_t^{\varphi_{\{e^{-n\beta}\}}}\}$-invariant, strongly positive linear functional on $\mathcal{M}$. We simply put $\Omega_\beta = \Omega_{\{e^{-n\beta}\}}$ and $\varphi_\beta = \varphi_{\{e^{-n\beta}\}}$. Suppose $\Omega_\beta^{-1}\rho$ is densely defined, $\overline{\rho^2\Omega_\beta^{-1}} \in \mathcal{S} \otimes \overline{L^2}$ and φ_ρ is $\{\sigma_t^{\varphi_\beta}\}$-invariant. We put

$$H_0 = (\Omega_\beta^{-1}\rho)(\Omega_\beta^{-1}\rho)^*.$$

Then $\pi'(H_0)$ is a positive self-adjoint operator in $L^2(\mathbb{R})$ affiliated with $\pi'(\mathcal{B}(L^2(\mathbb{R})))$. Since $\overline{\rho^2\Omega_\beta^{-1}} \in \mathcal{S} \otimes \overline{L^2}$ it follows that

$$\Omega_\beta \in \mathcal{D}(\pi'(H_0)) \text{ and } \pi'(H_0)\Omega_\beta = \overline{\rho^2\Omega_\beta^{-1}} \in \mathcal{S} \otimes \overline{L^2}, \qquad (3.5.17)$$

and hence

$$\begin{aligned} &\pi(\mathcal{M})\Omega_\beta \subset \mathcal{D}(\pi'(H_0)), \; \pi'(H_0)\pi(X)\Omega_\beta = \pi(X)\pi'(H_0)\Omega_\beta, \\ &\varphi_\rho(X) = < \pi(X)\pi'(H_0)\Omega_\beta|\Omega_\beta >, \; {}^\forall X \in \mathcal{M}. \end{aligned}$$

Since φ_ρ is $\{\sigma_t^{\varphi_\beta}\}$-invariant, it follows that

$$\begin{aligned} \varphi_\rho(Y^\dagger\sigma_t^{\varphi_\beta}(X)) &= < \pi(Y^\dagger\sigma_t^{\varphi_\beta}(X))\pi'(H_0)\Omega_\beta|\Omega_\beta > \\ &= < \pi'(H_0)\Delta^{it}_{\Omega_\beta}\pi(X)\Omega_\beta|\pi(Y)\Omega_\beta > \\ \varphi_\rho(Y^\dagger\sigma_t^{\varphi_\beta}(X)) &= \varphi_\rho(\sigma_t^{\varphi_\beta}(Y^\dagger)X) \\ &= < \Delta^{it}_{\Omega_\beta}\pi'(H_0)\pi(X)\Omega_\beta|\pi(Y)\Omega_\beta > \end{aligned}$$

for all $X, Y \in \mathcal{M}$ and $t \in \mathbb{R}$, which implies since $\pi''(\mathcal{B}(L^2(\mathbb{R})))\Omega_\beta \subset \mathcal{D}(\pi'(H_0))$ that

$$\begin{aligned} < \pi(X)\Omega_\beta | \Delta^{it}_{\Omega_\beta} \pi'(H_0)\pi''(A)\Omega_\beta > &=< \pi'(H_0)\Delta^{-it}_{\Omega_\beta} \pi(X)\Omega_\beta | \pi''(A)\Omega_\beta > \\ &=< \Delta^{-it}_{\Omega_\beta} \pi'(H_0)\pi(X)\Omega_\beta | \pi''(A)\Omega_\beta > \\ &=< \pi(X)\Omega_\beta | \pi'(H_0)\Delta^{it}_{\Omega_\beta} \pi''(A)\Omega_\beta > \end{aligned}$$

for all $A \in \mathcal{B}(L^2(\mathbb{R}))$, $X \in \mathcal{M}$ and $t \in \mathbb{R}$. Hence it follows since $\Delta^{it}_{\Omega_\beta} = \pi'(\Omega_\beta^{-2it})\pi''(\Omega_\beta^{2it})$, $^\forall t \in \mathbb{R}$ that

$$\pi''(\Omega_\beta^{2it})\pi'(\Omega_\beta^{-2it})\pi'(H_0)\pi''(A)\Omega_\beta = \pi'(H_0)\pi''(\Omega_\beta^{2it})\pi'(\Omega_\beta^{-2it})\pi''(A)\Omega_\beta$$

for all $A \in \mathcal{B}(L^2(\mathbb{R}))$ and $t \in \mathbb{R}$. We here put $A = f_n \otimes \overline{f_n}$, $n \in \mathbb{N} \cup \{0\}$. Then, since $\{f_k\} \subset \mathcal{D}(H_0)$, it follows that

$$\begin{aligned} e^{-2k\beta it}(H_0 f_k | f_n) f_n &= (f_n \otimes \overline{f_n}) H_0 \Omega_\beta^{-2it} f_k \\ &= (f_n \otimes \overline{f_n}) \Omega_\beta^{-2it} H_0 f_k \\ &= e^{-2n\beta it}(H_0 f_k | f_n) f_n, \end{aligned}$$

which implies

$$H_0 f_n = (H_0 f_n | f_n) f_n, \quad n \in \mathbb{N} \cup \{0\}.$$

By (3.5.17) we have

$$\{\alpha_n \equiv e^{-n\beta}(H_0 f_n | f_n)^{\frac{1}{2}}\} \in s_+ \text{ and } \varphi_\rho = \varphi_{\{\alpha_n\}}.$$

Suppose φ_ρ is φ_β-dominated. By Theorem 3.5.2, φ_ρ is represented as

$$\varphi_\rho(X) =< \pi(X)\pi'(H_0)\Omega_\beta | \pi'(H_0)\Omega_\beta >, \; X \in \mathcal{M}$$

for some positive self-adjoint operator H_0 in $\mathcal{B}(L^2(\mathbb{R}))$, and

$$\begin{aligned} \Omega_\beta^{-1}\pi'(H_0)\Omega_\beta &= H_0 \in \mathcal{B}(L^2(\mathbb{R})), \\ \overline{(\pi'(H_0)\Omega_\beta)^2 \Omega_\beta^{-1}} &= (\pi'(H_0)\Omega_\beta)H_0 \in \mathcal{S} \otimes \overline{L^2}. \end{aligned}$$

Hence, taking the above ρ to $\pi'(H_0)\Omega_\beta$, we can show that $\varphi_\rho = \varphi_{\pi'(H_0)\Omega_\beta} = \varphi_{\{\alpha_n\}}$ for some $\{\alpha_n\} \in s_+$. The statement (3) follows from Theorem 30 in Gudder-Hudson [1].

We finally give concreate examples of $\varphi_{\{e^{-n\beta}\}}$-singular positive linear functionals on $\mathcal{L}^\dagger(\mathcal{S}(\mathbb{R}))$ and of $\varphi_{\{e^{-n\beta}\}}$-absolutely continuous positive linear functionals on $\mathcal{L}^\dagger(\mathcal{S}(\mathbb{R}))$, and characterize $\{\sigma_t^{\varphi_{\{e^{-n\beta}\}}}\}$-invariant positive linear functionals on $\mathcal{L}^\dagger(\mathcal{S}(\mathbb{R}))$.

Example 3.5.15. (1) We put

$$f_\infty = \sum_{n=0}^{\infty} e^{-n\beta} f_n, \quad f'_\infty = 2f_0 - f_\infty.$$

Then $\varphi_{f_\infty \otimes \overline{f_\infty}}$ and $\varphi_{f'_\infty \otimes \overline{f'_\infty}}$ are $\varphi_{\{e^{-n\beta}\}}$-singular positive linear functionals on $\mathcal{L}^\dagger(\mathcal{S}(\mathbb{R}))$ and $\varphi_{f_\infty \otimes \overline{f_\infty}} + \varphi_{f'_\infty \otimes \overline{f'_\infty}}$ is not a $\varphi_{\{e^{-n\beta}\}}$-singular positive linear functional on $\mathcal{L}^\dagger(\mathcal{S}(\mathbb{R}))$.

(2) The $\varphi_{\{e^{-n\beta}\}}$-absolutely continuous positive linear functional $\varphi_{\{e^{-n\frac{\beta}{2}}\}}$ on $\mathcal{L}^\dagger(\mathcal{S}(\mathbb{R}))$ dominates a positive linear functional ψ on $\mathcal{L}^\dagger(\mathcal{S}(\mathbb{R}))$ which is not $\varphi_{\{e^{-n\beta}\}}$-absolutely continuous.

(3) The Lebesgue decomposition of $\varphi_{\{e^{-n\frac{\beta}{2}}\}}$ is not unique.

(4) Every $\varphi_{\{e^{-n\beta}\}}$-absolutely continuous and $\{\sigma_t^{\varphi_{\{e^{-n\beta}\}}}\}$-invariant, strongly positive linear functional φ on $\mathcal{L}^\dagger(\mathcal{S}(\mathbb{R}))$ is represented as $\varphi = \varphi_{\{\alpha_n\}}$ for some $\{\alpha_n\} \in s_+$.

(5) Every $\{\sigma_t^{\varphi_{\{e^{-n\beta}\}}}\}$-KMS strongly positive linear functional φ on $\mathcal{L}^\dagger(\mathcal{S}(\mathbb{R}))$ is represented as $\varphi = \gamma\varphi_{\{e^{-n\beta}\}}$ for some constant $\gamma > 0$.

We show the statement (1). We put $\Omega_\beta = \Omega_{\{e^{-n\beta}\}}$ and $\varphi_\beta = \varphi_{\{e^{-n\beta}\}}$. For any $X \in \mathcal{L}^\dagger(\mathcal{S}(\mathbb{R}))$ we put

$$X_m = \frac{1}{\log m}\sum_{k=1}^{m}\frac{1}{k}e^{\beta k}(Xf_\infty \otimes \overline{f_k}), \quad m = 2,3\cdots.$$

Then we have

$$X_m \in \mathcal{S} \otimes \overline{L^2}, \ \pi(X_m)\Omega_\beta = \frac{1}{\log m}\sum_{k=1}^{m}\frac{1}{k}(Xf_\infty|f_k),$$

$$\pi(X_m)(f_\infty \otimes \overline{f_\infty}) = (\frac{1}{\log m}\sum_{k=1}^{m}\frac{1}{k})\pi(X)(f_\infty \otimes \overline{f_\infty}), \ m = 2,3,\cdots.$$

Hence it follows that

$$\lim_{m\to\infty}\pi(X_m)\Omega_\beta = 0 \text{ and } \lim_{m\to\infty}\pi(X_m)(f_\infty \otimes \overline{f_\infty}) = \pi(X)(f_\infty \otimes \overline{f_\infty})$$

for each $X \in \mathcal{L}^\dagger(\mathcal{S}(\mathbb{R}))$, which means that $\varphi_{f_\infty \otimes \overline{f_\infty}}$ is φ_β-singular. Similarly, we can show that $\varphi_{f'_\infty \otimes \overline{f'_\infty}}$ is φ_β-singular. Since

$$(f_\infty \otimes \overline{f_\infty})^2 + (f'_\infty \otimes \overline{f'_\infty})^2 = \frac{e^{2\beta}}{e^{2\beta}-1}(f_\infty \otimes \overline{f_\infty} + f'_\infty \otimes \overline{f'_\infty}),$$

$$((f_\infty \otimes \overline{f_\infty})^2 + (f'_\infty \otimes \overline{f'_\infty})^2)(f_\infty + f'_\infty) = \frac{2e^{2\beta}}{e^{2\beta}-1}(f_\infty + f'_\infty),$$

$$((f_\infty \otimes \overline{f_\infty})^2 + (f'_\infty \otimes \overline{f'_\infty})^2)(f_\infty - f'_\infty) = \frac{2e^{2\beta}}{(e^{2\beta}-1)^2}(f_\infty - f'_\infty),$$

it follows that $f_\infty + f'_\infty = 2f_0$ and $f_\infty - f'_\infty$ are eigenvectors for $((f_\infty \otimes \overline{f_\infty})^2 + (f'_\infty \otimes \overline{f'_\infty})^2)$ with eigenvalues $2e^{2\beta}/e^{2\beta} - 1$ and $2e^{2\beta}/(e^{2\beta}-1)^2$, respectively, which implies

$$((f_\infty \otimes \overline{f_\infty})^2 + (f'_\infty \otimes \overline{f'_\infty})^2) \geq \frac{2e^{2\beta}}{e^{2\beta}-1}(f_0 \otimes \overline{f_0}).$$

Hence we have

$$\begin{aligned}(\varphi_{f_\infty \otimes \overline{f_\infty}} + \varphi_{f'_\infty \otimes \overline{f'_\infty}})(X^\dagger X) &= tr((f_\infty \otimes \overline{f_\infty})^2 + (f'_\infty \otimes \overline{f'_\infty})^2)X^\dagger X \\ &\geq \frac{2e^{2\beta}}{e^{2\beta}-1} tr(f_0 \otimes \overline{f_0})X^\dagger X \\ &= \frac{2e^{2\beta}}{e^{2\beta}-1}\varphi_{f_0 \otimes \overline{f_0}}(X^\dagger X)\end{aligned}$$

for all $X \in \mathcal{L}^\dagger(\mathcal{S}(\mathbb{R}))$, and hence $(2e^{2\beta}/e^{2\beta}-1)\varphi_{f_0 \otimes \overline{f_0}}$ is a non-zero positive linear functional on $\mathcal{L}^\dagger(\mathcal{S}(\mathbb{R}))$ which is dominated by both φ_β and $(\varphi_{f_\infty \otimes \overline{f_\infty}} + \varphi_{f'_\infty \otimes \overline{f'_\infty}})$, so that by Proposition 3.5.4 that $(\varphi_{f_\infty \otimes \overline{f_\infty}} + \varphi_{f'_\infty \otimes \overline{f'_\infty}})$ is not φ_β-singular.

We next show the statement (2). Let $\mathcal{H}_1$ be the closed subspace of $L^2(\mathbb{R})$ generated by $\{f_1, f_3, \cdots, f_{2n+1}, \cdots\}$ and P the projection of $L^2(\mathbb{R})$ onto $\mathcal{H}_1$. Since $\Omega_{\{e^{-n\frac{\beta}{2}}\}} P = P\Omega_{\{e^{-n\frac{\beta}{2}}\}}$ and it is a non-singular compact operator on $\mathcal{H}_1$, it follows from Lemma 8.8 in Kosaki [1] that there exists a unitary operator $\widetilde{U}$ on $\mathcal{H}_1$ such that

$$\text{Range}(\Omega_{\{e^{-n\frac{\beta}{2}}\}} P) \cap \widetilde{U}\text{Range}(\Omega_{\{e^{-n\frac{\beta}{2}}\}} P) = \{0\}.$$

We here put

$$\begin{aligned}&\rho = \Omega_{\{e^{-n\frac{\beta}{2}}\}} U \Omega_{\{e^{-n\frac{\beta}{2}}\}}, \text{ where } U = \widetilde{U}P + (1-P), \\ &\psi(X) = tr\rho\rho^* X, \quad X \in \mathcal{L}^\dagger(\mathcal{S}(\mathbb{R})).\end{aligned}$$

Since

$$\begin{aligned}\psi(X^\dagger X) &= \|\pi(X)\Omega_{\{e^{-n\frac{\beta}{2}}\}} U \Omega_{\{e^{-n\frac{\beta}{2}}\}}\|_2^2 \\ &= \|\pi'(U\Omega_{\{e^{-n\frac{\beta}{2}}\}})\pi(X)\Omega_{\{e^{-n\frac{\beta}{2}}\}}\|_2^2 \\ &\leq \|\pi'(U\Omega_{\{e^{-n\frac{\beta}{2}}\}})\|^2 \|\pi(X)\Omega_{\{e^{-n\frac{\beta}{2}}\}}\|_2^2 \\ &\leq \varphi_{\{e^{-n\frac{\beta}{2}}\}}(X^\dagger X)\end{aligned}$$

for all $X \in \mathcal{L}^\dagger(\mathcal{S}(\mathbb{R}))$, it follows that ψ is $\varphi_{\{e^{-n\frac{\beta}{2}}\}}$-dominated. Suppose ψ is φ-absolutely continuous. By Theorem 3.5.11 ψ is represented as

$$\psi(X) = < \pi(X)H'_\psi \Omega_\beta | H'_\psi \Omega_\beta >, \quad X \in \mathcal{L}^\dagger(\mathcal{S}(\mathbb{R})).$$

Hence, the positive linear functional ψ'' on $\mathcal{B}(L^2(\mathbb{R}))$ defined by

$$\psi''(A) = < \pi''(A)H'_\psi \Omega_\beta | H'_\psi \Omega_\beta >, \ A \in \mathcal{B}(L^2(\mathbb{R}))$$

is faithful and φ''_β-absolutely continuous, so that by Corollary 7.3 in Kosaki [1] that

$$\pi'(\mathcal{B}(L^2(\mathbb{R})))\Omega_\beta \cap \pi'(\mathcal{B}(L^2(\mathbb{R})))\rho \text{ is dense in } L^2 \otimes \overline{L^2}. \qquad (3.5.18)$$

Take an arbitrary $T \in \pi'(\mathcal{B}(L^2(\mathbb{R})))\Omega_\beta \cap \pi'(\mathcal{B}(L^2(\mathbb{R})))\rho$. Since $T = \pi'(A)\rho = \pi'(B)\Omega_\beta$ for $A, B \in \mathcal{B}(L^2(\mathbb{R}))$), we have

$$U\Omega_{\{e^{-n\frac{\beta}{2}}\}} A\xi = \Omega_{\{e^{-n\frac{\beta}{2}}\}} B\xi, \quad {}^\forall \xi \in L^2(\mathbb{R}),$$

which implies

$$\begin{aligned} P\Omega_{\{e^{-n\frac{\beta}{2}}\}} B\xi &= \tilde{U} P\Omega_{\{e^{-n\frac{\beta}{2}}\}} A\xi \\ &\in \mathrm{Range}(P\Omega_{\{e^{-n\frac{\beta}{2}}\}}) \cap \tilde{U}\mathrm{Range}(P\Omega_{\{e^{-n\frac{\beta}{2}}\}}) = \{0\}. \end{aligned}$$

Hence we have

$$PT\xi = P\Omega_\beta B\xi = \Omega_{\{e^{-n\frac{\beta}{2}}\}} P\Omega_{\{e^{-n\frac{\beta}{2}}\}} B\xi = 0$$

for each $\xi \in L^2(\mathbb{R})$, and so Range$(T) \subset (I - P)L^2(\mathbb{R})$, which contradicts (3.5.18). Hence ψ is not φ_β-absolutely continuous.

We show the statement (3). The φ_β-absolutely continuous positive linear functional $\varphi_{\{e^{-n\frac{\beta}{2}}\}}$ on $\mathcal{L}^\dagger(\mathcal{S}(\mathbb{R}))$ is decomposed into

$$\begin{aligned} \varphi_{\{e^{-n\frac{\beta}{2}}\}} &= \varphi_{\{e^{-n\frac{\beta}{2}}\}} + 0 \\ &= \{(\varphi_{\{e^{-n\frac{\beta}{2}}\}} - \psi) + \psi_c^{\varphi_\beta}\} + \psi_s^{\varphi_\beta}, \end{aligned}$$

where ψ is in (3). Since $\varphi_{\{e^{-n\frac{\beta}{2}}\}} - \psi \leq \varphi_\beta$ and $\psi_s^{\varphi_\beta} \neq 0$, it follows that $((\varphi_{\{e^{-n\frac{\beta}{2}}\}} - \psi) + \psi_c^{\varphi_\beta})$ is $\varphi_{\{e^{-n\frac{\beta}{2}}\}}$-dominated and φ_β-absolutely continuous, and $\psi_s^{\varphi_\beta}$ is non-zero and φ_β-singular, which shows that the Lebesque decomposition of $\varphi_{\{e^{-n\frac{\beta}{2}}\}}$ is not unique.

We show the statement (4). By Proposition 3.5.12,(2) φ is represented as

$$\varphi(X) = < \pi(X)H\Omega_\beta | H\Omega_\beta >, \ X \in \mathcal{L}^\dagger(\mathcal{S}(\mathbb{R}))$$

for some positive self-adjoint operator H in $L^2 \otimes \overline{L^2}$ affiliated with $\pi''(\mathcal{B}(L^2(\mathbb{R})))^{\sigma^{\varphi_\beta}}$ such that $\Omega_\beta \in \mathcal{D}(H)$ and $H\Omega_\beta \in \mathcal{S} \otimes \overline{L^2}$. It is easily shown that

$$\pi''(\mathcal{B}(L^2(\mathbb{R})))^{\sigma^{\varphi_\beta}} = \{\pi''(A);\ A = \sum_{n=0}^{\infty} \alpha_n f_n \otimes \overline{f_n} \in \mathcal{B}(L^2(\mathbb{R}))\}.$$

Hence we have

$$H_n = \sum_{k=0}^{\infty} \beta_k^{(n)} f_k \otimes \overline{f_k} \in \mathcal{B}(L^2(\mathbb{R})),\ n \in \mathbb{N} \text{ and } \lim_{n\to\infty} \pi''(H_n)\Omega_\beta = H\Omega_\beta,$$

which implies

$$\lim_{n\to\infty} \beta_k^{(n)} e^{-k\beta} = \alpha_k,\ k = 0, 1, 2, \cdots \text{ and } H\Omega_\beta = \sum_{k=0}^{\infty} \alpha_k f_k \otimes \overline{f_k} \in \mathcal{S} \otimes \overline{L^2}.$$

Hence we have $\{\alpha_k\} \in s_+$ and $\varphi = \varphi_{\{\alpha_k\}}$. We finally show the statement (5). By Proposition 3.5.12 φ is represented as

$$\varphi(X) = <\pi(X)H\Omega_\beta | H\Omega_\beta>,\ X \in \mathcal{L}^\dagger(\mathcal{S}(\mathbb{R}))$$

for some positive self-adjoint operator H affiliated with $\pi''(\mathcal{B}(L^2(\mathbb{R}))) \cap \pi'(\mathcal{B}(L^2(\mathbb{R})))$ such that $\Omega_\beta \in \mathcal{D}(H)$ and $H\Omega_\beta \in \mathcal{S} \otimes \overline{L^2}$. It is easily shown that $\pi''(\mathcal{B}(L^2(\mathbb{R}))) \cap \pi'(\mathcal{B}(L^2(\mathbb{R}))) = \mathbb{C}I$, which implies $\varphi = \gamma\varphi_\beta$ for some constant $\gamma > 0$.

3.6 Standard weights by vectors in Hilbert spaces

We first investigate the regularity, the singularity and the standardness of the quasi-weights ω_ξ defined by elements ξ of Hilbert space. Let $\mathcal{M}$ be a closed O*-algebra on $\mathcal{D}$ in $\mathcal{H}$ and put

$$\mathcal{D}^*(\mathcal{M}) = \bigcap_{X\in\mathcal{M}} \mathcal{D}(X^*) \text{ and } \mathcal{D}^{**}(\mathcal{M}) = \bigcap_{X\in\mathcal{M}} \mathcal{D}((X^* \lceil \mathcal{D}^*(\mathcal{M}))^*).$$

Suppose $\xi \in \mathcal{D}^{**}(\mathcal{M})$ and put

$$\omega_\xi(X) = (X^{\dagger *}\xi \mid \xi), \qquad X \in \mathcal{M}.$$

Then ω_ξ is a positive linear functional on $\mathcal{M}$. If $\xi \in \mathcal{D}^*(\mathcal{M}) \backslash \mathcal{D}^{**}(\mathcal{M})$, then ω_ξ is a linear functional on $\mathcal{M}$, but it is not necessarily positive. If $\xi \notin \mathcal{D}^*(\mathcal{M})$, then ω_ξ is not defined, and so we regard ω_ξ as the quasi-weight on $\mathcal{P}(\mathcal{M})$ as follows:

$$\begin{cases} \mathfrak{N}_{\omega_\xi} = \{X \in \mathcal{M}\ ;\ \xi \in \mathcal{D}(X^{\dagger *}) \text{ and } X^{\dagger *}\xi \in \mathcal{D}\}, \\ \omega_\xi(X^\dagger X) = \| X^{\dagger *}\xi \|^2, \qquad X \in \mathfrak{N}_{\omega_\xi}. \end{cases}$$

This coincides the quasi-weights φ_{λ_ξ} on $\mathcal{P}(\mathcal{M})$ induced by the generalized vector λ_ξ. We here investigate such quasi-weigts ω_ξ on $\mathcal{P}(\mathcal{M})$ in detail. We first investigate when a quasi-weight ω_ξ can be extended to a weight.

Proposition 3.6.1. Let $\mathcal{M}$ be a commutative integrable O*-algebra on $\mathcal{D}$ in $\mathcal{H}$ and $\xi \in \mathcal{H} \setminus \mathcal{D}$. We put

$$\widetilde{\omega_\xi}(\sum_k X_k^\dagger X_k) = \begin{cases} (\overline{\sum_k X_k^\dagger X_k}\,\xi|\xi) & \text{if } \xi \in \mathcal{D}(\overline{\sum_k X_k^\dagger X_k}) \\ \infty & \text{if otherwise.} \end{cases}$$

Then $\widetilde{\omega_\xi}$ is a weight on $\mathcal{P}(\mathcal{M})$ which is an extension of ω_ξ such that

$$\begin{aligned} \mathfrak{N}^0_{\widetilde{\omega_\xi}} &\equiv \{X \in \mathcal{M}\ ;\ \widetilde{\omega_\xi}(X^\dagger X) < \infty\} = \{X \in \mathcal{M}\ ;\ \xi \in \mathcal{D}(X^*\overline{X})\}, \\ \mathfrak{N}_{\widetilde{\omega_\xi}} &\equiv \{X \in \mathcal{M}\ ;\ AX \in \mathfrak{N}^0_{\widetilde{\omega_\xi}},\ {}^\forall A \in \mathcal{M}\} \\ &= \mathfrak{N}_{\omega_\xi}. \end{aligned}$$

Proof. Since $\mathcal{M}$ is commutative and integrable, it follows that $\xi \in \mathcal{D}(\overline{\sum_k X_k^\dagger X_k})$ if and only if there exists a sequence $\{\xi_n\}$ in $\mathcal{D}$ such that $\xi_n \to \xi$ and both $\{X_k\xi_n\}$ and $\{X_k^\dagger X_k\xi_n\}$ are Cauchy sequences in $\mathcal{H}$ for each k if and only if $\xi \in \mathcal{D}(\overline{X_k^\dagger X_k}) = \mathcal{D}(X_k^*\overline{X_k})$ for each k, and then $\overline{\sum_k X_k^\dagger X_k}\xi = \sum_k X_k^*\overline{X}_k\xi$, which implies that $\widetilde{\omega_\xi}$ is a weight on $\mathcal{P}(\mathcal{M})$. It is easy to show that $\mathfrak{N}_{\widetilde{\omega_\xi}} = \mathfrak{N}_{\omega_\xi}$ and $\widetilde{\omega_\xi}$ is an extension of ω_ξ.

Example 3.6.2. Let H be a positive self-adjoint unbounded operator in $\mathcal{H}$, $\mathcal{D}^\infty(H) \equiv \bigcap_{n\in\mathbb{N}} \mathcal{D}(H^n)$ and $H_0 \equiv H\lceil\mathcal{D}^\infty(H)$.

(1) The polynomial algebra $\mathcal{M} \equiv \mathcal{P}(H_0)$ is a commutative integrable O*-algebra on $\mathcal{D}^\infty(H)$ in $\mathcal{H}$ and the following statements hold:

(i) If $\xi \notin \mathcal{D}(H^2)$, then $\mathfrak{N}^0_{\widetilde{\omega_\xi}} = \mathbb{C}I$ and $\mathfrak{N}_{\widetilde{\omega_\xi}} = \mathfrak{N}_{\omega_\xi} = \{0\}$.

(ii) If $\xi \in \mathcal{D}(H^{2n}) \setminus \mathcal{D}(H^{2n+2})$ $(n \in \mathbb{N})$, then

$$\mathfrak{N}^0_{\widetilde{\omega_\xi}} = \{p(H_0)\ ;\ p \text{ is a polynomial with the degree } \leq n\},$$

$$\mathfrak{N}_{\widetilde{\omega_\xi}} = \mathfrak{N}_{\omega_\xi} = \{0\}.$$

(2) Let $\mathcal{M}$ be a commutative EW*-algebra on $\mathcal{D}^\infty(H)$ containing H_0 and $\xi \in \mathcal{H}\setminus\mathcal{D}^\infty(H)$. Suppose ξ is a cyclic vector for $(\mathcal{M}'_{\mathrm{w}})'$. Then $\mathfrak{N}^{\dagger *}_{\omega_\xi}\xi$ is dense in $\mathcal{D}^\infty(H)[t_\mathcal{M}]$.

As seen in Example 3.6.2, (1), there are the following cases:

(i) Even if $\xi \neq 0$, $\quad \mathfrak{N}_{\omega_\xi} = \{0\}$, that is, $\omega_\xi = 0$.

(ii) Even if $\widetilde{\omega_\xi} \neq 0$, $\quad (\widetilde{\omega_\xi})_q = \omega_\xi = 0$.

It is meaningless to consider such (quasi-)weights.

We investigate the regularity and the singularity of quasi-weights ω_ξ. For the singularity of ω_ξ we have the following

Proposition 3.6.3. Let $\xi \notin \mathcal{D}^*(\mathcal{M})$. Suppose that $\mathcal{M}'_{\rm w} = \mathbb{C}I$, $\mathfrak{N}^\dagger_{\omega_\xi}\mathcal{D}$ is dense in $\mathcal{H}$ and $\mathfrak{N}^{\dagger *}_{\omega_\xi}\xi$ is dense in $\mathcal{D}[t_\mathcal{M}]$. Then ω_ξ is singular.

Proof. Since $\mathfrak{N}^{\dagger *}_{\omega_\xi}\xi$ is dense in $\mathcal{D}[t_\mathcal{M}]$, $\pi_{\omega_\xi}(\mathcal{M})$ is unitarily equivalent to $\mathcal{M}$, that is, there exists a unitary operator U of $\mathcal{H}_{\omega_\xi}$ onto $\mathcal{H}$ such that $U\lambda_{\omega_\xi}(X) = X^{\dagger *}\xi$ for all $X \in \mathfrak{N}_{\omega_\xi}$ and $U\pi_{\omega_\xi}(A)U^* = A$ for all $A \in \mathcal{M}$. Take an arbitrary $K \in T(\omega_\xi)'_\delta$. Then there is a constant $\alpha \in \mathbb{C}$ such that $\alpha X^{\dagger *}\xi = X^{\dagger *}U\lambda'(K)$ for all $X \in \mathfrak{N}_{\omega_\xi}$. Since $\mathfrak{N}^\dagger_{\omega_\xi}\mathcal{D}$ is dense in $\mathcal{H}$, we have $\alpha\xi = U\lambda'(K) \in \mathcal{D}^*(\mathcal{M})$, and so $\alpha = 0$. Hence $K = 0$, which implies by Lemma 3.2.3 that ω_ξ is singular.

Since $\mathcal{L}^\dagger(\mathcal{D})$ satisfies all conditions of Proposition 3.6.3, we have the following

Corollary 3.6.4. For every $\xi \in \mathcal{H} \setminus \mathcal{D}^*(\mathcal{L}^\dagger(\mathcal{D}))$ ω_ξ is a singular quasi-weight on $\mathcal{P}(\mathcal{L}^\dagger(\mathcal{D}))$.

Proposition 3.6.5. Let $\mathcal{M}$ be a self-adjoint O*-algebra on $\mathcal{D}$ in $\mathcal{H}$ and $\xi \in \mathcal{H} \setminus \mathcal{D}$. Suppose $\mathfrak{N}^{\dagger *}_{\omega_\xi}\xi$ is dense in $\mathcal{D}[t_\mathcal{M}]$ and put

$$\mathcal{C}_\xi \equiv \{C \in \mathcal{M}'_{\rm w}; C\xi, C^*\xi \in \mathcal{D}\} = \mathcal{D}(\lambda_\xi^{\mathcal{C}})^* \cap \mathcal{D}(\lambda_\xi^{\mathcal{C}}),$$
$$P'_\xi = \text{Proj}\ \overline{\mathcal{C}_\xi\mathcal{H}},$$
$$\xi_r = P'_\xi\xi, \qquad \xi_s = (I - P'_\xi)\xi.$$

Then the following statements hold:

(1) $\omega_\xi \subset \omega_{\xi_r} + \omega_{\xi_s}$, ω_{ξ_r} is a regular quasi-weight on $\mathcal{P}(\mathcal{M})$ and ω_{ξ_s} is a singular quasi-weight on $\mathcal{P}(\mathcal{M})$. Further, $\pi_{\omega_{\xi_r}}$ (resp. $\pi_{\omega_{\xi_s}}$) is unitarily equivalent to $\pi_{\omega_\xi^{(r)}}$ (resp. $\pi_{\omega_\xi^{(s)}}$), where $\omega_\xi^{(r)}$ is the regular part of ω_ξ and $\omega_\xi^{(s)}$ is the singular part of ω_ξ in Theorem 3.2.7.

(2) ω_ξ is singular if and only if $\mathcal{C}_\xi = \{0\}$ if and only if $\xi_r = 0$, and ω_ξ is regular if and only if $\mathcal{C}_\xi$ is a nondegenerate $*$-subalgebra of $\mathcal{M}'_{\rm w}$ if and only if $\xi_s = 0$.

Proof. (1) By Theorem 3.2.7, the quasi-weight ω_ξ on $\mathcal{P}(\mathcal{M})$ is decomposed into $\omega_\xi = \omega_\xi^{(r)} + \omega_\xi^{(s)}$, where $\omega_\xi^{(r)}$ is a regular quasi-weight on $\mathcal{P}(\mathcal{M})$ and $\omega_\xi^{(s)}$ is a singular quasi-weight on $\mathcal{P}(\mathcal{M})$ with $\mathfrak{N}_{\omega_\xi^{(r)}} = \mathfrak{N}_{\omega_\xi^{(s)}} = \mathfrak{N}_{\omega_\xi}$ defined by

$$\omega_\xi^{(r)}(X^\dagger X) = \| P'_\xi X^{\dagger *}\xi \|^2, \quad \omega_\xi^{(s)}(X^\dagger X) = \| (I - P'_\xi)X^{\dagger *}\xi \|^2, \quad X \in \mathfrak{N}_{\omega_\xi}.$$

We have the relation that the quasi-weights ω_{ξ_r} and $\omega_\xi^{(r)}$ are equivalent ($\omega_{\xi_r} \sim \omega_\xi^{(r)}$), that is, $\pi_{\omega_{\xi_r}}$ and $\pi_{\omega_\xi^{(r)}}$ are unitarily equivalent. In fact, it is clear that

$\omega_\xi^{(r)} \subset \omega_{\xi_r}$, that is, $\mathfrak{N}_{\omega_\xi^{(r)}} (= \mathfrak{N}_{\omega_\xi}) \subset \mathfrak{N}_{\omega_{\xi_r}}$ and $\omega_\xi^{(r)}(X^\dagger X) = \omega_{\xi_r}(X^\dagger X)$ for all $X \in \mathfrak{N}_{\omega_\xi^{(r)}}$, and so $\pi_{\omega_\xi^{(r)}} \subset \pi_{\omega_{\xi_r}}$ unitarily and $\pi_{\omega_\xi^{(r)}}$ is self-adjoint. Hence $\pi_{\omega_\xi^{(r)}}$ is unitarily equivalent to $\pi_{\omega_{\xi_r}}$. Similarly, we have $\omega_{\xi^{(s)}} \sim \omega_{\xi_s}$. Thus ω_{ξ_r} is a regular quasi-weight on $\mathcal{P}(\mathcal{M})$ and ω_{ξ_s} is a singular quasi-weight on $\mathcal{P}(\mathcal{M})$ and $\omega_\xi = \omega_{\xi_r} + \omega_{\xi_s}$ on $\mathcal{P}(\mathfrak{N}_{\omega_\xi})$.

(2) This follows from Theorem 3.2.5.

Hence, we call ξ_r and ξ_s the *regular part* and the *singular part* of ξ, respectively.

Corollary 3.6.6. Let H be a positive self-adjoint unbounded operator in $\mathcal{H}$, $\mathcal{M}$ an O*-algebra on $\mathcal{D}^\infty(H)$ containing

$$\{f(H)\lceil \mathcal{D}^\infty(H); f \text{ is a measurable function on } \mathbb{R}_+ \text{ such that } |f(t)| \leq p(t),\ t \in \mathbb{R}_+ \text{ for some polynomial } p\},$$

and $\xi \in \mathcal{H} \setminus \mathcal{D}^\infty(H)$. Suppose $\mathfrak{N}_{\omega_\xi}^{\dagger *}\xi$ is dense in $\mathcal{H}$. Then the following statements hold:

(1) Suppose $\mathcal{M}'_{\rm w} = \mathbb{C}I$. Then ω_ξ is singular.

(2) Suppose $\mathcal{M}$ is commutative. Then ω_ξ is regular.

Proof. It is easily shown that $\mathcal{M}$ is self-adjoint and $\mathfrak{N}_{\omega_\xi}^{\dagger *}\xi$ is dense in $\mathcal{D}^\infty(H)[t_\mathcal{M}]$ using the spectral resolution of H.

(1) Since $\mathcal{M}'_{\rm w} = \mathbb{C}I$, we have $\mathcal{C}_\xi = \{0\}$ and hence by Theorem 3.6.5 ω_ξ is singular.

(2) Let $H = \int_0^\infty t\,dE_H(t)$ be the spectral resolution of H. Since $\mathcal{M}$ is commutative, we have $\{E_H(n); n \in \mathbb{N}\} \subset \mathcal{C}_\xi$, and hence $\mathcal{C}_\xi$ is a nondegenerate $*$-subalgebra of $\mathcal{M}'_{\rm w}$. Therefore, by Theorem 3.6.5 ω_ξ is regular.

Example 3.6.7. Let $\mathcal{A}$ be the unbounded CCR-algebra for one degree of freedom and π_0 the Schrödinger representation of $\mathcal{A}$. Then $\pi_0(\mathcal{A})$ is a self-adjoint O*-algebra on $\mathcal{S}(\mathbb{R})$ satisfying $\pi_0(\mathcal{A})'_{\rm w} = \mathbb{C}I$. Let $\mathcal{M}$ be the O*-algebra on $\mathcal{S}(\mathbb{R})$ generated by $\pi_0(\mathcal{A})$ and $\{f(N)\ ;\ f$ is a real-valued continuous function on $\mathbb{R}_+$ such that $|f(t)| \leq p(t)$ $(t \in \mathbb{R}_+)$ for some polynomial $p\}$, and $\xi \in L^2(\mathbb{R}) \setminus \mathcal{S}(\mathbb{R})$. Suppose $\mathfrak{N}_{\omega_\xi}^{\dagger *}\xi$ is dense in $L^2(\mathbb{R})$. Then $\mathcal{S}(\mathbb{R}) = \bigcap_{k=1}^{\infty} \mathcal{D}(N^k)$, where N is the number operator, and $\mathcal{M}$ and ξ satisfy all of the conditions of Corollary 3.6.6, (1). Hence ω_ξ is singular.

We investigate the standardness of the quasi-weights ω_ξ. The standardness of the quasi-weight ω_ξ and the generalized vector λ_ξ is equivalent, and so by Proposition 2.4.1 we have the following

Proposition 3.6.8. Suppose

$(S)_1$ $\{YX^{\dagger *}\xi \ ; \ X, Y \in \mathfrak{N}^{\dagger}_{\omega_\xi} \cap \mathfrak{N}_{\omega_\xi}\}$ is total in $\mathcal{H}$,

$(S)_2$ $\mathcal{C}_\xi \xi$ is dense in $\mathcal{H}$.

Then ω_ξ is a faithful regular quasi-weight on $\mathcal{P}(\mathcal{M})$ and ξ is a cyclic and separating vector for the von Neumann algebra $(\mathcal{M}'_{\mathrm{w}})'$ and denote by Δ''_ξ the modular operator for the left Hilbert algebra $(\mathcal{M}'_{\mathrm{w}})'\xi$. Further, we have the following results:

(1) ω_ξ is quasi-standard if and only if the above $(S)_1$, $(S)_2$ and the following condition $(S)_3$ hold:

$(S)_3$ $\Delta''^{it}_{\xi}\mathcal{D} \subset \mathcal{D}$ for each $t \in \mathbb{R}$.

(2) ω_ξ is standard if and only if the above conditions $(S)_1$, $(S)_2$ and $(S)_3$ and the following condition $(S)_4$ hold:

$(S)_4$ $\Delta''^{it}_{\xi}\mathcal{M}\Delta''^{-it}_{\xi} = \mathcal{M}$ for each $t \in \mathbb{R}$.

We next give some examples of regular quasi-weights , singular quasi-weights and standard quasi-weights defined in the Hilbert space of Hilbert-Schmidt operators, which are important for the quantum physics.

Let $\mathcal{M}$ be a self-adjoint O*-algebra on $\mathcal{D}$ in $\mathcal{H}$ such that $\mathcal{M}'_{\mathrm{w}} = \mathbb{C}I$. We put

$$\begin{cases} \mathfrak{S}_2(\mathcal{M}) = \{T \in \mathcal{H} \otimes \overline{\mathcal{H}} \ ; \ T\mathcal{H} \subset \mathcal{D} \text{ and } XT \in \mathcal{H} \otimes \overline{\mathcal{H}}, \ {}^\forall X \in \mathcal{M}\}, \\ \pi(X)T = XT, \quad X \in \mathcal{M}, \quad T \in \mathfrak{S}_2(\mathcal{M}). \end{cases}$$

Then π is a self-adjoint representation of $\mathcal{M}$ on $\mathfrak{S}_2(\mathcal{M})$ in $\mathcal{H} \otimes \overline{\mathcal{H}}$ such that $\pi(\mathcal{M})'_{\mathrm{w}} = \pi'(\mathcal{B}(\mathcal{H}))$ and $(\pi(\mathcal{M})'_{\mathrm{w}})' = \pi''(\mathcal{B}(\mathcal{H}))$, where $\pi'(A)T = TA$ and $\pi''(A)T = AT$ for $A \in \mathcal{B}(\mathcal{H})$ and $T \in \mathcal{H} \otimes \overline{\mathcal{H}}$ (Lemma 2.4.14). Let $\Omega \in \mathcal{H} \otimes \overline{\mathcal{H}}$ and $\Omega \geq 0$. We define the quasi-weight φ_Ω on $\mathcal{P}(\mathcal{M})$ by

$$\begin{cases} \mathfrak{N}_{\varphi_\Omega} = \{X \in \mathcal{M}; \ \pi(X) \in \mathfrak{N}_{\omega_\Omega}\}, \\ \varphi_\Omega(X^\dagger X) = tr(X^{\dagger *}\Omega)^*(X^{\dagger *}\Omega) = \omega_\Omega(\pi(X)^\dagger \pi(X)), \quad X \in \mathfrak{N}_{\varphi_\Omega}. \end{cases}$$

Then we have

$$\begin{cases} \mathfrak{N}_{\omega_\Omega} = \pi(\mathfrak{N}_{\varphi_\Omega}) \\ \varphi_\Omega(X^\dagger X) = \omega_\Omega(\pi(X)^\dagger \pi(X)), \quad X \in \mathfrak{N}_{\varphi_\Omega}. \end{cases}$$

Hence, by Proposition 3.6.5 we have the following

Proposition 3.6.9. Suppose $\mathfrak{N}^{\dagger *}_{\varphi_\Omega}\Omega$ is dense in $\mathfrak{S}_2(\mathcal{M})[t_{\pi(\mathcal{M})}]$. Then the following statements hold:

(1) φ_Ω is singular if and only if

$$\mathcal{C}_\Omega \equiv \{\pi'(K); K \in \mathcal{B}(\mathcal{H}) \text{ and } \Omega K, \Omega K^* \in \mathfrak{S}_2(\mathcal{M})\} = \{0\}.$$

(2) φ_Ω is regular if and only if $\mathcal{C}_\Omega$ is a nondegenerate $*$-subalgebra of $\pi'(\mathcal{B}(\mathcal{H}))$.

(3) Ω is decomposed into $\Omega = \Omega_r + \Omega_s$, where Ω_r is the regular part of Ω and Ω_s is the singular part of Ω. Hence, φ_{Ω_r} is a regular quasi-weight on $\mathcal{P}(\mathcal{M})$, φ_{Ω_s} is a singular quasi-weight on $\mathcal{P}(\mathcal{M})$ and $\varphi_\Omega = \varphi_{\Omega_r} + \varphi_{\Omega_s}$ on $\mathcal{P}(\mathfrak{N}_{\varphi_\Omega})$.

By Theorem 2.4.18 we have the following

Proposition 3.6.10. Suppose there exists a dense subspace $\mathcal{E}$ in $\mathcal{D}[t_\mathcal{M}]$ such that
(i) $\mathcal{M} \supset \{\xi \otimes \overline{\eta}\ ;\ \xi, \eta \in \mathcal{E}\}$,
(ii) $\Omega\mathcal{E} \subset \mathcal{D}$ and $\Omega\mathcal{E}$ is dense in $\mathcal{H}$.
(iii) Ω^{-1} is densely defined,
(iv) $\Omega^{it}\mathcal{D} \subset \mathcal{D}$ for each $t \in \mathbb{R}$.
Then φ_Ω is a quasi-standard quasi-weight on $\mathcal{P}(\mathcal{M})$. Further, suppose
(v) $\Omega^{it}\mathcal{M}\Omega^{-it} = \mathcal{M}$ for each $t \in \mathbb{R}$.
Then φ_Ω is a standard quasi-weight on $\mathcal{P}(\mathcal{M})$.

Let Ω be a positive self-adjoint unbounded operator in $\mathcal{H}$. We define a quasi-weight on $\mathcal{P}(\mathcal{M})$ by

$$\begin{cases} \mathfrak{N}_{\varphi_\Omega} = \{X \in \mathcal{M}; \overline{X^{\dagger *}\Omega} \in \mathfrak{S}_2(\mathcal{M})\}, \\ \varphi_\Omega(X^\dagger X) = tr(X^{\dagger *}\Omega)^*(\overline{X^{\dagger *}\Omega}), \quad X \in \mathfrak{N}_{\varphi_\Omega}. \end{cases}$$

By Theorem 2.4.23 we have the following

Proposition 3.6.11. Suppose there exists a subspace $\mathcal{E}$ of $\mathcal{D} \cap \mathcal{D}(\Omega)$ such that
(i) $\mathcal{E}$ is dense in $\mathcal{D}[t_\mathcal{M}]$,
(ii) $\mathcal{M} \supset \{\xi \otimes \overline{\eta}; \xi, \eta \in \mathcal{E}\}$, (iii) $\Omega\mathcal{E} \subset \mathcal{D}$ and $\Omega\mathcal{E}$ is dense in $\mathcal{H}$,
(iv) Ω^{-1} is densely defined and $\mathcal{D} \cap \mathcal{D}(\Omega^{-1})$ is a core for Ω^{-1},
(v) Suppose $\Omega^{it}\mathcal{D} \subset \mathcal{D}$ for all $t \in \mathbb{R}$.
Then φ_Ω is a quasi-standard quasi-weight on $\mathcal{P}(\mathcal{M})$. Further, suppose
(vi) Suppose $\Omega^{it}\mathcal{M}\Omega^{-it} = \mathcal{M}$ for all $t \in \mathbb{R}$.
Then φ_Ω is a standard quasi-weight on $\mathcal{P}(\mathcal{M})$.

Notes

3.1, 3.2. The works of these sections are due to Inoue-Ogi [1].
3.3, 3.4. These are due to Inoue-Karwowski-Ogi [1].

3.5. Definition 3.5.1 $\sim$ Proposition 3.5.12 in the first part of this section are generalization of the results obtained in Inoue [10] for positive linear functionals to (quasi-)weights. Examples 3.5.13, 3.5.14, 3.5.15 in the larst part are due to §6 in Inoue [10].

3.6. The works of this section are due to Inoue-Ogi [1].

4. Physical Applications

In this chapter we apply the Tomita-Takesaki theory in O*-algebras studied in Chapter II, III to quantum statistical mechanics and the Wightman quantum field theory.

Given a physical system, the first task of quantum statistical mechanics is to try and construct equilibrium states of the system. In the traditional algebraic formulation, the system is characterized by the algebra $\mathcal{A}$ of its observables, usually taken as an algebra of bounded operators. The latter in turn may be obtained by applying the well-known GNS construction defined by a state on some abstract $*$-algebra. Then the standard treatment of the basic problem consists in applying to $\mathcal{A}$ the Tomita-Takesaki theory of modular automorphisms, which yields states on $\mathcal{A}$ that satisfy the KMS condition. The latter is a characteristic of equilibrium: Gibbs states satisfy the KMS condition. For finite systems, the converse is also true, whereas, for infinite systems, the KMS condition characterizes only the local thermodynamical stability. For many models, the equality between the sets of KMS states and Gibbs equilibrium states persists also after the thermodynamical limit. This fact suggests the general interpretation of KMS states as equilibrium states in the Gibbs formulation, at least if the system is described as a C*- or W*-dynamical system. However, there are systems for which the standard approach fails, typically spin systems with long range interactions such as the BCS model of superconductivity and its relatives. For such systems, indeed, the thermodynamic limit does not converge in any norm topology. An elegant way of circumventing the difficulty consists in taking for observable algebra, namely an O*-algebra on some dense invariant domain $\mathcal{D}$ in the Hilbert space $\mathcal{H}$. The same technique may be applied when unbounded observables are considered, such as position and momentum in the CCR algebra (then $\mathcal{H} = L^2(\mathbb{R}^3)$ and $\mathcal{D}$ is Schwartz space $\mathcal{S}(\mathbb{R}^3)$). Since the examples presented in this chapter are of that nature, we will adopt the O*-approach. This means that the observables of the system (either local or in the thermodynamical limit) are represented by the elements of an O*-algebra $\mathcal{M}$. Thus we are facing the same question as before: how does one construct KMS states on an O*-algebra? It is important to consider a Tomita-Takesaki theory in O*-algebras. The Tomita-Takesaki theory may be derived for $\mathcal{M}$ if, among other conditions, $\mathcal{M}$ possesses a strongly cyclic vector $\xi_o \in \mathcal{D}$. In that case, one obtains states on $\mathcal{M}$ (in the usual sense) that satisfy the KMS condition. However, the existence of the cyclic vector is a rather restrictive condition as seen in late physical examples, that we want to avoid.

An interesting possibility is to consider it as a *generalized vector*. The main advantage of this interpretation is that generalized vectors (vectors in $\mathcal{H} \setminus \mathcal{D}$ are only the simplest case) are also closely related to the concept of *weights* and *quasi-weights* on O*-algebras. For a system whose observable algebra is assumed to be an O*-algebra $\mathcal{M}$, we will be able to show the existence of quasi-weights on $\mathcal{M}$ satisfying the KMS condition. In view of the discussion above, in a generalized setup where physical states of the system would be represented by quasi-weights on the algebra of obsevables, it is plausible that these KMS quasi-weights would represent equilibrium states.

This chapter is organized as follows:

In Section 4.1 we consider the quantum moment problem for states on an O*-algebras. Many important examples of states f in quantum physics are trace functionals, that is, they are of the form $f(X) = \operatorname{tr}\overline{TX}$ with a certain trace operator $T \in \mathfrak{S}_1(\mathcal{M})$. Hence we study the following quantum moment problem: Under what conditions is every strongly positive linear functional on an O*-algebra a trace functional? The results of this section are applied to Section 4.2, 4.3 and 4.4.

In Section 4.2 we extend some results obtained for states in Section 4.1 to (quasi-) weights.

In Section 4.3 we first study standard systems in unbounded CCR-algebras in one degree of freedom. Let $\mathcal{A}$ be a $*$-algebra generated by identity 1 and hermitian elements p and q satisfying the Heisenberg commutation relation: $[p,q] = -i1$. For a self-adjoint representation π of $\mathcal{A}$ satisfying $\pi(p)$ and $\pi(q)$ are essentially self-adjoint, Von Neumann [1] had the result that the strongly continous unitary groups $U(s) \equiv e^{is\overline{\pi(p)}}$ and $V(t) \equiv e^{it\overline{\pi(q)}}$ satisfy the Weyl commutation relation: $U(s)V(t) = e^{ist}V(t)U(s)$ for each $s,t \in \mathbb{R}$ if and only if π is a direct sum $\bigoplus_{\alpha \in I_0} \pi_\alpha$ of $*$-representations π_α which are unitary equivalent to the Schrödinger representation π_0. We call such a $*$-representation the Weyl representation of the cardinal I_0. Powers [1] defined the notion of strong positivity of self-adjoint representations of $\mathcal{A}$ and using it he characterized the Weyl representations. Here we introduce the Powers results. Further, we show that a Weyl representation of countable cardinal is unitarily equivalent to the self-adjoint representation $\pi_\otimes$ of $\mathcal{A}$ defined by $\mathcal{D}(\pi_\otimes) = \mathcal{S}(\mathbb{R}) \otimes \overline{L^2(\mathbb{R})}$ and $\pi_\otimes(a)T = \pi_0(a)T$ for $a \in \mathcal{A}$ and $T \in \mathcal{D}(\pi_\otimes)$, and $\Omega_\beta \equiv \sum_{n=0}^{\infty} e^{-\frac{n\beta}{2}} f_n \otimes \overline{f_n}$ $(\beta > 0)$ is a standard vector for $\pi_\otimes(\mathcal{A})$, where $\{f_n\}_{n=0,1,\cdots}$ is an ONB in $L^2(\mathbb{R})$ consisting of the normalized Hermite functions. We consider more general $\Omega_{\{\alpha_n\}} \equiv \sum_{n=0}^{\infty} \alpha_n f_n \otimes \overline{f_n}$ $(\alpha_n > 0, n = 0,1,2,\cdots)$. Let $\mathcal{M}$ be an O*-algebra on $\mathcal{S}(\mathbb{R})$ generated by $\pi_0(\mathcal{A})$

and $f_0 \otimes \overline{f_0}$. Then the positive self-adjoint operator $\Omega_{\{\alpha_n\}}$ defines a quasi-standard generalized vector $\lambda_{\Omega_{\{\alpha_n\}}}$ for the self-adjoint O*-algebra $\pi(\mathcal{M})$ on $\mathcal{S}(\mathbb{R}) \otimes \overline{L^2(\mathbb{R})}$ defined by $\pi(X)T = XT$ for $X \in \mathcal{M}$ and $T \in \mathcal{S}(\mathbb{R}) \otimes \overline{L^2(\mathbb{R})}$. Further $\lambda_{\Omega_{\{\alpha_n\}}}$ is standard if and only if $\alpha_n = e^{\beta n}, n \in \mathbb{N} \cup \{0\}$ for some $\beta \in \mathbb{R}$. We next give a standard generalized vector and a modular generalized vector in an interacting Boson model.

In Section 4.4 we study standard systems in the BCS-Bogoluvov model. In case of the BCS model, a rigorous algebraic description, in the quasi-spin formulation, was given long ago by Thirring-Wehrl [1,2]. Using this formulation Lassner [2,3] solved the problem of the thermodynamical limit discussed above by constructing a rather complicated topological quasi *-algebra. We show here that the existence of KMS quasi-weights may be obtained with a much simpler O*-algebra, provided one uses appropriate generalized vectors, as described in Chapter II, III.

In Section 4.5 we study standard systems in the Wightman quantum field theory. The general theory of quantum fields has been developed along two main lines: One is based on the Wightman axioms and makes use of unbounded field operators, and the other is the theory of local nets of bounded observables initiated by Haag-Kastler [1] and Araki [1]. Here we characterize the passage from a Wightman field to a local net of von Neumann algebras by the existence of standard systems obtained from the right wedge-region in Minkowski space.

4.1 Quantum moment problem I

In this section we consider under what conditions every strongly positive linear functional on an O*-algebra is a trace functional. This problem is closely related to the so-called problem of moments, and so we call it the *quantum moment problem* .

Throughout this section let $\mathcal{M}$ be a closed O*-algebra on $\mathcal{D}$ in $\mathcal{H}$ with the identity operator I. We denote by $\mathcal{F}(\mathcal{H})$ the $*$-invariant subspace of $\mathcal{B}(\mathcal{H})$ consisting of finite dimensional operators on $\mathcal{H}$, and by $\mathcal{M}_{\mathcal{F}(\mathcal{H})}$ the linear span of $\mathcal{M}$ and $\mathcal{F}(\mathcal{H})\lceil\mathcal{D}$ regarded as operators on $\mathcal{D}$. We first investigate under which conditions a *continuous* linear functional f on $\mathcal{M}$ is a trace functional, that is, $f(X) = \text{tr}\ XT, X \in \mathcal{M}$ for some $T \in \mathfrak{S}_1(\mathcal{M})$.

We prepare the following lemma:

Lemma 4.1.1. Suppose f is a strongly positive linear functional on $\mathcal{M}_{\mathcal{F}(\mathcal{H})}$. Then there exists an element T of ${}_1\mathfrak{S}(\mathcal{M})_+$ such that $f(A) = \text{tr}\ TA$ for all $A \in \mathcal{F}(\mathcal{H})$.

Proof. Since $x \otimes \overline{x} \leq \|x\|^2 I$ for each $x \in \mathcal{H}$, it follows from the strong positivity of f that $f(x \otimes \overline{x}) \leq \|x\|^2 f(I)$ for all $x \in \mathcal{H}$, so that by the Schwartz inequality

$$\begin{aligned}|f(x \otimes \overline{y})|^2 &\leq f(I) f((x \otimes \overline{y})^\dagger (x \otimes \overline{y})) \\ &= f(I)\|x\|^2 f(y \otimes \overline{y}) \\ &\leq f(I)^2 \|x\|^2 \|y\|^2\end{aligned}$$

for all $x, y \in \mathcal{H}$. Hence, $(x, y) \to f(x \otimes \overline{y})$ is a continuous positive sesquilinear form on $\mathcal{H} \oplus \mathcal{H}$, and so there exists a positive bounded operator T on $\mathcal{H}$ such that

$$f(x \otimes \overline{y}) = (Tx|y) \tag{4.1.1}$$

for all $x, y \in \mathcal{H}$. Since

$$\begin{aligned}|(X\xi|Ty)|^2 &= |f(X(\xi \otimes \overline{y}))|^2 \\ &\leq f(XX^\dagger) f((\xi \otimes \overline{y})^\dagger (\xi \otimes \overline{y})) \\ &\leq f(XX^\dagger)\|\xi\|^2 f(y \otimes \overline{y}) \\ &\leq f(XX^\dagger) f(I)\|y\|^2 \|\xi\|^2\end{aligned}$$

for all $X \in \mathcal{M}, \xi \in \mathcal{D}$ and $y \in \mathcal{H}$, we have $T\mathcal{H} \subset \mathcal{D}^*(\mathcal{M})$, and so it follows from the closed graph theorem that $X^*T \in \mathcal{B}(\mathcal{H})$ for all $X \in \mathcal{M}$. We show that $X^*T \in \mathfrak{S}_1(\mathcal{H})$ for all $X \in \mathcal{M}$. Let $\{x_i\}_{i \in F}$ be an orthonormal subset in $\mathcal{H}$ and $\{i_1, i_2, \cdots, i_n\}$ any finite subset F. For each $k \in \{1, 2, \cdots, n\}$ there exists $\xi_{i_k} \in \mathcal{D}$ such that

$$\|x_{i_k} - \xi_{i_k}\| < \frac{1}{2^k \|X^*T\|}. \tag{4.1.2}$$

Further, we take a number $\alpha_k \in \mathbb{C}$ such that $|\alpha_k| = 1$ and $|(X^*T\xi_{i_k}|\xi_{i_k})| = \alpha_k(X^*T\xi_{i_k}|\xi_{i_k})$. Then we have

$$\begin{aligned}\|\sum_{k=1}^n \alpha_k(\xi_{i_k} \otimes \overline{\xi_{i_k}}) - \sum_{k=1}^n \alpha_k(x_{i_k} \otimes \overline{x_{i_k}})\| &\leq \sum_{k=1}^n \|\xi_{i_k} \otimes \overline{\xi_{i_k}} - x_{i_k} \otimes \overline{x_{i_k}}\| \\ &\leq \sum_{k=1}^n \|\xi_{i_k} - x_{i_k}\|(\|\xi_{i_k}\| + 1) \\ &\leq \sum_{k=1}^n \frac{1}{2^k\|X^*T\|}(2 + \frac{1}{2^k\|X^*T\|}) \\ &\leq \frac{7}{3\|X^*T\|},\end{aligned}$$

and since $\{x_{i_k}\}$ is an orthonormal system, it follows that

$$\begin{aligned}\|\sum_{k=1}^n \alpha_k(\xi_{i_k} \otimes \overline{\xi_{i_k}})\| &\leq \frac{7}{3\|X^*T\|} + \|\sum_{k=1}^n \alpha_k(x_{i_k} \otimes \overline{x_{i_k}})\| \\ &\leq (\frac{7}{3\|X^*T\|} + 1).\end{aligned} \tag{4.1.3}$$

We put

$$A = \sum_{k=1}^{n} \alpha_k \xi_{i_k} \otimes \overline{X^\dagger \xi_{i_k}}.$$

By (4.1.1) and (4.1.3) we have

$$\begin{aligned}
\sum_{k=1}^{n} \alpha_k (X^* T\xi_{i_k} | \xi_{i_k}) &= f(A) = |f(A)| \\
&= |f((\sum_{k=1}^{n} \alpha_k \xi_{i_k} \otimes \overline{\xi_{i_k}})X)| \\
&\le f((\sum_{k=1}^{n} \alpha_k \xi_{i_k} \otimes \overline{\xi_{i_k}})(\sum_{k=1}^{n} \alpha_k \xi_{i_k} \otimes \overline{\xi_{i_k}})^\dagger)^{1/2} f(X^\dagger X)^{1/2} \\
&\le \| \sum_{k=1}^{n} \alpha_k \xi_{i_k} \otimes \overline{\xi_{i_k}} \| f(I)^{1/2} f(X^\dagger X)^{1/2} \\
&\le (\frac{7}{3\|X^* T\|} + 1) f(I)^{1/2} f(X^\dagger X)^{1/2}.
\end{aligned}$$

Hence we have

$$\begin{aligned}
&\sum_{k=1}^{n} |(X^* T x_{i_k} | x_{i_k})| \\
&\quad = \sum_{k=1}^{n} |(X^* T(x_{i_k} - \xi_{i_k}) | x_{i_k}) + (X^* T\xi_{i_k} | x_{i_k} - \xi_{i_k}) + (X^* T\xi_{i_k} | \xi_{i_k})| \\
&\quad = \sum_{k=1}^{n} (\frac{1}{2^k} + \frac{1}{2^{k-1}} + |(X^* T\xi_{i_k} | \xi_{i_k})|) \\
&\quad \le 3 + \sum_{k=1}^{n} \alpha_k (X^* T\xi_{i_k} | \xi_{i_k}) \\
&\quad \le 3 + (\frac{7}{3\|X^* T\|} + 1) f(I)^{1/2} f(X^\dagger X)^{1/2},
\end{aligned}$$

which implies

$$\sum_{i \in F} |(XT^* x_i | x_i)| \le 3 + (\frac{7}{3\|X^* T\|} + 1) f(I)^{1/2} f(X^\dagger X)^{1/2}.$$

Hence we have $X^* T \in \mathfrak{S}_1(\mathcal{H})$. Thus we have $T \in {}_1\mathfrak{S}(\mathcal{M})$, and by (4.1.1) $f(A) = \text{ tr } AT$ for each $A \in \mathcal{F}(\mathcal{H})$. This completes the proof.

We study the trace representation of continuous linear functionals on $\mathcal{M}$.

Theorem 4.1.2. Suppose f is a continuous linear functional on $\mathcal{M}[\tau_c]$. Then there exists an element T of $\mathfrak{S}_1(\mathcal{M})$ such that $f(X) = \text{ tr } XT$ for all $X \in \mathcal{M}$.

Proof. Since f is τ_c-continuous, there is a relatively compact subset $\mathfrak{M}$ of $\mathcal{D}[t_{\mathcal{M}}]$ such that

$$|f(X)| \leq \sup_{\xi \in \mathfrak{M}} |(X\xi|\xi)|, \quad X \in \mathcal{M}. \tag{4.1.4}$$

Without loss of generality we may assume that $\mathfrak{M}[t_{\mathcal{M}}]$ is a compact Hausdorff space. Let $\mathcal{C}(\mathfrak{M})$ be the C*-algebra of all continuous functions on the compact Hausdorff space $\mathfrak{M}$. For each $X \in \mathcal{M}$ we define a continuous function φ_X on $\mathfrak{M}[t_{\mathcal{M}}]$ by

$$\varphi_X(\xi) = (X\xi|\xi), \quad \xi \in \mathfrak{M}$$

and denote by $\mathcal{F}$ the subspace of $\mathcal{C}(\mathfrak{M})$ generated by $\{\varphi_X; X \in \mathcal{M}\}$. By (4.1.4) we can define a continuous linear functional F on $\mathcal{F}$ by

$$F(\varphi_X) = f(X), \quad X \in \mathcal{M}. \tag{4.1.5}$$

By the Hahn-Banach theorem F can be extended to a continuous linear functional on the C*-algebra $\mathcal{C}(\mathfrak{M})$ and also denoted by F, and F can be written as a linear combination $F = F_1 - F_2 + i(F_3 - F_4)$ of positive linear functionals $F_1, \cdots, F_4$ on $\mathcal{C}(\mathfrak{M})$. For any $B = \sum_k x_k \otimes \overline{y_k} \in \mathcal{F}(\mathcal{H})$ we define a continuous function φ_B on $\mathfrak{M}$ by

$$\varphi_B(\xi) = (B\xi|\xi), \quad \xi \in \mathfrak{M},$$

and so $\{\varphi_B; B \in \mathcal{F}(\mathcal{H})\} \subset \mathcal{C}(\mathfrak{M})$. Hence we have $\{\varphi_A; A \in \mathcal{M}_{\mathcal{F}(\mathcal{H})}\} \subset \mathcal{C}(\mathfrak{M})$. Therefore $F_j (j = 1, \cdots, 4)$ induces a strongly positive linear functional f_j on $\mathcal{M}_{\mathcal{F}(\mathcal{H})}$ by

$$f_j(A) = F_j(\varphi_A), \quad A \in \mathcal{M}_{\mathcal{F}(\mathcal{H})}.$$

By the Riesz representation theorem there exists a positive Borel measure μ_j on the compact space $\mathfrak{M}$ such that

$$F_j(\varphi) = \int_{\mathfrak{M}} \varphi(\xi) d\mu_j(\xi)$$

for all $\varphi \in \mathcal{C}(\mathfrak{M})$. In particular, for the functions $\varphi_A, A \in \mathcal{M}_{\mathcal{F}(\mathcal{H})}$, it means that

$$f_j(A) = F_j(\varphi_A) = \int_{\mathfrak{M}} (A\xi|\xi) d\mu_j(\xi). \tag{4.1.6}$$

On the other hand, by Lemma 4.1.1 there is an element T_j of ${}_1\mathfrak{S}(\mathcal{M})_+$ such that

$$f_j(B) = \operatorname{tr} T_j B, \qquad B \in \mathcal{F}(\mathcal{H}). \tag{4.1.7}$$

Take arbitrary $\eta \in \mathcal{D}(X^*)$ and $x \in \mathcal{H}$. Since $\varphi_{x\otimes\overline{X^*\eta}} \in \mathcal{C}(\mathfrak{M})$, it follows from (4.1.6) and (4.1.7) that

$$\begin{aligned}
|(T_j x|X^*\eta)| &= |\operatorname{tr} T_j(x \otimes \overline{X^*\eta})| \\
&= |f_j(x \otimes \overline{X^*\eta})| \\
&= \left| \int_{\mathfrak{M}} (X\xi|\eta)(x|\xi) d\mu_j(\xi) \right| \\
&\leq \|\eta\|\|x\| \int_{\mathfrak{M}} \|X\xi\|\|\xi\| d\mu_j(\xi) \\
&\leq \left(\|x\| \int_{\mathfrak{M}} ((I + X^\dagger X)\xi|\xi) d\mu_j(\xi) \right) \|\eta\|,
\end{aligned}$$

which implies $T_j\mathcal{H} \subset \mathcal{D}$. Therefore, $T_j \in \mathfrak{S}_1(\mathcal{M})_+$. We show that $f_j(X) = \operatorname{tr} \overline{T_j X}$ for all $X \in \mathcal{M}$. Let $T_j = \sum_n t_n \zeta_n \otimes \overline{\zeta_n}$ be a canonical representation of T_j. Since $T_j \in \mathfrak{S}_1(\mathcal{M})_+$, we have $t_n \geq 0$ for each $n \in \mathbb{N}$ and $\{\zeta_n\} \subset \mathcal{D}$, and further since $\mathfrak{M}$ is a compact subset of $\mathcal{D}[t_\mathcal{M}]$, there exists an orthonormal set $\{\xi_n\}$ in $\mathcal{H}$ such that $\{\xi_n\} \subset \mathcal{D}$ and $\mathfrak{M} \cup \{\zeta_n\} \subset$ closed linear span of $\{\xi_n\}$. Since

$$\sum_{n=1}^{\infty} |(X\xi|\xi_n)(\xi_n|\xi)| \leq \|X\xi\|\|\xi\| \leq ((I + X^\dagger X)\xi|\xi)$$

and $\varphi_{I+X^\dagger X}(\xi) = ((I + X^\dagger X)\xi|\xi)$ is μ_j-integrable, it follows from the Lebesgue theorem that

$$
\begin{aligned}
\text{tr } \overline{T_j X} = \text{ tr } XT_j &= \sum_{n=1}^{\infty}(T_j\xi_n|X^\dagger\xi_n) \\
&= \sum_{n=1}^{\infty} \text{ tr } T_j(\xi_n \otimes \overline{X^\dagger\xi_n}) \\
&= \sum_{n=1}^{\infty} f_j(\xi_n \otimes \overline{X^\dagger\xi_n}) && \text{(by 4.1.7)} \\
&= \sum_{n=1}^{\infty} \int_{\mathfrak{M}} ((\xi_n \otimes \overline{X^\dagger\xi_n})\xi|\xi)d\mu_j(\xi) && \text{(by 4.1.6)} \\
&= \sum_{n=1}^{\infty} \int_{\mathfrak{M}} (X\xi|\xi_n)(\xi_n|\xi)d\mu_j(\xi) \\
&= \int_{\mathfrak{M}} \sum_{n=1}^{\infty} (X\xi|\xi_n)(\xi_n|\xi)d\mu_j(\xi) \\
&= \int_{\mathfrak{M}} (X\xi|\xi)d\mu_j(\xi) && \text{(by 4.1.6)} \\
&= f_j(X).
\end{aligned}
$$

Thus we have

$$
T_j \in \mathfrak{S}_1(\mathcal{M})_+ \text{ and } f_j(X) = \text{ tr } \overline{T_j X} = \text{ tr } XT_j, \quad X \in \mathcal{M}. \qquad (4.1.8)
$$

Here we put $T = T_1 - T_2 + i(T_3 - T_4)$. Then it follows from (4.1.8) and (4.1.5) that $T \in \mathfrak{S}_1(\mathcal{M})$ and $f(X) = \text{ tr } \overline{TX} = \text{ tr } XT$ for each $X \in \mathcal{M}$. This completes the proof.

Corollary 4.1.3. Suppose $\mathcal{D}[t_{\mathcal{M}}]$ is a Fréchet space and f is a linear functional on $\mathcal{M}$. The following statements are equivalent:
(i) f is continuous on $\mathcal{M}[\tau_c]$.
(ii) $f = f_T$ for some $T \in \mathfrak{S}_1(\mathcal{M})$.

Proof. This follows from Proposition 1.9.11 and Theorem 4.1.2.

Corollary 4.1.4. Suppose $\mathcal{D}[t_{\mathcal{M}}]$ is a semi-Montel space and f is a continuous linear functional on $\mathcal{M}[\tau_u]$. Then $f = f_T$ for some $T \in \mathfrak{S}_1(\mathcal{M})$.

Proof. In a complete semi-Montel space, a set is precompact if and only if it is bounded. Hence we have $\tau_u = \tau_c$, and so by Theorem 4.1.2 $f = f_T$ for some $T \in \mathfrak{S}_1(\mathcal{M})$.

Example 4.1.5. Let $\mathcal{M}$ be the O*-algebra generated by the position and momentum operators Q_j, P_j $(j = 1, 2, \cdots, n)$ on the Schwartz space $\mathcal{S}(\mathbb{R}^n)$. Then every linear functional f on $\mathcal{M}$ is a trace functional, that is, $f = f_T$

for some $T \in \mathfrak{S}_1(\mathcal{M})$. In fact, since the graph topology $t_{\mathcal{M}}$ is the usual topology of the space $\mathcal{S}(\mathbb{R}^n)$ and the Schwartz space is a Fréchet Montel space, it follows that $\mathcal{D}[t_{\mathcal{M}}]$ is a Fréchet Montel space. Further, it follows from Example 4.5.7 in Schmüdgen [21] that the uniform topology τ_u on $\mathcal{M}$ equals the strongest locally convex topology τ_{st} on $\mathcal{M}$, which implies that every linear funtional f on $\mathcal{M}$ is τ_u-continuous, so that by Corollary 4.1.4 $f = f_T$ for some $T \in \mathfrak{S}_1(\mathcal{M})$.

We consider the quantum moment problem: Under what conditions is every strongly positive linear functional f on an O*-algebra $\mathcal{M}$ a trace functional f_T, $T \in \mathfrak{S}_1(\mathcal{M})_+$? If this is true, then an O*-algebra $\mathcal{M}$ is said to be *QMP-solvable* .

We first consider an ordered $*$-vector space $\mathcal{L}(\mathcal{D}, \mathcal{D}^\dagger)$. Let $\mathcal{D}$ be a dense subspace of a Hilbert space $\mathcal{H}$. We denote by $\mathcal{D}^\dagger$ (or $\mathcal{V}(\mathcal{D})$) the algebraic conjugate dual of $\mathcal{D}$, that is, the set of all conjugate linear functionals on $\mathcal{D}$. The set $\mathcal{D}^\dagger$ is a vector space under the following operations:

$$< v_1 + v_2, \xi > = < v_1, \xi > + < v_2, \xi >, \qquad < \alpha v, \xi > = \alpha < v, \xi >, \qquad \xi \in \mathcal{D},$$

where $< v, \xi >$ is the value of $v \in \mathcal{D}^\dagger$ at $\xi \in \mathcal{D}$. Tomita [2] has called an element of $\mathcal{D}^\dagger$ an *unbounded vector* in $\mathcal{H}$. We denote by $\mathcal{L}(\mathcal{D}, \mathcal{D}^\dagger)$ the set of all linear maps from $\mathcal{D}$ to $\mathcal{D}^\dagger$. Then $\mathcal{L}(\mathcal{D}, \mathcal{D}^\dagger)$ is a $*$-vector space under the usual operations: $S + T, \lambda T$ and the involution $T \to T^\dagger$ ($< T^\dagger \xi, \eta > = \overline{< T\eta . \xi >}$, $\xi, \eta \in \mathcal{D}$). Furthermore, $\mathcal{L}(\mathcal{D}, \mathcal{D}^\dagger)_h \equiv \{T \in \mathcal{L}(\mathcal{D}, \mathcal{D}^\dagger); T^\dagger = T\}$ is an ordered set under the order $S \leq T$ ($< S\xi, \xi > \leq < T\xi, \xi >$, ${}^\forall \xi \in \mathcal{D}$). We remark that any linear operator X defined on $\mathcal{D}$ is regarded as an element of $\mathcal{L}(\mathcal{D}, \mathcal{D}^\dagger)$ by $< X\xi, \eta > = (X\xi|\eta)$, $\xi, \eta \in \mathcal{D}$. In particular, $\mathcal{L}^\dagger(\mathcal{D})$ and $\mathcal{B}(\mathcal{H})$ are regarded as ordered $*$-subspaces of $\mathcal{L}(\mathcal{D}, \mathcal{D}^\dagger)$. For any element $X, Y \in \mathcal{L}^\dagger(\mathcal{D})$ and $A \in \mathcal{B}(\mathcal{H})$ we define a multiplication $Y^\dagger \circ A \circ X$ by

$$< (Y^\dagger \circ A \circ X)\xi, \eta > = (AX\xi|\eta), \qquad \xi, \eta \in \mathcal{D}.$$

Then we have

$$Y^\dagger \circ A \circ X \in \mathcal{L}(\mathcal{D}, \mathcal{D}^\dagger), {}^\forall A \in \mathcal{B}(\mathcal{H}), {}^\forall X, Y \in \mathcal{L}^\dagger(\mathcal{D})$$

and

$$X^\dagger \circ A \circ X \leq \|A\| X^\dagger X, {}^\forall A^* = A \in \mathcal{B}(\mathcal{H}), {}^\forall X \in \mathcal{L}^\dagger(\mathcal{D}). \qquad (4.1.9)$$

Theorem 4.1.6. Suppose $\mathcal{M}$ is self-adjoint and there exists an element C of $\mathcal{M}$ such that $(I + C^*\overline{C})^{-1}$ is a compact operator on $\mathcal{H}$. Then $\mathcal{M}$ is QMP-solvable.

Proof. Let f be a strongly positive linear functional on $\mathcal{M}$. Let $\mathcal{L}_0$ denote the subspace of $\mathcal{L}(\mathcal{D}, \mathcal{D}^\dagger)$ generated by $\{X^\dagger \circ A \circ X; X \in \mathcal{M}, A \in \mathcal{B}(\mathcal{H})_+\}$. By (4.1.9) $\mathcal{M}$ is cofinal in the ordered $*$-vector space $\mathcal{L} \equiv \mathcal{M} + \mathcal{L}_0$, and so the functional f can be extended to a strongly positive linear functional on $\mathcal{L}$ and denote it by the same f. By Lemma 4.1.1 and the self-adjointness of $\mathcal{M}$ there exists an element T of $\mathfrak{S}_1(\mathcal{M})_+$ such that

$$f(A) = \text{ tr } TA = \text{ tr } AT \tag{4.1.10}$$

for all $A \in \mathcal{F}(\mathcal{H})$. We show that $f(X) = f_T(X)$ for all $X \in \mathcal{M}$. Let $\{\lambda_n\}$ be the eigenvalues of the compact operator $(I+C^*\overline{C})^{-1}$ and $\{\xi_n\}$ an orthonormal basis of the corresponding eigenvectors. Without loss of generality we may assume that $\lambda_n \geq \lambda_{n+1}$ for all $n \in \mathbb{N}$. Take an arbitrary $X \in \mathcal{M}$. We put

$$X_n = \sum_{k=1}^{n} (\xi_k \otimes \overline{\xi_k})X, \quad n \in \mathbb{N}.$$

Then X_n is a finite dimensional operator, but not bounded in general. Hence, we regard X_n as an element of $\mathcal{L}_0$. Since

$$X = \sum_{k=1}^{\infty} \lambda_k (\xi_k \otimes \overline{\xi_k})(I + C^\dagger C)X,$$

$$X_n = \sum_{k=1}^{n} \lambda_k (\xi_k \otimes \overline{\xi_k})(I + C^\dagger C)X,$$

we have

$$\begin{aligned}
|((X - X_n)\xi|\xi)| &= |\sum_{k=n+1}^{\infty} \lambda_k ((I + C^\dagger C)X\xi|\xi_k)(\xi_k|\xi)| \\
&\leq (\sup_{k \geq n+1} \lambda_k)\{\sum_{k=n+1}^{\infty} |((I + C^\dagger C)X\xi|\xi_k)||(\xi_k|\xi)|\} \\
&\leq \lambda_{n+1}\{\sum_{k=1}^{\infty} |((I + C^\dagger C)X\xi|\xi_k)|^2\}^{\frac{1}{2}}\{\sum_{k=1}^{\infty} |(\xi_k|\xi)|^2\}^{\frac{1}{2}} \\
&= \lambda_{n+1}\|(I + C^\dagger C)X\xi\|\|\xi\| \\
&\leq \lambda_{n+1}\{\|(I + C^\dagger C)X\xi\|^2 + \|\xi\|^2\} \\
&= \lambda_{n+1}((I + X^\dagger(I + C^\dagger C)^2 X)\xi|\xi)
\end{aligned}$$

for each $\xi \in \mathcal{D}$ and $n \in \mathbb{N}$, which implies that

$$\begin{aligned}
(X - X_n)_1 &\leq \lambda_{n+1}(I + X^\dagger(I + C^\dagger C)^2 X), \\
(X - X_n)_2 &\leq \lambda_{n+1}(I + X^\dagger(I + C^\dagger C)^2 X),
\end{aligned}$$

where $(X - X_n)_1$ and $(X - X_n)_2$ are the real part and the imaginary part of $(X - X_n) \in \mathcal{L}$, respectively. By the strong positivity of f we have

$$\begin{aligned}
|f((X - X_n)_1)| &\leq \lambda_{n+1} f(I + X^\dagger(I + C^\dagger C)^2 X), \\
|f((X - X_n)_2)| &\leq \lambda_{n+1} f(I + X^\dagger(I + C^\dagger C)^2 X),
\end{aligned}$$

so that

$$|f(X)-f(X_n)| \leq \sqrt{2}\lambda_{n+1} f(I+X^\dagger(I+C^\dagger C)^2 X) \qquad (4.1.11)$$

for each $n \in \mathbb{N}$. Since $T \in \mathfrak{S}_1(\mathcal{M})_+$, it follows from Lemma 1.9.7 that f_T is a strongly positive linear functional on $\mathcal{M}$. Hence, f_T can similarly be extended to a strongly positive linear functional on $\mathcal{L}$ and denoted by the same f_T; and so similarly to (4.1.11) we have

$$|f_T(X)-f_T(X_n)| \leq \sqrt{2}\lambda_{n+1} f_T(I+X^\dagger(I+C^\dagger C)^2 X) \qquad (4.1.12)$$

for each $n \in \mathbb{N}$. By (4.1.10), (4.1.11) and (4.1.12) we have

$$|f(X)-f_T(X)| \leq \sqrt{2}\lambda_{n+1}\{f(I+X^\dagger(I+C^\dagger C)^2 X)+f_T(I+X^\dagger(I+C^\dagger C)^2 X)\}$$

for each $n \in \mathbb{N}$. Since $\lim_{n\to\infty} \lambda_{n+1} = 0$, we have $f(X) = f_T(X)$. This completes the proof.

Theorem 4.1.7. Suppose $\mathcal{D}[t_\mathcal{M}]$ is a Fréchet Montel space. Then $\mathcal{M}$ is QMP-solvable and every strongly positive linear functional is continuous with respect to the topology τ_u.

Proof. Let f be a strongly positive linear functional on $\mathcal{M}$. Since $\mathcal{D}[t_\mathcal{M}]$ is a Fréchet space, it follows from the closed graph theorem that $t_\mathcal{M} = t_{\mathcal{L}^\dagger(\mathcal{D})}$, which implies $\mathcal{M}_h$ is a cofinal in $\mathcal{L}^\dagger(\mathcal{D})_h$. Hence f can be extended to a strongly positive linear functional on $\mathcal{L}^\dagger(\mathcal{D})$ and denote it by the same f. It follows from Corollary 4.4.3 in Schmüdgen [21] that $\tau_u = \tau_{ord}$ on $\mathcal{L}^\dagger(\mathcal{D})$, which implies that f is continuous on $\mathcal{L}^\dagger(\mathcal{D})[\tau_u]$. By Corollary 4.1.4 there exists an element T of $\mathfrak{S}_1(\mathcal{L}^\dagger(\mathcal{D}))$ such that $f(X) = \operatorname{tr} XT$ for all $X \in \mathcal{L}^\dagger(\mathcal{D})$. Then since

$$0 \leq f(\xi \otimes \bar{\xi}) = \operatorname{tr}(\xi \otimes \bar{\xi})T = (T\xi|\xi)$$

for all $\xi \in \mathcal{D}$, we have $T \geq 0$. This completes the proof.

Corollary 4.1.8. Suppose $\mathcal{M}$ is a countably generated closed O*-algebra containing the restriction N of the inverse of a compact operator. Then $\mathcal{M}$ is QMP-solvable.

Proof. Since $\mathcal{D}[t_\mathcal{M}]$ is a Fréchet Montel space, it follows from Theorem 4.1.7 that $\mathcal{M}$ is QMP-solvable.

Example 4.1.9. The O*-algebra $\mathcal{M}$ generated by the position and momentum operators $Q_k, P_k, k = 1, 2, \cdots, n$, on the Schwartz space $\mathcal{S}(\mathbb{R}^n)$ is QMP-solvable. In fact, the graph topology $t_\mathcal{M}$ is the usual topology of the space $\mathcal{S}(\mathbb{R}^n)$ and the Schwartz space $\mathcal{S}(\mathbb{R}^n)$ is a Fréchet Montel space. Hence it follows from Corollary 4.1.8 that $\mathcal{M}$ is QMP-solvable.

We remark that every positive linear functional on $\mathcal{M}$ is not a trace functional. For example, the positive linear functional f in Example 1.9.2 is not a trace functional because

$$\mathrm{tr}((A^\dagger A - I)(A^\dagger A - 2I)T) = \sum_{n=0}^{\infty}(n-1)(n-2)(\xi_n|T\xi_n) \geq 0$$

for each $T \in \mathfrak{S}_1(\mathcal{M})_+$.

Theorem 4.1.10. Suppose $\mathcal{D}[t_{\mathcal{L}^\dagger(\mathcal{D})}]$ is a Fréchet space. Then the following statements are equivalent:

(i) $\mathcal{D}[t_{\mathcal{L}^\dagger(\mathcal{D})}]$ is a Montel space.

(ii) $\mathcal{L}^\dagger(\mathcal{D})$ is QMP-solvable.

(iii) Every O*-algebra $\mathcal{M}$ on $\mathcal{D}$ with $t_\mathcal{M} = t_{\mathcal{L}^\dagger(\mathcal{D})}$ is QMP-solvable.

Proof. (i) $\Rightarrow$ (iii) This follows from Theorem 4.1.7.

(iii) $\Rightarrow$ (ii) This is trivial.

(ii) $\Rightarrow$ (i) Assume that $\mathcal{D}[t_{\mathcal{L}^\dagger(\mathcal{D})}]$ is not a Montel space. Then there exists a bounded set $\mathfrak{M}$ in $\mathcal{D}[t_{\mathcal{L}^\dagger(\mathcal{D})}]$ which is not relatively compact in $\mathcal{D}[t_{\mathcal{L}^\dagger(\mathcal{D})}]$. Hence $\mathfrak{M}$ contains a sequence $\{\zeta_n\}_{n\in\mathbb{N}}$ which has no cluster point in $\mathcal{D}[t_{\mathcal{L}^\dagger(\mathcal{D})}]$. Since $\{\zeta_n\}$ is bounded in the Hilbert space norm, it has a weakly convergent subsequence in $\mathcal{H}$. For simplicity suppose that $\{\zeta_n\}$ is weakly convergent to $\zeta_0 \in \mathcal{H}$. The set $K \equiv \{\zeta_n; n \in \mathbb{N}\}$ endowed with the induced topology by $\mathcal{D}[t_{\mathcal{L}^\dagger(\mathcal{D})}]$ is a Tychonoff space. Hence there exists the Stone-Czech compactfication $\beta(K)$ of K (Kelly [1], p.153). For $A \in \mathcal{L}^\dagger(\mathcal{D})$ the function $g_A(\xi) \equiv (A\xi|\xi)$ is a continuous bounded function on the topological space K, and so it can be extended uniquely to a continuous function on $\beta(K)$ and denote it by the same g_A. Since K is not compact, there is an element ξ_0 of $\beta(K) \setminus K$. We define a linear functional f on $\mathcal{L}^\dagger(\mathcal{D})$ by

$$f(A) = g_A(\xi_0), \quad A \in \mathcal{L}^\dagger(\mathcal{D}).$$

Since K is dense in $\beta(K)$, ξ_0 is a cluster point of K and since g_A is continuous on $\beta(K)$, we have

$$f(A) = \lim_{n\to\infty}(A\zeta_n|\zeta_n), \quad A \in \mathcal{L}^\dagger(\mathcal{D}). \tag{4.1.13}$$

Hence f is a strongly positive linear functional on $\mathcal{L}^\dagger(\mathcal{D})$. Since $\mathcal{L}^\dagger(\mathcal{D})$ is QMP-solvable, there exists an element T of $\mathfrak{S}_1(\mathcal{L}^\dagger(\mathcal{D}))_+$ such that

$$f(A) = \mathrm{tr}\; AT, \quad A \in \mathcal{L}^\dagger(\mathcal{D}). \tag{4.1.14}$$

By (4.1.13) and (4.1.14) we have

$$\begin{aligned}(T\xi|\eta) = \text{ tr } T(\xi\otimes\overline{\eta}) &= f(\xi\otimes\overline{\eta})\\ &= \lim_{n\to\infty}((\xi\otimes\overline{\eta})\zeta_n|\zeta_n)\\ &= \lim_{n\to\infty}(\zeta_n|\eta)(\xi|\zeta_n)\\ &= (\zeta_0|\eta)(\xi|\zeta_0)\\ &= ((\zeta_0\otimes\overline{\zeta_0})\xi|\eta)\end{aligned}$$

for all $\xi,\eta\in\mathcal{D}$, and so $T=\zeta_0\otimes\overline{\zeta_0}$. Since $T\in\mathfrak{S}_1(\mathcal{L}^\dagger(\mathcal{D}))_+$, we have $\zeta_0\in\mathcal{D}$. Further, we have

$$\begin{aligned}\lim_{n\to\infty}(A\zeta_n|\zeta_n) = f(A) = \text{ tr } A(\zeta_0\otimes\overline{\zeta_0})\\ = (A\zeta_0|\zeta_0)\end{aligned}$$

for all $A\in\mathcal{L}^\dagger(\mathcal{D})$, which implies that

$$\begin{aligned}&\lim_{n\to\infty}\|A\zeta_n - A\zeta_0\|^2\\ &= \lim_{n\to\infty}\{(A^\dagger A\zeta_n|\zeta_n)-(\zeta_n|A^\dagger A\zeta_0)-(A^\dagger A\zeta_0|\zeta_n)+\|A\zeta_0\|^2\}\\ &= (A^\dagger A\zeta_0|\zeta_0)-(\zeta_0|A^\dagger A\zeta_0)-(A^\dagger A\zeta_0|\zeta_0)+\|A\zeta_0\|^2\\ &= 0\end{aligned}$$

for each $A\in\mathcal{L}^\dagger(\mathcal{D})$. This means that ζ_0 is a cluster point in $\mathcal{D}[t_{\mathcal{L}^\dagger(\mathcal{D})}]$. This is a contradiction. This completes the proof.

We finally consider trace representation of *positive* linear functionals. For the relation of the trace representability and the σ-weak continuity of positive linear functionals on $\mathcal{M}$ we have the following

Proposition 4.1.11. Let f be a positive linear functional on $\mathcal{M}$. Consider the following statements:

(i) f is a trace functional on $\mathcal{M}$, that is, $f=f_T$ for some $T\in\mathfrak{S}_1(\mathcal{M})_+$.

(ii) $f=\sum_{n=1}^{\infty}\omega_{\xi_n}$ for some $\{\xi_n\}\in\mathcal{D}^\infty(\mathcal{M})$.

(iii) f is σ-weakly continuous.

Then the following implications hold:

$$\begin{array}{ccc}\text{(i)} & \Leftrightarrow & \text{(ii)}\\ & \Downarrow & \\ & \text{(iii)}. & \end{array}$$

Proof. The equivalence of (i) and (ii) follows from Lemma 1.9.4. The implication (ii) $\Rightarrow$ (iii) is trivial.

To consider when the above implication (iii) $\Rightarrow$ (ii) holds, we first define an O*-algebra $[\mathcal{M}]$ of diagonal operators by

$$[X]\{\xi_k\} = \{X\xi_k\}, \quad X \in \mathcal{M}, \{\xi_k\} \in \mathcal{D}^\infty(\mathcal{M});$$
$$[\mathcal{M}] = \{[X]; X \in \mathcal{M}\}.$$

Then we have the following

Lemma 4.1.12. $[\mathcal{M}]$ is an O*-algebra on $\mathcal{D}^\infty(\mathcal{M})$ in $\mathcal{H}^\infty$ satisfying
(i) $[\mathcal{M}]$ is closed if and only if $\mathcal{M}$ is closed;
(ii) $[\mathcal{M}]'_{\mathrm{w}}$ is a von Neumann algebra if and only if $\mathcal{M}'_{\mathrm{w}}$ is a von Neumann algebra;
(iii) $[\mathcal{M}]'_{\mathrm{w}}\mathcal{D}^\infty(\mathcal{M}) = \mathcal{D}^\infty(\mathcal{M})$ if and only if $\mathcal{M}'_{\mathrm{w}}\mathcal{D} = \mathcal{D}$;
(iv) $[\mathcal{M}]$ is self-adjoint if and only if $\mathcal{M}$ is self-adjoint;
(v) the topology τ_{w} (resp. τ_s, τ_s^*) on $[\mathcal{M}]$ is the topology $\tau_{\sigma\mathrm{w}}$ (resp. $\tau_{\sigma\mathrm{s}}, \tau_{\sigma\mathrm{s}}^*$) on $\mathcal{M}$.

Theorem 4.1.13. Let $\mathcal{M}$ be a closed O*-algebra on $\mathcal{D}$ in $\mathcal{H}$ such that $\mathcal{M}'_{\mathrm{w}}\mathcal{D} \subset \mathcal{D}$. Then the following statements hold:
(1) Suppose $\overline{\mathcal{M}}^{\tau_s^*} = \mathcal{M}''_{\mathrm{w}\sigma}$. Then f is a weakly continuous positive linear functional on $\mathcal{M}$ if and only if $f = \sum_{k=1}^{n} \omega_{\xi_k}$ for some $\{\xi_k\}_{k=1,\cdots,n}$ in $\mathcal{D}$.
(2) Suppose $\overline{\mathcal{M}}^{\tau_{\sigma s}^*} = \mathcal{M}''_{\mathrm{w}\sigma}$. Then f is a σ-weakly continuous positive linear functional on $\mathcal{M}$ if and only if $f = \sum_{k=1}^{\infty} \omega_{\xi_k}$ for some $\{\xi_k\} \in \mathcal{D}^\infty(\mathcal{M})$.

Proof. We first prove the statement (2). Suppose f is a σ-weakly continuous positive linear functional on $\mathcal{M}$. Then there exists an element $\{\eta_k\}$ of $\mathcal{D}^\infty(\mathcal{M})$ such that

$$|f(X)| \leq |\sum_{k=1}^{\infty}(X\eta_k|\eta_k)|, \quad X \in \mathcal{M},$$

so that

$$\|\lambda_f(X)\| \leq \|[X]\{\eta_k\}\|, \quad X \in \mathcal{M}. \tag{4.1.15}$$

We put

$$C_0[X]\{\eta_k\} = \lambda_f(X), \quad X \in \mathcal{M}.$$

By (4.1.15) C_0 can be extended to a continuous linear map $\overline{C_0}$ of $\overline{[\mathcal{M}]\{\eta_k\}}$ into $\mathcal{H}_f$. Since $[\mathcal{M}]'_{\mathrm{w}}\mathcal{D}^\infty(\mathcal{M}) \subset \mathcal{D}^\infty(\mathcal{M})$ and $\overline{[\mathcal{M}]}^{\tau_s^*} = [\mathcal{M}]''_{\mathrm{w}\sigma}$ by Lemma 4.1.12, it follows from Proposition 1.2.3 that $\overline{[\mathcal{M}]\{\eta_k\}} = \overline{([\mathcal{M}]'_{\mathrm{w}})'\{\eta_k\}}$, which implies that the projection P from $\mathcal{H}^\infty$ to $\overline{[\mathcal{M}]\{\eta_k\}}$ belongs to $[\mathcal{M}]'_{\mathrm{w}}$. We now show

$$C \equiv (\overline{C_0}P)^*(\overline{C_0}P) \in [\mathcal{M}]'_{\mathrm{w}}. \qquad (4.1.16)$$

Take arbitrary $A \in (\mathcal{M}'_{\mathrm{w}})'$ and $x, y \in \mathcal{H}^\infty$. Since

$$[(\mathcal{M}'_{\mathrm{w}})'] = ([\mathcal{M}]'_{\mathrm{w}})' \subset [\mathcal{M}]''_{\mathrm{w}\sigma} = \overline{[\mathcal{M}]}^{\tau_s^*},$$

there exists a net $\{X_\alpha\}$ in $\mathcal{M}$ such that

$$\lim_\alpha \sum_{k=1}^\infty \|X_\alpha \zeta_k - A\zeta_k\|^2 = \lim_\alpha \sum_{k=1}^\infty \|X_\alpha^\dagger \zeta_k - A^*\zeta_k\|^2 = 0$$

for each $\{\zeta_k\} \in \mathcal{D}^\infty(\mathcal{M})$. Since $Px, Py \in \overline{[\mathcal{M}]\{\eta_k\}}$, there exists sequences $\{Y_n\}$ and $\{Z_n\}$ in $\mathcal{M}$ such that $\lim_{n\to\infty} [Y_n]\{\eta_k\} = Px$ and $\lim_{n\to\infty} [Z_n]\{\eta_k\} = Py$. Then we have

$$\begin{aligned}
(C[A]x|y) &= (\overline{C_0}P[A]x|\overline{C_0}Py) \\
&= (\overline{C_0}[A]Px|\overline{C_0}Py) \\
&= \lim_{n\to\infty} (\overline{C_0}[A][Y_n]\{\eta_k\}|\overline{C_0}[Z_n]\{\eta_k\}) \\
&= \lim_{n\to\infty} \lim_\alpha (\overline{C_0}[X_\alpha Y_n]\{\eta_k\}|\overline{C_0}[Z_n]\{\eta_k\}) \\
&= \lim_{n\to\infty} \lim_\alpha (\lambda_f(X_\alpha Y_n)|\lambda_f(Z_n)) \\
&= \lim_{n\to\infty} \lim_\alpha (\lambda_f(Y_n)|\lambda_f(X_\alpha^\dagger Z_n)) \\
&= \lim_{n\to\infty} \lim_\alpha (\overline{C_0}[Y_n]\{\eta_k\}|\overline{C_0}[X_\alpha]^\dagger[Z_n]\{\eta_k\}) \\
&= \lim_{n\to\infty} (\overline{C_0}[Y_n]\{\eta_k\}|\overline{C_0}[A]^*[Z_n]\{\eta_k\}) \\
&= (\overline{C_0}Px|\overline{C_0}[A]^*Py) \\
&= (Cx|[A]^*y).
\end{aligned}$$

Hence $C \in [(\mathcal{M}'_{\mathrm{w}})']' = [\mathcal{M}]'_{\mathrm{w}}$. Since $[\mathcal{M}]'_{\mathrm{w}}\mathcal{D}^\infty(\mathcal{M}) \subset \mathcal{D}^\infty(\mathcal{M})$, it follows that $C^{1/2} \in [\mathcal{M}]'_{\mathrm{w}}$ and $\{\xi_k\} \equiv C^{1/2}\{\eta_k\} \in \mathcal{D}^\infty(\mathcal{M})$, which implies that

$$\begin{aligned}
f(X) = (\lambda_f(X)|\lambda_f(I)) &= (C[X]\{\eta_k\}|\{\eta_k\}) \\
&= ([X]C^{1/2}\{\eta_k\}|C^{1/2}\{\eta_k\}) \\
&= \sum_{k=1}^\infty (X\xi_k|\xi_k)
\end{aligned}$$

for all $X \in \mathcal{M}$. The converse is trivial. The proof of the statement (1) is similar to (2). This completes the proof.

We next consider when a positive linear functional f on $\mathcal{M}$ is represented as $f = \omega_\xi$ for some $\xi \in \mathcal{D}$.

Theorem 4.1.14. Suppose $\mathcal{M}$ is a closed O*-algebra on $\mathcal{D}$ in $\mathcal{H}$ such that $\mathcal{M}'_{\rm w}\mathcal{D} \subset \mathcal{D}$ and $\overline{\mathcal{M}}^{\tau^*_{\sigma s}} = \mathcal{M}''_{\rm w\sigma}$. Then the following statements hold:

(1) If $(\mathcal{M}'_{\rm w})'$ has a separating vector, then every σ-weakly continuous positive linear functional f on $\mathcal{M}$ is represented as $f = \omega_\xi$ for some $\xi \in \mathcal{D}$.

(2) If there is a vector ξ_0 in $\mathcal{D}$ which is strongly cyclic for $\mathcal{M}$ and separating for $(\mathcal{M}'_{\rm w})'$, then every σ-weakly continuous positive linear functional f on $\mathcal{M}$ is represented as $f = \omega_{\xi_f}$ for a unique element ξ_f of $\mathcal{P}^\#_{\xi_0} \equiv \mathcal{P}^\#_{(\mathcal{M}'_{\rm w})'\xi_0} \cap \mathcal{D}$, where $\mathcal{P}^\#_{(\mathcal{M}'_{\rm w})'\xi_0}$ is the natural positive cone associated with the left Hilbert algebra $(\mathcal{M}'_{\rm w})'\xi_0$ (Bratteli-Robinson[1]). Further, $f = \omega_{H\xi_0}$ for some positive self-adjoint operator H affiliated with $(\mathcal{M}'_{\rm w})'$ such that $\xi_0 \in \mathcal{D}(H)$ and $H\xi_0 \in \mathcal{D}$.

Proof. (1) By Proposition 4.1.13, (2) there exists an element $\{\xi_k\}$ of $\mathcal{D}^\infty(\mathcal{M})$ such that $f = \sum_{k=1}^{\infty} \omega_{\xi_k}$. We put

$$E_1 = \text{proj}\overline{[\mathcal{M}]\{\xi_k\}}, \qquad E_2 = \begin{pmatrix} 1 & 0 & \cdots \\ 0 & 0 & \cdots \\ \vdots & \vdots & \ddots \end{pmatrix}.$$

It is proved in the same way as the projection P in the proof of Proposition 4.1.13 belongs to $[\mathcal{M}]'_{\rm w}$ that $E_1 \in [\mathcal{M}]'_{\rm w}$. It is clear that $E_2 \in [\mathcal{M}]'_{\rm w}$ and $Z(E_1) \leq Z(E_2) = I$, where $Z(E_i)$ is the central support of E_i, $i = 1, 2$. Further, we have

$$\{\xi_k\} \in E_1\mathcal{H}^\infty \text{ and } \overline{([\mathcal{M}]'_{\rm w})'E_1\{\xi_k\}} = \overline{[\mathcal{M}]\{\xi_k\}} = E_1\mathcal{H}^\infty.$$

Let η_0 be a separating vector for $(\mathcal{M}'_{\rm w})'$ and put

$$\tilde{\eta}_0 = \begin{pmatrix} \eta_0 \\ 0 \\ \vdots \end{pmatrix}.$$

Then we have

$$\tilde{\eta}_0 \in E_2\mathcal{H}^\infty \text{ and } \overline{E_2[\mathcal{M}]'_{\rm w}E_2\tilde{\eta}_0} = \begin{pmatrix} \overline{\mathcal{M}'_{\rm w}\eta_0} \\ 0 \\ \vdots \end{pmatrix} = E_2\mathcal{H}^\infty.$$

By Dixmier [2] (Part III, Chap. I, Lemma 4) there exists an operator V in $[\mathcal{M}]'_{\rm w}$ such that $V^*V = E_1$ and $VV^* \leq E_2$. Then we have

$$VV^*V\{\xi_k\} = VE_1\{\xi_k\} = V\{\xi_k\},$$

and so $V\{\xi_k\} \in E_2\mathcal{H}^\infty$. Hence it follows from $[\mathcal{M}]'_{\rm w}\mathcal{D}^\infty(\mathcal{M}) \subset \mathcal{D}^\infty(\mathcal{M})$ that

$$V\{\xi_k\} = \begin{pmatrix} \xi \\ 0 \\ \vdots \end{pmatrix} \qquad \text{for some } \xi \in \mathcal{D}.$$

Thus we have

$$\begin{aligned} f(X) = ([X]\{\xi_k\}|\{\xi_k\}) &= (V^*V[X]\{\xi_k\}|\{\xi_k\}) \\ &= ([X]V\{\xi_k\}|V\{\xi_k\}) \\ &= (X\xi|\xi) \end{aligned}$$

for all $X \in \mathcal{M}$.

(2) By (1) $f = \omega_\xi$ for some $\xi \in \mathcal{D}$. It follows from Lemma 10.9 in Stratila-Zsido [1] that

$$(A\xi|\xi) = (A\xi_f|\xi_f), \quad A \in (\mathcal{M}'_{\mathrm{w}})' \tag{4.1.17}$$

for a unique vector ξ_f in $\mathcal{P}^{\#}_{(\mathcal{M}'_{\mathrm{w}})'\xi_0}$. Take an arbitrary $X \in \mathcal{M}$. Let $|\overline{X}| = \int_0^\infty t dE(t)$ be the spectral resolution of $|\overline{X}|$ and $E_n = \int_0^n dE(t)$, $n \in \mathbb{N}$. Since $\mathcal{M}'_{\mathrm{w}}\mathcal{D} \subset \mathcal{D}$, it follows that $E_n, \overline{X}E_n \in (\mathcal{M}'_{\mathrm{w}})'$ for $n \in \mathbb{N}$, so that by (4.1.17)

$$\begin{aligned} &\lim_{n\to\infty} E_n\xi_f = \xi_f, \\ &\lim_{m,n\to\infty} \|\overline{X}E_m\xi_f - \overline{X}E_n\xi_f\| = \lim_{m,n\to\infty} \|\overline{X}E_m\xi - \overline{X}E_n\xi\| = 0. \end{aligned}$$

Hence, $\xi_f \in \mathcal{D}$ and $f = \omega_{\xi_f}$. Suppose $f = \omega_{\xi_1} = \omega_{\xi_2}$ for $\xi_1, \xi_2 \in \mathcal{P}^{\#}_{\xi_0}$. Then it follows from $\overline{[\mathcal{M}]}^{\tau_s^*} = [\mathcal{M}]''_{\mathrm{w}\sigma}$ and (4.1.17) that $\xi_1 = \xi_2$, which shows the uniqueness of ξ_f. We put

$$H_0 C\xi_0 = C\xi_f, \quad C \in \mathcal{M}'_{\mathrm{w}}.$$

Since $\xi_f \in \mathcal{P}^{\#}_{(\mathcal{M}'_{\mathrm{w}})'\xi_0}$, it follows that H_0 is a positive operator whose closure $\overline{H_0}$ is affiliated with $(\mathcal{M}'_{\mathrm{w}})'$. The Fiedrich extension H of $\overline{H_0}$ fulfills our assertions. This completes the proof.

Applying the Randon-Nikodym theorem in Section 3.5 to O*-algebras, we have the following

Proposition 4.1.15. Let $\mathcal{M}$ be a closed O*-algebra on $\mathcal{D}$ in $\mathcal{H}$ such that $\mathcal{M}'_{\mathrm{w}}\mathcal{D} \subset \mathcal{D}$ and ξ_0 a strongly cyclic vector for $\mathcal{M}$. Suppose f is a positive linear functional on $\mathcal{M}$ such that $\pi_{f+\omega_{\xi_0}}(\mathcal{M})'_{\mathrm{w}}$ is a von Neumann algebra. Then the following statements hold:

(1) f is ω_{ξ_0}-absolutely continuous if and only if $f = \omega_{H'\xi_0}$ for some positive self-adjoint operator H' affiliated with $\mathcal{M}'_{\mathrm{w}}$ such that $\mathcal{D}(H') \supset \mathcal{M}\xi_0$.

(2) Further, suppose ξ_0 is a separating vector for $(\mathcal{M}'_{\rm w})'$ and f is ω_{ξ_0}-absolutely continuous. Then $f = \omega_{\xi_f}$ for a unique element ξ_f of $\mathcal{P}^{\#}_{\xi_0}$. Further, $f = \omega_{H\xi_0}$ for some positive self-adjoint operator H affiliated with $(\mathcal{M}'_{\rm w})'$ such that $\xi_0 \in \mathcal{D}(H)$ and $H\xi_0 \in \mathcal{D}$.

Proof. (1) This follows from Theorem 3.5.2..
(2) This follows from (1) and Proposition 4.1.14.

Remark 4.1.16. We may change the condition "$\mathcal{M}'_{\rm w}\mathcal{D} \subset \mathcal{D}$" in Proposition 4.1.13, 4.1.14, 4.1.15 to a weaker condition "$\mathcal{M}'_{\rm w}$ is a von Neumann algebra". For, when $\mathcal{M}'_{\rm w}$ is a von Neumann algebra, by Theorem 1.4.2 there exists a closed O*-algebra $\widehat{\mathcal{M}}$ on $\hat{\mathcal{D}}$ in $\mathcal{H}$ which is an extension of $\mathcal{M}$ satisfying $(\widehat{\mathcal{M}})'_{\rm w} = \mathcal{M}'_{\rm w}$ and $(\widehat{\mathcal{M}})'_{\rm w}\hat{\mathcal{D}} \subset \hat{\mathcal{D}}$. Hence we have only to consider the O*-algebra $\widehat{\mathcal{M}}$ instead of the O*-algebra $\mathcal{M}$.

4.2 Quantum moment problem II

Let f be a trace functional on a closed O*-algebra $\mathcal{M}$, that is,

$$f(X) = trXT = tr\overline{TX}, \quad X \in \mathcal{M}$$

for some $T \in \mathfrak{S}_1(\mathcal{M})_+$. Then, by Lemma 1.9.13 $\Omega \equiv T^{1/2}$ is a positive Hilbert-Schmidt operator on $\mathcal{H}$ such that $\Omega\mathcal{H} \subset \mathcal{D}$ and $X\Omega$ is a Hilbert-Schmidt operator for all $X \in \mathcal{M}$, and the f is represented as

$$f(X^\dagger X) = tr(X\Omega)^*(X\Omega), \quad X \in \mathcal{M}, \tag{4.2.1}$$

and so the GNS-representation π_f for f is unitarily equivalent to a $*$-subrepresentation of the $*$-representation π of $\mathcal{M}$ on the Hilbert space $\mathcal{H} \otimes \overline{\mathcal{H}}$ of Hilbert-Schmidt operators on $\mathcal{H}$. This result is useful for the unbounded Tomita-Takesaki theory and applicable to the quantum physics in later sections. In this section we consider under what conditions a weight on $\mathcal{M}_+$ is represented as the similar form to (4.2.1); that is, the weight φ is of the form

$$\varphi(X^\dagger X) = tr(X^{\dagger *}\Omega)^*\overline{X^{\dagger *}\Omega} \text{ whenever } X \in \mathfrak{N}^0_\varphi \equiv \{X \in \mathcal{M};\ \ \varphi(X^\dagger X) < \infty\} \tag{4.2.2}$$

for some positive self-adjoint operator Ω in $\mathcal{H}$. In this case we can define the generalized vector λ_Ω for the O*-algebra $\pi(\mathcal{M})$ in $\mathcal{H} \otimes \overline{\mathcal{H}}$ and develop the Tomita-Takesaki theory in the O*-algebra $\pi(\mathcal{M})$ using the theory of generalized vectors in Section 2.

Let $\mathcal{M}$ be an O*-algebra on $\mathcal{D}$ in $\mathcal{H}$ with identity operator I. For $T \in \mathfrak{S}_1(\mathcal{M})_+$ we define a strongly positive linear functional f_T on $\mathcal{M}$ by

$$f_T(X) \equiv tr\,\overline{TX} = tr\,XT, \qquad X \in \mathcal{M}.$$

Then we have

$$f_T(X^\dagger X) = tr(T^{1/2}X^\dagger)^*\overline{T^{1/2}X^\dagger} = tr\,\overline{XTX^\dagger}, \quad X \in \mathcal{M}. \qquad (4.2.3)$$

In fact, given $T \in \mathfrak{S}_1(\mathcal{M})_+$ and $X \in \mathcal{M}$, we consider orthonormal sequences $\{\xi_n\}$ and $\{\eta_n\}$ in $\mathcal{D}$ such that $\{\xi_n\}$ contains a complete set of eigenvectors with non-zero eigenvalues of T and that the closed linear span of $\{\eta_n\}$ in $\mathcal{H}$ contains $XT\mathcal{H} \cup T\mathcal{H}$. Then, the equation (4.2.3) follows from

$$\begin{aligned} f_T(X^\dagger X) &= \sum_n (XT\xi_n | X^\dagger \xi_n) \\ &= \sum_n \sum_m (XT\xi_n|\eta_m)(\eta_m|X\xi_n) \\ &= \sum_m (X^\dagger \eta_m | TX^\dagger \eta_m) \\ &= \sum_m \|T^{1/2}X^\dagger \eta_m\|^2 \\ &= \sum_m (XTX^\dagger \eta_m|\eta_m). \end{aligned}$$

Theorem 4.2.1. Suppose $\mathcal{M}$ is a QMP-solvable O*-algebra on $\mathcal{D}$ in $\mathcal{H}$ and φ is a regular weight on $\mathcal{M}_+$ such that $(\mathfrak{N}_\varphi^0)^\dagger\mathcal{D}$ is total in $\mathcal{H}$. Suppose further $\mathcal{F}(\mathcal{D}) \equiv$ linear span of $\{\xi \otimes \overline{\eta}; \xi, \eta \in \mathcal{D}\} \subset \mathcal{M}$ or φ is sequentially m-regular, that is, $\varphi = \sup_n f_n$ for some increasing sequence $\{f_n\}$ of strongly positive linear functionals on $\mathcal{M}$. Then there exists a positive self-adjoint operator Ω in $\mathcal{H}$ such that

$$\begin{aligned} \mathfrak{N}_\varphi^0 &= \{X \in \mathcal{M};\ \overline{\Omega X^\dagger} \in \mathcal{H} \otimes \overline{\mathcal{H}}\} \\ &\subset \{X \in \mathcal{M};\ \overline{X^{\dagger *}\Omega} \in \mathcal{H} \otimes \overline{\mathcal{H}}\}, \end{aligned}$$

$$\varphi(X^\dagger X) = tr(\Omega X^\dagger)^*\overline{\Omega X^\dagger} = tr(X^{\dagger *}\Omega)^*\overline{X^{\dagger *}\Omega}, \quad X \in \mathfrak{N}_\varphi^0.$$

For the proof of Theorem 4.2.1 we prepare some notations and one lemma. Suppose $\mathcal{M}$ is a QMP-solvable O*-algebra on $\mathcal{D}$ in $\mathcal{H}$ and φ is a regular weight on $\mathcal{M}_+$, then there exists a net $\{T_\alpha\}$ in $\mathfrak{S}_1(\mathcal{M})_+$ such that

$$\varphi(X) = \varphi_{\{T_\alpha\}}(X) \equiv \sup_\alpha tr XT_\alpha, \quad X \in \mathcal{M}_+. \qquad (4.2.4)$$

Hence we first investigate when such a weight $\varphi_{\{T_\alpha\}}$ is of the form (4.2.2). We denote by $\mathcal{T}(\mathcal{M})$ the set of all nets $\{T_\alpha\}$ in $\mathfrak{S}_1(\mathcal{M})_+$ for which the formula

$$\varphi_{\{T_\alpha\}}(X) = \sup_\alpha tr XT_\alpha, \quad X \in \mathcal{M}_+$$

defines a weight on $\mathcal{M}_+$, and denote by $\mathcal{T}_i(\mathcal{M})$ (resp. $\mathcal{T}_i^c(\mathcal{M})$) the set of all increasing (resp. increasing and mutually commuting) nets in $\mathfrak{S}_1(\mathcal{M})_+$.

Then, $\mathcal{T}_i^c(\mathcal{M}) \subset \mathcal{T}_i(\mathcal{M}) \subset \mathcal{T}(\mathcal{M})$. For trace representation of the weight $\varphi_{\{T_\alpha\}}$ we have the following

Lemma 4.2.2. Let $\mathcal{M}$ be an O*-algebra on $\mathcal{D}$ in $\mathcal{H}$. Let $\{T_\alpha\} \in \mathcal{T}(\mathcal{M})$ be given such that $(\mathfrak{N}^0_{\varphi_{\{T_\alpha\}}})^\dagger \mathcal{D}$ is total in $\mathcal{H}$. Suppose $\{T_\alpha\} \in \mathcal{T}_i(\mathcal{M})$ or that $\mathcal{F}(\mathcal{D}) \subset \mathcal{M}$. Then the following statements hold:

(1) There exists a positive self-adjoint operator Ω in $\mathcal{H}$ such that

$$\begin{aligned}\mathfrak{N}^0_{\varphi_{\{T_\alpha\}}} &= \{X \in \mathcal{M};\ \overline{\Omega X^\dagger} \in \mathcal{H} \otimes \overline{\mathcal{H}}\} \\ &\subset \{X \in \mathcal{M};\ \overline{X^{\dagger *}\Omega} \in \mathcal{H} \otimes \overline{\mathcal{H}}\}\end{aligned}$$

and

$$\begin{aligned}\varphi_{\{T_\alpha\}}(X^\dagger X) &= tr\,(\Omega X^\dagger)^* \overline{\Omega X^\dagger} \\ &= tr\,(X^{\dagger *}\Omega)^* \overline{X^{\dagger *}\Omega}, \quad X \in \mathfrak{N}^0_{\varphi_{\{T_\alpha\}}}.\end{aligned}$$

(2) If $\{T_\alpha\} \in \mathcal{T}_i^c(\mathcal{M})$, then there exist a set $\{\mu_\beta\}_{\beta \in B}$ of positive numbers and an orthonormal system $\{\xi_\beta\}_{\beta \in B}$ in $\mathcal{H}$ contained in $\mathcal{D}$ such that

$$\varphi_{\{T_\alpha\}}(X) = \sum_{\beta \in B} \mu_\beta (X\xi_\beta|\xi_\beta), \qquad X \in \mathcal{M}_+,$$

so that $\Omega \equiv \sum_{\beta \in B} \mu_\beta^{1/2} \xi_\beta \otimes \overline{\xi_\beta}$ is a positive self-adjoint operator in $\mathcal{H}$ and

$$\begin{aligned}\mathfrak{N}^0_{\varphi_{\{T_\alpha\}}} &= \{X \in \mathcal{M}; \overline{X^{\dagger *}\Omega} \in \mathcal{H} \otimes \overline{\mathcal{H}}\} \\ &= \{X \in \mathcal{M}; \overline{\Omega X^\dagger} \in \mathcal{H} \otimes \overline{\mathcal{H}}\}, \\ \varphi_{\{T_\alpha\}}(X^\dagger X) &= tr(X^{\dagger *}\Omega)^* \overline{X^{\dagger *}\Omega}, \qquad X \in \mathfrak{N}^0_{\varphi_{\{T_\alpha\}}}.\end{aligned}$$

Proof. (1) We define a form θ by

$$\begin{aligned}\mathcal{D}_\theta &= \{\xi \in \mathcal{D};\ \sup_\alpha (T_\alpha \xi|\xi) < \infty\}, \\ \theta(\xi, \xi) &= \sup_\alpha (T_\alpha \xi|\xi), \quad \xi \in \mathcal{D}_\theta.\end{aligned}$$

We show that $\mathcal{D}_\theta$ is a dense subspace of $\mathcal{H}$ and that θ can be extended to a closable sesquilinear form on $\mathcal{D}_\theta \times \mathcal{D}_\theta$. Since

$$(T_\alpha(\xi + \eta)|\xi + \eta) \le 2\{(T_\alpha \xi|\xi) + (T_\alpha \eta|\eta)\}$$

for each $\xi, \eta \in \mathcal{D}_\theta$, it follows that $\mathcal{D}_\theta$ is a subspace of $\mathcal{H}$. Since

$$\begin{aligned}\sup_\alpha (T_\alpha X^\dagger \xi|X^\dagger \xi) &\le \sup_\alpha tr\, \overline{X T_\alpha X^\dagger} \\ &= \sup_\alpha tr\, \overline{T_\alpha X^\dagger X} \qquad \text{(by 4.2.3)} \\ &= \varphi_{\{T_\alpha\}}(X^\dagger X)\end{aligned}$$

for each $X \in \mathfrak{N}_\varphi^0$ and $\xi \in \mathcal{D}$, we have $(\mathfrak{N}^0_{\varphi_{\{T_\alpha\}}})^\dagger \mathcal{D} \subset \mathcal{D}_\theta$, and so $\mathcal{D}_\theta$ is dense in $\mathcal{H}$. If $\{T_\alpha\}$ is an increasing net, θ extends to a positive sesquilinear form on $\mathcal{D}_\theta \times \mathcal{D}_\theta$ (also denoted by the same θ) by

$$\theta(\xi,\eta) = \lim_\alpha (T_\alpha \xi | \eta) = \frac{1}{4}\sum_{k=0}^{3} i^k \lim_\alpha (T_\alpha(\xi + i^k\eta)|\xi + i^k\eta), \qquad \xi,\eta \in \mathcal{D}_\theta.$$

Suppose that $\mathcal{F}(\mathcal{D}) \subset \mathcal{M}$. Then $\xi \in \mathcal{D}_\theta$ if and only if $\varphi_{\{T_\alpha\}}(\xi \otimes \overline{\xi}) < \infty$. There exists a positive linear functional $\dot{\varphi}$ defined on the linear span of $\{X \in \mathcal{M}_+; \varphi(X) < \infty\}$ such that $\dot{\varphi}(X) = \varphi(X)$ for each $X \in \mathcal{M}_+$ with $\varphi(X) < \infty$. Consequently, θ extends to a positive sesquilinear form on $\mathcal{D}_\theta \times \mathcal{D}_\theta$ defined by

$$\begin{aligned}\theta(\xi,\eta) &= \dot{\varphi}(\xi \otimes \overline{\eta}) \\ &= \frac{1}{4}\sum_{k=0}^{3} i^k \dot{\varphi}((\xi + i^k\eta) \otimes \overline{(\xi + i^k\eta)}), \qquad \xi,\eta \in \mathcal{D}_\theta.\end{aligned}$$

We show that θ is closable. In fact, take an arbitrary $\{\xi_n\}$ in $\mathcal{D}_\theta$ such that $\lim_{n\to\infty} \xi_n = 0$ and $\lim_{n,m\to\infty} \theta(\xi_n - \xi_m,\ \xi_n - \xi_m) = 0$. For any $\varepsilon > 0$ there exists a natural number N_ε such that

$$\|\xi_n\| < \varepsilon \ \text{ and } \ \theta(\xi_n - \xi_m,\ \xi_n - \xi_m) < \varepsilon \qquad \text{for all } n,m \geq N_\varepsilon,$$

and so

$$(T_\alpha(\xi_n - \xi_m) \mid \xi_n - \xi_m) < \varepsilon \qquad \text{for all } n,m \geq N_\varepsilon \text{ and } \alpha.$$

Taking $m \to \infty$, we have

$$(T_\alpha \xi_n \mid \xi_n) \leq \varepsilon \qquad \text{for all } n \geq N_\varepsilon \text{ and } \alpha,$$

and $\sup_\alpha (T_\alpha \xi_n | \xi_n) \leq \varepsilon$ for all $n \geq N_\varepsilon$. Hence, $\lim_{n\to\infty} \theta(\xi_n,\ \xi_n) = 0$. Thus, θ is closable. Let $\overline{\theta}$ be the closure of θ. By Theorem 2.1 in Faris [1] there exists a positive self-adjoint operator Ω in $\mathcal{H}$ with $\mathcal{D}(\Omega) = \mathcal{D}_{\overline{\theta}}$ and $(\Omega\xi \mid \Omega\eta) = \overline{\theta}(\xi,\eta)$ for each $\xi,\eta \in \mathcal{D}_{\overline{\theta}}$. We next show $\mathfrak{N}^0_{\varphi_{\{T_\alpha\}}} = \{X \in \mathcal{M}; \overline{\Omega X^\dagger} \in \mathcal{H} \otimes \overline{\mathcal{H}}\}$. Suppose that $X \in \mathfrak{N}^0_{\varphi_{\{T_\alpha\}}}$. Let $S = \{\{\xi_\beta\};\ \{\xi_\beta\}$ is an nonempty orthonormal set in $\mathcal{H}$ contained in $\mathcal{D}\}$. It follows that

$$\begin{aligned}\varphi_{\{T_\alpha\}}(X^\dagger X) &= \sup_\alpha tr\, \overline{T_\alpha X^\dagger X} \\ &= \sup_\alpha tr\, \overline{X T_\alpha X^\dagger} \qquad \text{(by 4.2.3)} \\ &= \sup_\alpha \sup_{\{\xi_\beta\}\in S} \sum_\beta (T_\alpha X^\dagger \xi_\beta \mid X^\dagger \xi_\beta) \\ &= \sup_{\{\xi_\beta\}\in S} \sum_\beta \|\Omega X^\dagger \xi_\beta\|^2 \\ &= tr\,(\Omega X^\dagger)^* \overline{\Omega X^\dagger},\end{aligned}$$

which implies that $\overline{\Omega X^\dagger}$ is a Hilbert-Schmidt operator. Conversely, if $\overline{\Omega X^\dagger}$ is a Hilbert-Schmidt operator, we can write down the same equations with $S' = \{\{\eta_\beta\}\,;\,\{\eta_\beta\}$ is a nonempty orthonormal set in $\mathcal{H}$ contained in $\mathcal{D}(\Omega X^\dagger)\}$ in place of S. This implies $X \in \mathfrak{N}^0_{\varphi_{\{T_\alpha\}}}$. Finally, given $X \in \mathfrak{N}^0_{\varphi_{\{T_\alpha\}}}$, $\xi \in \mathcal{D}$ and $\eta \in \mathcal{D}(\Omega)$, we have

$$(X^\dagger\xi \mid \Omega\eta) = (\Omega X^\dagger\xi \mid \eta) = (\xi \mid (\Omega X^\dagger)^*\eta) = (\xi | X^{\dagger *}\Omega\eta),$$
$$\overline{X^{\dagger *}\Omega} = (\Omega X^\dagger)^* \in \mathcal{H} \otimes \overline{\mathcal{H}},$$
$$\varphi_{\{T_\alpha\}}(X^\dagger X) = tr(\Omega X^\dagger)^*\overline{\Omega X^\dagger} = tr\,(X^{\dagger *}\Omega)^*\overline{X^{\dagger *}\Omega}.$$

(2) Let $\{T_\alpha\} \in \mathcal{T}_i^c(\mathcal{M})$. Then there exist an orthonormal system $\{\xi_\beta\}_{\beta\in B}$ in $\mathcal{H}$ and nonnegative numbers $\mu_{\alpha,\beta}$ such that

$$T_\alpha = \sum_{\beta\in B}\mu_{\alpha,\beta}\xi_\beta \otimes \overline{\xi_\beta} \qquad \text{for each } \alpha. \tag{4.2.5}$$

In fact, using Zorn's lemma, one finds a maximal system $\{P_i\}_{i\in I}$ of non-zero finite rank orthogonal projections in $\mathcal{H}$ such that

(i) $P_iP_j = 0,\ i \neq j$;
(ii) for $\alpha \in A$ and $i \in I$ there exists $\lambda_{\alpha,i}$ such that $T_\alpha P_i = \lambda_{\alpha,i}P_i$;
(iii) for $i \in I$ there exists $\alpha \in A$ such that $T_\alpha P_i \neq 0$.

We show that

$$(\sum_{i\in I}P_i)T_\alpha = T_\alpha \qquad \text{for each } \alpha \in A. \tag{4.2.6}$$

Suppose that this is not true. Then, for some $\alpha_0 \in A$ $(\sum_{i\in I}P_i)T_{\alpha_0} \neq T_{\alpha_0}$. Let $T_{\alpha_0} = \sum_n \lambda_n^{(\alpha_0)}P_n^{(\alpha_0)}$ be the spectral decomposition $(\lambda_n^{(\alpha_0)} > 0)$. Then for some n, $P_n^{(\alpha_0)}$ is not a subprojection of $\sum_{i\in I}P_i$. This means that

$$Q \equiv P_n^{(\alpha_0)} - P_n^{(\alpha_0)}(\sum_{i\in I}P_i) = P_n^{(\alpha_0)} - \sum_{\{i;\lambda_{\alpha_0,i}=\lambda_n^{(\alpha_0)}\}} P_i \neq 0.$$

Since the $*$-algebra generated by $\{T_\alpha Q\}$ is a commutative C*-algebra of finite dimension, we can find a non-zero subprojection P of Q such that $T_\alpha P = \lambda_\alpha P$ for each α, which implies that $\{P_i, P\}$ satisfies the conditions (i) ~ (iii). This contradicts the maximality of $\{P_i\}$ and thus the statement (4.2.6) holds. Furthermore, we can choose in each $P_i(\mathcal{H})$ an orthonormal basis and collect all these basis vectors together.

By (4.2.4) we have

$$\{\xi_\beta\} \subset \mathcal{D} \text{ and } \varphi_{\{T_\alpha\}}(X) = \sum_\beta \mu_\beta (X\xi_\beta|\xi_\beta), \quad X \in \mathcal{M}_+, \tag{4.2.7}$$

where $\mu_\beta \equiv \sup_\alpha \mu_{\alpha,\beta}$ for each $\beta \in B$. Now suppose $\mu_{\beta_0} = \infty$ for some $\beta_0 \in B$. Then $X\xi_{\beta_0} = 0$ for each $X \in \mathfrak{N}^0_{\varphi_{\{T_\alpha\}}}$ and so $(\xi_{\beta_0} \mid X^\dagger \xi) = 0$ for each $X \in \mathfrak{N}^0_{\varphi_{\{T_\alpha\}}}$ and $\xi \in \mathcal{D}$. Since $(\mathfrak{N}^0_{\varphi_{\{T_\alpha\}}})^\dagger \mathcal{D}$ is total in $\mathcal{H}$, we have $\xi_{\beta_0} = 0$. This is a contradiction. Hence we have

$$\mu_\beta < \infty, \quad {}^\forall \beta \in B. \tag{4.2.8}$$

Hence, $\Omega \equiv \sum_{\beta \in B} \mu_\beta^{\frac{1}{2}} \xi_\beta \otimes \overline{\xi_\beta}$ is a positive self-adjoint operator in $\mathcal{H}$. Let $X \in \mathcal{M}$. Since $X^{\dagger *}\Omega = 0$ on $\{\xi_\beta\}^\perp$ and

$$\begin{aligned}
|(X^{\dagger *}\Omega \sum_{r \in F} \lambda_r \xi_r \mid \eta)|^2 &= |\sum_{r \in F} \lambda_r \mu_r^{\frac{1}{2}} (X\xi_r \mid \eta)|^2 \\
&\leq (\sum_{r \in F} |\lambda_r|^2) \sum_{\beta \in B} \mu_\beta |(X\xi_\beta \mid \eta)|^2 \\
&\leq \|\sum_{r \in F} \lambda_r \xi_r\|^2 \|\eta\|^2 (\sum_{\beta \in B} \mu_\beta \|X\xi_\beta\|^2) \\
&= \|\sum_{r \in F} \lambda_r \xi_r\|^2 \|\eta\|^2 \varphi_{\{T_\alpha\}}(X^\dagger X)
\end{aligned}$$

for each $\eta \in \mathcal{H}$ and each finite subset F of B, it follows that $(\Omega X^\dagger)^* = \overline{X^{\dagger *}\Omega}$ are bounded and

$$\begin{aligned}
tr((X^{\dagger *}\Omega)^* \overline{X^{\dagger *}\Omega} &= \sum_{\beta \in B} \|\overline{X^{\dagger *}\Omega}\xi_\beta\|^2 \\
&= \sum_{\beta \in B} \mu_\beta \|X\xi_\beta\|^2 \\
&= \varphi_{\{T_\alpha\}}(X^\dagger X),
\end{aligned}$$

which implies that $X \in \mathfrak{N}^0_{\varphi_{\{T_\alpha\}}}$ if and only if $\overline{X^{\dagger *}\Omega}$ is a Hilbert-Schmidt operator on $\mathcal{H}$ and that this is the case if and only if $\overline{\Omega X^\dagger}$ is a Hilbert-Schmidt operator on $\mathcal{H}$. This completes the proof of Lemma 4.2.2.

Proof of Theorem 4.2.1. : Suppose φ is sequentially m-regular, that is, $\varphi = \sup_n f_n$ for some increasing sequence $\{f_n\}$ of strongly positive linear functionals on $\mathcal{M}$. In this case it may be represented also as $\varphi = \sum_n g_n$, where $g_1 \equiv f_1$ and $g_{n+1} \equiv f_{n+1} - f_n$ are strongly positive. This implies that $\varphi = \varphi_{\{T_n\}}$ for some $\{T_n\} \in \mathcal{T}_i(\mathcal{M})$. Hence our assertions follow from (4.2.4) and Lemma 4.2.2.

Remark 4.2.3. In Lemma 4.2.2 we have showed

$$\{X \in \mathcal{M};\ \overline{\Omega X^\dagger} \in \mathcal{H} \otimes \overline{\mathcal{H}}\} \subset \{X \in \mathcal{M};\ \overline{X^{\dagger *}\Omega} \in \mathcal{H} \otimes \overline{\mathcal{H}}\}.$$

Do the above two sets coincide? Suppose $\overline{X^{\dagger *}\Omega} \in \mathcal{H} \otimes \overline{\mathcal{H}}$ and $\Omega X^\dagger$ is densely defined. Then $\overline{\Omega X^\dagger} \in \mathcal{H} \otimes \overline{\mathcal{H}}$ and $\overline{\Omega X^\dagger} = (X^{\dagger *}\Omega)^*$. Hence, in the following cases two sets coincide:

(i) $\mathcal{D} \subset \mathcal{D}(\Omega)$, that is, $\sup_\alpha (T_\alpha \xi|\xi) < \infty$ for each $\xi \in \mathcal{D}$.

(ii) Ω is bounded, that is, $\sup_\alpha trT_\alpha < \infty$.

By Theorem 4.2.1 and Remark 4.2.3 we have the following

Corollary 4.2.4. Suppose $\mathcal{M}$ is a QMP-solvable O*-algebra on $\mathcal{D}$ in $\mathcal{H}$. For every regular weight φ on $\mathcal{M}_+$ satisfying $\varphi(I) < \infty$ there exists a positive Hilbert-Schmidt operator Ω on $\mathcal{H}$ such that

$$\begin{aligned} \mathfrak{N}^0_\varphi &= \{X \in \mathcal{M};\ \overline{\Omega X^\dagger} \in \mathcal{H} \otimes \overline{\mathcal{H}}\} \\ &= \{X \in \mathcal{M};\ \overline{X^{\dagger *}\Omega} \in \mathcal{H} \otimes \overline{\mathcal{H}}\}, \\ \varphi(X^\dagger X) &= tr(\Omega X^\dagger)^* \overline{\Omega X^\dagger} = tr(X^{\dagger *}\Omega)^* \overline{X^{\dagger *}\Omega}, \quad X \in \mathfrak{N}^0_\varphi. \end{aligned}$$

Remark 4.2.5. In bounded case the condition $\varphi(I) < \infty$ in Corollary 4.2.4 implies φ is finite, that is, $\varphi(X) < \infty$ for each $X \in \mathcal{M}_+$. But, in unbounded case this does not necessarily hold as seen in next example. Let $\mathcal{M}$ be an O*-algebra on the Schwartz space $\mathcal{S}(\mathbb{R})$ generated by the momentum operator P and the position operator Q and $\{f_n\}_{n=0,1,\cdots} \subset \mathcal{S}(\mathbb{R})$ an ONB in $L^2(\mathbb{R})$ consisting of the Hermite functions. For $m \in \mathbb{N} \cup \{0\}$ we define a regular weight φ_m on $\mathcal{M}_+$ by

$$\varphi_m(X) = \sum_{n=1}^{\infty} \frac{1}{(n+1)^{2m}} (X f_n|f_n), \qquad X \in \mathcal{M}_+.$$

Then the following cases arise:

(i) If $m = 0$, then $\mathfrak{N}^0_{\varphi_m} = \{0\}$.

(ii) If $m \neq 0$, then $\{I, N, \cdots, N^m\} \subset \mathfrak{N}^0_{\varphi_m}$ but $N^k \notin \mathfrak{N}^0_{\varphi_m}$ for $k \geq m+1$, where $N \equiv \sum_{n=0}^{\infty}(n+1) f_n \otimes \overline{f_n}$ is the number operator.

We next consider trace representation of weights without the assumption of regularity. We generalize the Schmüdgen result (Theorem 4.1.6) for strongly positive linear functionals to weights. The proof is according to that of Theorem 4.1.6.

Theorem 4.2.6. Let $\mathcal{M}$ be an O*-algebra on $\mathcal{D}$ in $\mathcal{H}$ and φ a weight on $\mathcal{M}_+$. Suppose that there exists an elemen N of $\mathfrak{N}^0_\varphi$ which has a positive self-adjoint extension $\widetilde{N}$ such that $\widetilde{N}^{-1}$ is a bounded compact operator on $\mathcal{H}$. Then there exists a positive trace class operator T on $\mathcal{H}$ such that

(i) $\overline{T^{1/2}X^\dagger}$ is a Hilbert-Schmidt operator and $Y^{\dagger *}\overline{TX^\dagger}$ is a trace class operator on $\mathcal{H}$ for all $X, Y \in \mathfrak{N}^0_\varphi$;

(ii) $\varphi(X^\dagger X) = tr\,(T^{1/2}X^\dagger)^* \overline{T^{1/2}X^\dagger}$
$= tr\, X^{\dagger *}\overline{TX}$ for all $X \in \mathcal{M}$ such that $NX \in \mathfrak{N}^0_\varphi$.

(iii) $\varphi(X) = tr\,\overline{TX}$ for each positive operator X in $\mathfrak{N}^0_\varphi$.

Proof. Note first that $I \in \mathfrak{N}^0_\varphi$ since

$$N \in \mathfrak{N}^0_\varphi \text{ and } \|\widetilde{N}^{-1/2}\|^2(N\xi|\xi) \geq \|\xi\|^2, \quad \xi \in \mathcal{D}.$$

We put

$$\mathcal{D}(\dot{\varphi}) = \text{ linear span of } \{X \in \mathcal{M}_+; \varphi(X) < \infty\},$$
$$\dot{\varphi}(\sum_n \alpha_n X_n) = \sum_n \alpha_n \varphi(X_n), \quad \sum_n \alpha_n X_n \in \mathcal{D}(\dot{\varphi}).$$

Then $\mathcal{D}(\dot{\varphi})$ is a $*$-vector space with $I \in \mathcal{D}(\dot{\varphi})$ and it is not difficult to show that $\dot{\varphi}$ is a strongly positive linear functional on $\mathcal{D}(\dot{\varphi})$. Since

$$(X+Y)^\dagger(X+Y) \leq 2(X^\dagger X + Y^\dagger Y),$$
$$X = \frac{1}{4}\sum_{k=0}^{3} i^k (X + i^k I)^\dagger (X + i^k I),$$

it follows that $\mathfrak{N}^0_\varphi$ is a subspace of $\mathcal{D}(\dot{\varphi})$. Let $\mathcal{L}_0$ denote the subspace of $\mathcal{L}(\mathcal{D}, \mathcal{D}^\dagger)$ generated by $\{X^\dagger \circ A \circ X;\ X \in \mathfrak{N}^0_\varphi, A \in \mathcal{B}(\mathcal{H})_+\}$. By (4.1.9) $\mathcal{D}(\dot{\varphi})$ is cofinal in the ordered $*$-vector space $\mathcal{L} \equiv \mathcal{D}(\dot{\varphi}) + \mathcal{L}_0$, and so there exists a strongly positive extension of $\dot{\varphi}$ to $\mathcal{L}$, denoted also by $\dot{\varphi}$. Since $\mathcal{F}(\mathcal{H}) \equiv$ linear span of $\{x \otimes \overline{x}\lceil\mathcal{D}; x \in \mathcal{H}\} \subset \mathcal{L}_0$, it follows from (4.1.1) that there exists a unique positive trace operator T on $\mathcal{H}$ such that

$$\dot{\varphi}(A) = tr\, T\overline{A}, \quad A \in \mathcal{F}(\mathcal{H}). \tag{4.2.9}$$

We show the statement (i). Let $X, Y \in \mathfrak{N}^0_\varphi$. Since a linear functional $\dot{\varphi}_X$ on $\mathcal{F}(\mathcal{H})$ defined by $\dot{\varphi}_X(A) = \dot{\varphi}(X^\dagger \circ A \circ X)$, $A \in \mathcal{F}(\mathcal{H})$ is strongly positive, it follows also from (4.1.1) that

$$\dot{\varphi}_X(A) = tr\, T_X\overline{A}, \quad A \in \mathcal{F}(\mathcal{H}) \tag{4.2.10}$$

for a unique positive trace class operator T_X on $\mathcal{H}$. By (4.2.9) and (4.2.10) we have

$$\begin{aligned}(TX^\dagger\xi|X^\dagger\eta) &= tr\,T\overline{X^\dagger(\xi\otimes\overline{\eta})X}\\ &= \dot{\varphi}_X(\xi\otimes\overline{\eta})\\ &= tr\,T_X(\xi\otimes\overline{\eta})\\ &= (T_X\xi|\eta)\end{aligned}$$

for each $\xi,\eta\in\mathcal{D}$, and so

$$T_X = X^{\dagger *}\overline{TX^\dagger} = (T^{1/2}X^\dagger)^*\overline{T^{1/2}X^\dagger}. \tag{4.2.11}$$

Using the polarization formula, we get

$$\begin{aligned}(Y^{\dagger *}\overline{TX^\dagger}\xi|\eta) &= (\overline{TX^\dagger}\xi|Y^\dagger\eta)\\ &= (\frac{1}{4}\sum_{k=0}^{3} i^k T_{X+i^kY}\xi|\eta)\end{aligned}$$

for each $\xi,\eta\in\mathcal{D}$, which implies that $Y^{\dagger *}\overline{TX^\dagger}$ is a trace class operator and in particular, $Y^{\dagger *}T$ is a trace class operator and $(Y^{\dagger *}T)^* = \overline{TY^\dagger}$. Thus the statement (i) holds. We next show the statement (ii). Let $X\in\mathcal{M}$ such that $NX\in\mathfrak{N}_\varphi^0$. Let $\widetilde{N}^{-1}\equiv\sum_{n=1}^{\infty}\lambda_n\xi_n\otimes\overline{\xi_n}$, where $\{\xi_n\}$ is an orthonormal basis of $\mathcal{H}$ and $\lambda_n\downarrow 0$. We put

$$E_n = \sum_{k=1}^{n}\xi_k\otimes\overline{\xi_k}, \qquad n\in\mathbb{N}.$$

Since

$$(NX)^\dagger NX = X^\dagger N^2X \geq \|\widetilde{N}^{-1}\|^{-2}X^\dagger X,$$

we have $X\in\mathfrak{N}_\varphi^0$. Furthermore, since

$$\begin{aligned}&< (X^\dagger N)\circ(\widetilde{N}^{-1}(I-E_n)\widetilde{N}^{-1})\circ(NX)\xi,\eta >\\ &= ((I-E_n)X\xi|X\eta)\\ &= < (X^\dagger X - X^\dagger\circ E_n\circ X)\xi,\eta >\end{aligned}$$

for each $\xi,\eta\in\mathcal{D}$, it follows from (4.1.9) that

$$\begin{aligned}0\leq X^\dagger X - X^\dagger\circ E_n\circ X &= (X^\dagger N)\circ(\widetilde{N}^{-1}(I-E_n)\widetilde{N}^{-1})\circ(NX)\\ &\leq \|\widetilde{N}^{-2}(I-E_n)\|(NX)^\dagger NX,\end{aligned}$$

which implies

$$\begin{aligned}0 &\leq \dot{\varphi}(X^\dagger X - X^\dagger\circ E_n\circ X)\\ &\leq \|\widetilde{N}^{-2}(I-E_n)\|\varphi(X^\dagger N^2X)\\ &\leq \lambda_{n+1}^2\varphi(X^\dagger N^2X).\end{aligned} \tag{4.2.12}$$

Hence, it follows from (4.2.10) and (4.2.11) that

$$\begin{aligned}\varphi(X^\dagger X) &= \lim_{n\to\infty} \dot{\varphi}(X^\dagger \circ E_n \circ X)\\ &= \lim_{n\to\infty} \sum_{k=1}^{n} (T_X \xi_k | \xi_k)\\ &= \sum_{k=1}^{\infty} (X^{\dagger *}\overline{TX^\dagger}\xi_k | \xi_k)\\ &= \sum_{k=1}^{\infty} \|\overline{T^{1/2}X^\dagger}\xi_k\|^2\\ &= tr\, X^{\dagger *}\overline{TX^\dagger}\\ &= tr\,(T^{1/2}X^\dagger)^*\overline{T^{1/2}X^\dagger}.\end{aligned}$$

We finally show the statement (iii). Suppose now that $X \in \mathfrak{N}_\varphi^0$ (NX may belong to $\mathfrak{N}_\varphi^0$ or not). Then, given $\varepsilon > 0$ and $k \in \mathbb{Z}$, we have

$$\begin{aligned}0 &\le (\varepsilon X + \frac{1}{\varepsilon} i^k (I - E_n))^\dagger \circ (\varepsilon X + \frac{1}{\varepsilon} i^k (I - E_n))\\ &= \varepsilon^2 X^\dagger X + i^k X^\dagger \circ (I - E_n) + i^{-k}(I - E_n) \circ X + \frac{1}{\varepsilon^2}(I - E_n),\end{aligned}$$

which implies

$$\pi_\mu \dot{\varphi}(i^k X^\dagger \circ (I - E_n) + i^{-k}(I - E_n) \circ X) \le \varepsilon^2 \varphi(X^\dagger X) + \frac{1}{\varepsilon^2}\dot{\varphi}(I - E_n). \tag{4.2.13}$$

Since $NI = N \in \mathfrak{N}_\varphi^0$, it follows from (4.2.12) that $\lim_{n\to\infty} \dot{\varphi}(I - E_n) = 0$, which implies by (4.2.13) that $\lim_{n\to\infty} \dot{\varphi}((I - E_n) \circ X) = 0$. Moreover, it follows from (i), (4.2.10) and (4.2.11) that

$$\begin{aligned}\lim_{n\to\infty} \dot{\varphi}(E_n \circ X) &= \lim_{n\to\infty} \dot{\varphi}(\frac{1}{4}\sum_{k=0}^{3} i^k (X + i^k I)^\dagger \circ E_n \circ (X + i^k I))\\ &= \lim_{n\to\infty} \frac{1}{4}\sum_{k=0}^{3} i^k tr\,((X + i^k I)^{\dagger *}\overline{T(X + i^k I)^\dagger} E_n)\\ &= tr\, X^{\dagger *} T.\end{aligned}$$

This completes the proof.

Remark 4.2.7. The statement (ii) in Theorem 4.2.6 does not hold for *every* $X \in \mathfrak{N}_\varphi^0$ as seen next. Let $\mathcal{M}$, $\{f_n\}$ and N be as in Remark 4.2.5. We define a weight φ on $\mathcal{M}_+$ by

$$\varphi(X) = \lim_{n\to\infty} \frac{1}{n^2}(Xf_n|f_n), \quad X \in \mathcal{M}_+.$$

Then $\varphi(I) = 0$ and $\varphi(N^2) = 1$, and further the pair $(\mathcal{M}, \varphi)$ satisfies the assumptions of Theorem 4.2.6. Suppose now that there exists a positive trace class operator T on $\mathcal{H}$ such that $\varphi(X^\dagger X) = tr\,(T^{1/2}X^\dagger)^*\overline{T^{1/2}X^\dagger}$ for every $X \in \mathfrak{N}^0_\varphi$. Since $\varphi(I) = 0$, it follows that $T = 0$, which implies that $1 = \varphi(N^2) = tr\,(T^{1/2}N)^*\overline{T^{1/2}N} = 0$. This is a contradiction.

4.3 Unbounded CCR-algebras

A. *Unbounded CCR-algebras in finite degrees of freedom*

Let $\mathcal{A}$ be a $*$-algebra consisting of all polynomials in $\{p_i, q_i; i = 1, 2, \cdots, s\}$ satisfying

$$\begin{aligned} &p_i^* = p_i, \quad q_i^* = q_i, \\ &[p_i, q_j] \equiv p_i q_j - q_j p_i = -i\delta_{ij}\mathit{1}, \\ &[p_i, p_j] = [q_i, q_j] = 0 \quad (i, j = 1, 2, \cdots, s), \end{aligned}$$

and it is called the *canonical algebra for s-degree of freedom* . We treat with only the canonical algebra of one degree of freedom, that is, $\mathcal{A}$ is a $*$-algebra generated by identity $\mathit{1}$ and two hermitian elements p and q satisfying the Heisenberg commutation relation :

$$[p, q] \equiv pq - qp = -i\mathit{1}. \tag{4.3.1}$$

All of results obtained here hold for the canonical algebra for s-degree of freedom. It is not possible to find hermitian matrices P and Q satifsying the Heisenberg commutation relation (4.3.1), however Heisenberg found a solution in the form of infinite matrices:

$$P = \frac{-i}{\sqrt{2}}\begin{pmatrix} 0 & \sqrt{1} & 0 & 0 & 0 & 0 \cdots \\ -\sqrt{1} & 0 & \sqrt{2} & 0 & 0 & 0 \cdots \\ 0 & -\sqrt{2} & 0 & \sqrt{3} & 0 & 0 \cdots \\ 0 & 0 & -\sqrt{3} & 0 & \sqrt{4} & 0 \cdots \\ 0 & 0 & 0 & -\sqrt{4} & 0 & \cdots \\ \cdots \end{pmatrix},$$

$$Q = \frac{1}{\sqrt{2}}\begin{pmatrix} 0 & \sqrt{1} & 0 & 0 & 0 & 0 \cdots \\ \sqrt{1} & 0 & \sqrt{2} & 0 & 0 & 0 \cdots \\ 0 & \sqrt{2} & 0 & \sqrt{3} & 0 & 0 \cdots \\ 0 & 0 & \sqrt{3} & 0 & \sqrt{4} & 0 \cdots \\ 0 & 0 & 0 & \sqrt{4} & 0 & \cdots \\ \cdots \end{pmatrix}. \tag{4.3.2}$$

Schrödinger found another solution to the Heisenberg solution in his formulation of quantum mechanics. The operators P_0 and Q_0 in $L^2(\mathbb{R})$ defined by

$$(P_0\xi)(t) = -i\frac{d}{dt}\xi(t),$$
$$(Q_0\xi)(t) = t\xi(t), \qquad \xi \in \mathcal{S}(\mathbb{R})$$

are essentially self-adjoint operators satisfying the Heisenberg commutation relation. Sch-
rödinger showed that his formulation agreed with the Heisenberg formulation, that is, P_0 and Q_0 have the same matirix elements as given in equation (4.3.2) with respect to an orthonormal basis for $L^2(\mathbb{R})$ defined by

$$f_n(t) = \pi^{-\frac{1}{4}}(2^n n!)^{-\frac{1}{2}}(t - \frac{d}{dt})^n e^{-\frac{t^2}{2}}, \quad n \in \mathbb{N} \cup \{0\}.$$

We define a $*$-representation π_0 of $\mathcal{A}$ on $L^2(\mathbb{R})$ with $\mathcal{D}(\pi_0) = \mathcal{S}(\mathbb{R})$ by $\pi_0(p) = P_0$ and $\pi_0(q) = Q_0$, and it is called the *Schrödinger representation* of $\mathcal{A}$. We note some well-known results for the Schrödinger representation. Let $a^+ = 2^{-\frac{1}{2}}(q - ip)$ and $a^- = 2^{-\frac{1}{2}}(q + ip)$. Since $a^-a^+ - a^+a^- = 1$, $\mathcal{A}$ is generated by $\{1, a^+, a^-\}$. We have the following

Lemma 4.3.1.
(1) $\pi_0(a^+)f_n = \sqrt{n+1}f_{n+1}, \quad n = 0, 1, \cdots$

$$\pi_0(a^-)f_n = \begin{cases} 0 & , n = 0 \\ \sqrt{n}f_{n-1} & , n \geq 1. \end{cases}$$

(2) $\pi_0(a^-a^+)$ is essentially self-adjoint and $\overline{\pi_0(a^-a^+)} = \sum_{n=0}^{\infty}(n+1)f_n \otimes \overline{f_n}$.

(3) $\mathcal{D}(\pi_0) = \mathcal{S}(\mathbb{R}) = \bigcap_{n\in\mathbb{N}} \mathcal{D}(\overline{\pi_0(a^-a^+)}^n)$ and the graph topology $t_{\pi_0} =$ the Schwartz space topology on $\mathcal{S}(\mathbb{R})$ = the Fréchet topology defined by the sequence $\{\|\ \|_n; n \in \mathbb{N} \cup \{0\}\}$ of seminorms $\|\xi\|_n \equiv \|\pi_0(a^-a^+)^n\xi\|, \xi \in \mathcal{S}(\mathbb{R})$.

(4) f_0 is a strongly cyclic vector for π_0.

(5) π_0 is a faithful, self-adjoint representation of $\mathcal{A}$ such that $\pi_0(\mathcal{A})'_{\mathrm{w}} = \mathbb{C}I$.

This raised the question as to whether there were other self-adjoint representations π of $\mathcal{A}$ such that $\pi(p)$ and $\pi(q)$ are essentially self-adjoint. Weyl [1] proposed the question should be formulated in terms of the strongly continuous unitary groups $U(s) \equiv \exp is\overline{\pi(p)}$ and $V(t) \equiv \exp it\overline{\pi(q)}$. Suppose $U(s)\mathcal{D}(\pi) \subset \mathcal{D}(\pi)$ for all $s \in \mathbb{R}$. Then we have

$$e^{is\overline{\pi(p)}}\pi(q)e^{-is\overline{\pi(p)}}\xi = \pi(q)\xi + is[\pi(p), \pi(q)]\xi + \frac{(is)^2}{2!}[\pi(p), [\pi(p), \pi(q)]]\xi + \cdots$$
$$= (\pi(q) + sI)\xi \tag{4.3.3}$$

for all $\xi \in \mathcal{D}(\pi)$ and $s \in \mathbb{R}$, and so π is not bounded. Exponentiating both sides of equation (4.3.3) we have the *Weyl formulation of the commutation relations*:

$$U(s)V(t) = e^{ist}V(t)U(s), \quad {}^{\forall}s,t \in \mathbb{R}. \tag{4.3.4}$$

For the Schrödinger representation π_0 we have

$$(U(s)\xi)(x) = \xi(x+s),$$
$$(V(t)\xi)(x) = e^{itx}\xi(x)$$

for each $\xi \in L^2(\mathbb{R})$ and $s,t \in \mathbb{R}$, and so $\{U(s), V(t); s,t \in \mathbb{R}\}$ satisfies the Weyl commutation relations. Von Neumann [1] characterized self-adjoint representations π satisfying the Weyl formulation as follows:

Theorem 4.3.2. Let π be a self-adjoint representation of $\mathcal{A}$ such that $\pi(p)$ and $\pi(q)$ are essentially self-adjoint. Then the following statements are equivalent:

(i) $\{U(s), V(t); s,t \in \mathbb{R}\}$ satisfies the Weyl commutation relations.

(ii) π is a direct sum of $\bigoplus_{\alpha \in I_0} \pi_\alpha$ of $*$-representations π_α which are unitarily equivalent to the Schrödinger representation. We call such a $*$-representation the *Weyl representation of the cardinal* I_0.

(iii) $U(s)\mathcal{D}(\pi) \subset \mathcal{D}(\pi)$ for all $s \in \mathbb{R}$.

Fugledge [1] constructed a self-adjoint representation but not a Weyl representation.

Example 4.3.3. Let π be a $*$-representation of $\mathcal{A}$ on $L^2(\mathbb{R})$ defined by

$$\mathcal{D}(\pi) = \text{ linear span of } \{x^n e^{-rx^2+cx} \ ; \ n = 0,1,2,\cdots, r > 0, c \in \mathbb{C}\},$$
$$(\pi(p)\xi)(x) = -i\xi'(x) + e^{-\sqrt{2\pi}x}\xi(x),$$
$$(\pi(q)\xi)(x) = x\xi(x) + \xi(x + i\sqrt{2\pi}), \qquad \xi \in \mathcal{D}(\pi).$$

Then $\pi(p)$ and $\pi(q)$ are essentially adjoint and π^{**} is not a Weyl representation of $\mathcal{A}$.

The following problem arised again: under what conditions a $*$-representation π of $\mathcal{A}$ is a Weyl representation?

Dixmier [3] obtained the following

Theorem 4.3.4. Let π be a $*$-representation of $\mathcal{A}$. Then the following statements are equivalent:

(i) π is a Weyl representation.

(ii) $\pi(p^2 + q^2)$ is essentially self-adjoint.

Powers [1] defined the notion of strong positivity of self-adjoint representations of $\mathcal{A}$ and using it he characterized the Weyl representations.

Definition 4.3.5. Let ω be a positive linear functional on $\mathcal{A}$. If $\pi_0(a) \geq 0$ implies $\omega(a) \geq 0$, then ω is said to be *strongly positive* . A $*$-representation π of $\mathcal{A}$ is said to be *strongly positive* if $\pi_0(a) \geq 0$ implies $\pi(a) \geq 0$.

Lemma 4.3.6. Let π be a Weyl representation of $\mathcal{A}$. Suppose π_1 is a self-adjoint subrepresentation of π. Then π_1 is a Weyl representation of $\mathcal{A}$.

Proof. Since π is a Weyl representation of $\mathcal{A}$ on $\mathcal{H}$, we have $\pi = \bigoplus_{\alpha \in I_0} \pi_\alpha$ and $\mathcal{H} = \bigoplus_{\alpha \in I_0} \mathcal{H}_{\alpha_0}$, where $\pi_\alpha = \pi_0$ and $\mathcal{H}_\alpha = L^2(\mathbb{R})$ for all $\alpha \in I_0$. The vector $f_0(t) \equiv \pi^{-\frac{1}{4}} e^{-\frac{1}{2}t^2}$ is a strongly cyclic vector for π_0. It is easily shown that $\pi_0(ip+q)f_0 = 0$ and

$$\eta \in \mathcal{D}(\pi) \text{ and } \pi_0(ip+q)\eta = 0 \text{ implies } \eta = \alpha f_0 \text{ for some } \alpha \in \mathbb{C}. \quad (4.3.5)$$

Let $\mathfrak{M}_0$ be a closed linear span of $\{\overset{\circ}{f}_\alpha; \alpha \in I_0\}$ in the Hilbert space $\mathcal{H}$, where $f_\alpha = f_0$ and $\overset{\circ}{f}_\alpha = (\delta_{\alpha\beta} f_\beta)_{\beta \in I_0}$ for each $\alpha \in I_0$. Then we have

$$\mathfrak{M}_0 \subset \mathcal{D}(\pi)$$

and (4.3.6)

$$\xi \in \mathfrak{M}_0 \text{ iff } \xi \in \mathcal{D}(\pi) \text{ and } \pi(ip+q)\xi = 0.$$

In fact, let $\xi \in \mathfrak{M}_0$. There exists a sequence $\{\xi_n\}$ in $\mathcal{H}$ such that ξ_n is a finite linear combination $\sum_{\alpha \in I_0} c_n(\alpha) \overset{\circ}{f}_\alpha$ of $\overset{\circ}{f}_\alpha$ and $\lim_{n\to\infty} \xi_n = \xi$. Then since

$$\begin{aligned}
\|\pi(a)(\xi_n - \xi_m)\|^2 &= \| \sum_{\alpha \in I_0} (c_n(\alpha) - c_m(\alpha))\pi_0(a) \overset{\circ}{f}_\alpha \|^2 \\
&= \sum_{\alpha \in I_0} |c_n(\alpha) - c_m(\alpha)|^2 \|\pi_0(a) \overset{\circ}{f}_\alpha \|^2 \\
&= \|\xi_n - \xi_m\|^2 \|\pi_0(a) f_0\|^2 \\
&\to 0 \quad (n, m \to \infty)
\end{aligned}$$

for all $a \in \mathcal{A}$, it follows from the closedness of π that $\xi \in \mathcal{D}(\pi)$. Furthermore, since $\pi(ip+q)\xi_n = 0$ for all $n \in \mathbb{N}$, we have $\pi(ip+q)\xi = \lim_{n\to\infty} \pi(ip+q)\xi_n = 0$. Conversely, suppose $\xi = (\xi_\alpha)_{\alpha \in I_0} \in \mathcal{D}(\pi)$ and $\pi(ip+q)\xi = 0$. Then we have $\pi_0(ip+q)\xi_\alpha = 0$ for each $\alpha \in I_0$. By (4.3.5), $\xi_\alpha = c(\alpha) f_0$ for some $c(\alpha) \in \mathbb{C}$. Hence we have $\xi \in \mathfrak{M}_0$.

Since π_1 is a self-adjoint subrepresentation of π, it follows from Theorem 1.3.3 that $\pi_1 = \pi_E$ for some projection E in $\pi(\mathcal{A})'_{\mathrm{w}}$. Let $\mathfrak{N} = E\mathfrak{M}_0$. By (4.3.5) we have

$$\pi(ip+q)E\xi = E\pi(ip+q)\xi = 0$$

for each $\xi \in \mathfrak{M}_0$, and so $E\xi \in \mathfrak{M}_0$. Hence we have

$$\mathfrak{N} = E\mathfrak{M}_0 \subset \mathfrak{M}_0. \tag{4.3.7}$$

Let $\{\eta_\beta; \beta \in I_1\}$ be an orthonormal basis for $\mathfrak{N}$ and put $\mathfrak{N}_\beta = \pi(\mathcal{A})\eta_\beta$. By (4.3.7) we have $\eta_\beta, \eta_{\beta'} \in \mathfrak{M}_0$ and so $\eta_\beta = (c(\alpha)f_\alpha)_{\alpha\in I_0}$ and $\eta_{\beta'} = (c'(\alpha)f_\alpha)_{\alpha\in I_0}$. Then since

$$\begin{aligned}(\pi(a)\eta_\beta|\eta_{\beta'}) &= \sum_{\alpha\in I_0} c(\alpha)\overline{c'(\alpha)}(\pi_0(a)f_0|f_0)\\ &= (\eta_\beta|\eta_{\beta'})(\pi_0(a)f_0|f_0)\\ &= \delta_{\beta\beta'}(\pi_0(a)f_0|f_0)\end{aligned}$$

for all $a \in \mathcal{A}$, it follows that $\pi_\beta \equiv \pi_{\mathfrak{N}_\beta}$ is unitarily equivalent to π_0, so that $\pi_2 \equiv \bigoplus_{\beta\in I_0} \pi_\beta$ is a self-adjoint subrepresentation of π. By Theorem 1.3.3, $\pi_2 = \pi_F$ for some projection F in $\pi(\mathcal{A})'_{\mathrm{w}}$. We show $\pi_1 = \pi_2$. In fact, we have $\mathfrak{N} = E\mathfrak{M}_0 \subset E\mathcal{D}(\pi) = \mathcal{D}(\pi_1)$, and so $\eta_\beta \in \mathcal{D}(\pi_1)$ and $\mathfrak{N}_\beta = \pi(\mathcal{A})\eta_\beta = \pi_1(\mathcal{A})\eta_\beta \subset \mathcal{D}(\pi_1)$. Hence, $\pi_2 \subset \pi_1$, and so $F \leq E$. Take an arbitrary $\xi \in \mathcal{D}(\pi_1)$. Since $\pi = \bigoplus_{\alpha\in I_0} \pi_\alpha$ and f_0 is a strongly cyclic vector for $\pi_\alpha = \pi_0$, it follows that for any $\varepsilon > 0$ there exists a sequence $\{a_n\}$ in $\mathcal{A}$ such that $\|\xi - \sum_{k=1}^{n} \pi(a_k)\, \mathring{f}_k\,\| < \varepsilon$. Then we have

$$\begin{aligned}\|\xi - \sum_{k=1}^{n} \pi(a_k)E\, \mathring{f}_k\,\| &= \|E\xi - E\sum_{k=1}^{n} \pi(a_k)\, \mathring{f}_k\,\|\\ &\leq \|\xi - \sum_{k=1}^{n} \pi(a_k)\, \mathring{f}_k\,\|\\ &< \varepsilon.\end{aligned} \tag{4.3.8}$$

Further, since $E\, \mathring{f}_\alpha \in E\mathfrak{M}_0 = \mathfrak{N}$ for each $\alpha \in I_0$ and $\eta_\beta \in \mathcal{D}(\pi_2) \subset F\mathcal{H}$ for each $\beta \in I_1$, we have $E\, \mathring{f}_\alpha \in F\mathcal{H}$ for each $\alpha \in I_0$, which implies by (4.3.8) that $\xi \in F\mathcal{H}$. Hence, $E \leq F$. Thus we have $\pi_1 = \pi_2$. This completes the proof.

Theorem 4.3.7. Let π be a $*$-representation of $\mathcal{A}$ with a strongly cyclic vector ξ_0. Then the following statements are equivalent:

(i) π is a Weyl representation of at most countable cardinal.

(ii) π is self-adjoint and $\omega_{\xi_0} \circ \pi$ is strongly positive.

Proof. (i) $\Rightarrow$ (ii) This is trivial.

(ii) $\Rightarrow$ (i) By Example 4.1.9 there exists a sequence $\{\xi_n\}$ in $\mathcal{S}(\mathbb{R})$ such that

$$(\pi(a)\xi_0|\xi_0) = \sum_{n=1}^{\infty}(\pi_0(a)\xi_n|\xi_n) \tag{4.3.9}$$

for all $a \in \mathcal{A}$. Let $\pi_1 = \bigoplus_{n=1}^{\infty}\pi_0$ and $\xi \equiv \{\xi_n\}_{n\in\mathbb{N}}$. By (4.3.9) we have $\xi \in \mathcal{D}(\pi_1)$ and

$$(\pi(a)\xi_0|\xi_0) = (\pi_1(a)\xi|\xi), \qquad x \in \mathcal{A}. \tag{4.3.10}$$

We put

$$\mathfrak{M} = \pi_1(\mathcal{A})\xi \qquad \text{and} \qquad \pi_2 = (\pi_1)_{\mathfrak{M}}.$$

Then it follows from (4.3.10) that π_2 is unitary equivalent to π, and so π_2 is a self-adjoint subrepresentation of π_1. By Lemma 4.3.6 π_2 is a Weyl representation. Hence π is a Weyl representation of at most countable cardinal. This completes the proof.

Corollary 4.3.8. Let ω be a positive linear functional on $\mathcal{A}$. Then the following statements are equivalent:
(i) ω is strongly positive and π_ω is self-adjoint.
(ii) π_ω is a Weyl representation of at most countable cardinal.

Further, we show that a Weyl representation of countable cardinal is unitarily equivalent to the self-adjoint representation $\pi_\otimes$ of $\mathcal{A}$ on the Hilbert space of Hilbert -Schmidt operators on $L^2(\mathbb{R})$ defined by

$$\mathcal{D}(\pi_\otimes) = \mathcal{S}(\mathbb{R}) \otimes \overline{L^2(\mathbb{R})},$$
$$\pi_\otimes(a)T = \pi_0(a)T, \qquad a \in \mathcal{A},\ T \in \mathcal{D}(\pi_\otimes).$$

For any $\beta > 0$ we put

$$\Omega_\beta = \sum_{n=0}^{\infty} e^{-n\frac{\beta}{2}} f_n \otimes \overline{f_n}.$$

Gudder and Hudson [1] obtained the following

Lemma 4.3.9. Ω_β is a strongly cyclic vector for $\pi_\otimes$.

Theorem 4.3.10. Let π be a $*$-representation of $\mathcal{A}$. Then the following statements are equivalent:
(i) π is a Weyl representation of countable cardinal.
(ii) π is unitarily equivalent to $\pi_\otimes$.

Proof. We put

$$U\{\xi_n\} = \sum_{n=0}^{\infty} \xi_n \otimes \overline{f_n}, \qquad \{\xi_n\} \in \bigoplus_{n=0}^{\infty} L^2(\mathbb{R}).$$

Then it is easily shown that U is a unitary transformation of $\bigoplus_{n=0}^{\infty} L^2(\mathbb{R})$ onto $L^2(\mathbb{R}) \otimes \overline{L^2(\mathbb{R})}$.

(i) $\Rightarrow$ (ii) Let $\pi = \bigoplus_{n=0}^{\infty} \pi_n$, $\pi_n = \pi_0$ $(n = 0, 1, 2, \cdots)$. Since

$$\|\pi_\otimes(a)U\{\xi_n\}\|^2 = \|\pi_\otimes(a) \sum_{n=0}^{\infty} \xi_n \otimes \overline{f_n}\|_2^2 = \sum_{n=0}^{\infty} \|\pi_0(a)\xi_n\|^2$$

for each $\{\xi_n\} \in \mathcal{D}(\bigoplus_{n=0}^{\infty} \pi_n)$, we have $U\mathcal{D}(\bigoplus_{n=0}^{\infty} \pi_n) \subset \mathcal{S}(\mathbb{R}) \otimes \overline{L^2(\mathbb{R})}$. Take arbitrary $\xi \in \mathcal{S}(\mathbb{R})$ and $\eta \in L^2(\mathbb{R})$. Since $\eta = \sum_{n=0}^{\infty} \alpha_n f_n$ for some $\{\alpha_n\} \subset \mathbb{C}$, we have

$$\xi \otimes \overline{\eta} = \sum_{n=0}^{\infty} \overline{\alpha_n} \xi \otimes \overline{f_n},$$

$$\sum_{n=0}^{\infty} \|\overline{\alpha_n} \pi_0(a)\xi\|^2 = (\sum_{n=0}^{\infty} |\alpha_n|^2) \|\pi_0(a)\xi\|^2,$$

so that $\{\overline{\alpha_n}\xi\} \in \mathcal{D}(\bigoplus_{n=0}^{\infty} \pi_n)$ and $\xi \otimes \overline{\eta} \in U\mathcal{D}(\bigoplus_{n=0}^{\infty} \pi_n)$. Hence we have $U\mathcal{D}(\bigoplus_{n=0}^{\infty} \pi_n)$ $= \mathcal{S}(\mathbb{R}) \otimes \overline{L^2(\mathbb{R})}$. Further, since

$$\pi_\otimes(a)U\{\xi_n\} = \sum_{n=0}^{\infty} \pi_0(a)\xi_n \otimes \overline{f_n} = U\{\pi_0(a)\xi_n\} = U(\bigoplus_{n=0}^{\infty} \pi_n)(a)\{\xi_n\}$$

for each $a \in \mathcal{A}$ and $\{\xi_n\} \in \mathcal{D}(\bigoplus_{n=0}^{\infty} \pi_n)$, it follows that $\pi_\otimes$ is unitarily equivalent to $\bigoplus_{n=0}^{\infty} \pi_n$.

(ii) $\Rightarrow$ (i) By Lemma 4.3.9 Ω_β is a strongly cyclic vector for $\pi_\otimes$, and so $\omega_{\Omega_\beta} \circ \pi_\otimes$ is strongly positive. Hence it follows from Theorem 4.3.7 that $\pi_\otimes$ is a Weyl representation of at most countable cardinal. Further, since U is a unitary transformation of $\bigoplus_{n=0}^{\infty} L^2(\mathbb{R})$ onto $L^2(\mathbb{R}) \otimes \overline{L^2(\mathbb{R})}$, it follows that $\pi_\otimes$ is a Weyl representation of countable cardinal. This completes the proof.

By Theorem 4.3.10 it is useful to consider the self-adjoint O*-algebra $\pi_\otimes(\mathcal{A})$ on $\mathcal{S}(\mathbb{R}) \otimes \overline{L^2(\mathbb{R})}$ for the study of Tomita-Takesaki theory for CCR-algebras, and apply the results of Section 2.3, D to it. We put

$$\Omega_\beta (\equiv \Omega_{\{e^{-n\beta/2}\}}) = \sum_{n=0}^{\infty} e^{-n\frac{\beta}{2}} f_n \otimes \overline{f_n}, \quad \beta \in \mathbb{R}.$$

Then Ω_β is a non-singular positive self-adjoint operator in $L^2(\mathbb{R})$ and if $\beta > 0$ then $\Omega_\beta \in \mathcal{S}(\mathbb{R}) \otimes \overline{L^2(\mathbb{R})}$.

Lemma 4.3.11. $\Omega_\beta^{it}\pi_0(\mathcal{A})\Omega_\beta^{-it} = \pi_0(\mathcal{A})$ for each $t \in \mathbb{R}$.

Proof. For each $\xi \in \mathcal{S}(\mathbb{R})$ and $t \in \mathbb{R}$ we have

$$\begin{aligned}
\pi_0(a^+)\Omega_\beta^{it}\xi &= \sum_{n=0}^{\infty} e^{-\beta n \frac{it}{2}}(\xi|f_n)\pi_0(a^+)f_n \\
&= \sum_{n=0}^{\infty} e^{-\beta n \frac{it}{2}}\sqrt{n+1}(\xi|f_n)f_{n+1} \\
&= \sum_{n=0}^{\infty} e^{-\beta n \frac{it}{2}}(\xi|\pi_0(a^-)f_{n+1})f_{n+1} \\
&= e^{\beta\frac{it}{2}}\sum_{n=0}^{\infty} e^{-\beta(n+1) \frac{it}{2}}(\xi|\pi_0(a^-)f_{n+1})f_{n+1} \\
&= e^{\beta\frac{it}{2}}\sum_{n=1}^{\infty} e^{-\beta n \frac{it}{2}}(\xi|\pi_0(a^-)f_n)f_n \\
&= e^{\beta\frac{it}{2}}\sum_{n=0}^{\infty} e^{-\beta n \frac{it}{2}}(\pi_0(a^+)\xi|f_n)f_n \\
&= e^{\beta\frac{it}{2}}\Omega_\beta^{it}\pi_0(a^+)\xi.
\end{aligned}$$

Similarly, we have

$$\pi_0(a^{-m})\pi_0(a^{+n})\Omega_\beta^{it}\xi = e^{-(m-n)\beta\frac{it}{2}}\Omega_\beta^{it}\pi_0(a^{-m})\pi_0(a^{+n})\xi$$

for each $m, n \in \mathbb{N}\cup\{0\}$, $\xi \in \mathcal{S}(\mathbb{R})$ and $t \in \mathbb{R}$. Hence, we have $\Omega_\beta^{it}\pi_0(\mathcal{A})\Omega_\beta^{-it} = \pi_0(\mathcal{A})$ for each $t \in \mathbb{R}$.

Theorem 4.3.12. Let $\beta > 0$. Then the following statements hold:
(1) Ω_β is a standard vector for $\pi_\otimes(\mathcal{A})$.
(2) $\pi_\otimes(\mathcal{A})\Omega_\beta \subset \bigcap_{\alpha\in\mathbb{C}} \mathcal{D}(\Delta_{\Omega_\beta}^\alpha)$ and it is a core for each $\Delta_{\Omega_\beta}^\alpha$.
(3) There exists a complex one-parameter group $\{\Delta_\beta(\alpha); \alpha \in \mathbb{C}\}$ of automorphisms of $\mathcal{A}$ such that $\Delta_{\Omega_\beta}^\alpha\pi_\otimes(x)\Omega_\beta = \pi_\otimes(\Delta_\beta(\alpha)x)\Omega_\beta$ for all $x \in \mathcal{A}$ and $\alpha \in \mathbb{C}$.

Proof. (1) This follows from Lemma 4.3.9, 4.3.10 and Theorem 2.4.18.
(2) By Theorem 2.4.18 we have

$$\Delta_{\Omega_\beta} = \pi'(\Omega_\beta^{-2})\pi''(\Omega_\beta^2),$$

and so for each $\alpha \in \mathbb{C}$

$$\begin{aligned}
\Delta^{\alpha}_{\Omega_\beta}\pi_\otimes(a^+)\Omega_\beta &= \pi'(\Omega_\beta^{-2\alpha})\pi''(\Omega_\beta^{2\alpha})\pi_\otimes(a^+)\Omega_\beta \\
&= \sum_{n=0}^{\infty} e^{(-\frac{1}{2}+\alpha)\beta n}\sqrt{n+1}\pi''(\Omega_\beta^{2\alpha})f_{n+1}\otimes\overline{f_n} \\
&= \sum_{n=0}^{\infty} e^{(-\frac{1}{2}+\alpha)\beta n}\sqrt{n+1}(\sum_{k-0}^{\infty} e^{-\beta\alpha k} f_k\otimes\overline{f_k})(f_{n+1}\otimes\overline{f_n}) \\
&= e^{-\beta\alpha}\sum_{n=0}^{\infty} e^{-\beta\frac{n}{2}}\pi_\otimes(a^+)f_n\otimes\overline{f_n} \\
&= e^{-\beta\alpha}\pi_\otimes(a^+)\Omega_\beta, \qquad (4.3.11)
\end{aligned}$$

and

$$\Delta^{\alpha}_{\Omega_\beta}\pi_\otimes(a^-)\Omega_\beta = e^{\beta\alpha}\pi_\otimes(a^-)\Omega_\beta. \qquad (4.3.12)$$

Similarly we have

$$\Delta^{\alpha}_{\Omega_\beta}\pi_\otimes(a^{+j}a^{-k})\Omega_\beta = e^{-(j-k)\beta\alpha}\pi_\otimes(a^{+j}a^{-k})\Omega_\beta$$

for $j, k = 0, 1, 2, \cdots$, and so since $\pi_\otimes(\mathcal{A})\Omega_\beta \subset \mathcal{D}(\Delta^{\alpha}_{\Omega_\beta})$ and $(I + \Delta^{\mathrm{Re}\,\alpha}_{\Omega_\beta})$ $\pi_\otimes(\mathcal{A})\ \Omega_\beta(= \pi_\otimes(\mathcal{A})\Omega_\beta)$ is dense in $L^2(\mathbb{R})\otimes\overline{L^2(\mathbb{R})}$, it follows that $\pi_\otimes(\mathcal{A})\Omega_\beta$ is a core for $\Delta^{\alpha}_{\Omega_\beta}$.

(3) By (4.3.11) and (4.3.12) we have

$$\begin{aligned}
\Delta^{\alpha}_{\Omega_\beta}\pi_\otimes(p)\Omega_\beta &= \pi_\otimes(p^{\Delta_\beta(\alpha)})\Omega_\beta, \\
\Delta^{\alpha}_{\Omega_\beta}\pi_\otimes(q)\Omega_\beta &= \pi_\otimes(q^{\Delta_\beta(\alpha)})\Omega_\beta,
\end{aligned}$$

where

$$\begin{aligned}
p^{\Delta_\beta(\alpha)} &= \cosh(\beta\alpha)p - i\sinh(\beta\alpha)q, \\
q^{\Delta_\beta(\alpha)} &= i\sinh(\beta\alpha)p + \cosh(\beta\alpha)q.
\end{aligned}$$

Since $p^{\Delta_\beta(\alpha)}q^{\Delta_\beta(\alpha)} - q^{\Delta_\beta(\alpha)}p^{\Delta_\beta(\alpha)} = -i1$, there is a unique automorphism $\Delta_\beta(\alpha)$ of $\mathcal{A}$ under which $p^{\Delta_\beta(\alpha)}$ and $q^{\Delta_\beta(\alpha)}$ are the images of p and q, respectively, and

$$\Delta^{\alpha}_{\Omega_\beta}\pi_\otimes(x)\Omega_\beta = \pi_\otimes(\Delta_\beta(\alpha)x)\Omega_\beta, \qquad x\in\mathcal{A}, \alpha\in\mathbb{C}.$$

Further, it is shown that $\{\Delta_\beta(\alpha); \alpha\in\mathbb{C}\}$ is a one-parameter group of automorphisms of $\mathcal{A}$. This completes the proof.

Theorem 4.3.12 shows that $\pi_\otimes(\mathcal{A})\Omega_\beta$ is an *unbounded Tomita algebra* in $L^2(\mathbb{R})\otimes\overline{L^2(\mathbb{R})}$, that is, it has the properties of a Tomita algebra (Takesaki [1]), with the exception of the continuity of left multiplication.

We consider generalized vectors Ω_β ($\beta \le 0$). Let $\mathcal{M}$ be an O*-algebra on $\mathcal{S}(\mathbb{R})$ generated by $\pi_0(\mathcal{A})$ and $f_0 \otimes \overline{f_0}$. Since

$$f_n \otimes \overline{f_m} = \frac{1}{\sqrt{n!m!}} \pi_0(a^{+n})(f_0 \otimes \overline{f_0})\pi_0(a^{-m})$$

for $n, m \in \mathbb{N} \cup \{0\}$, we have

$$\{f_n \otimes \overline{f_m}; n, m \in \mathbb{N} \cup \{0\}\} \subset \mathcal{M}. \tag{4.3.13}$$

Let $\alpha_n > 0$, $n = 0, 1, 2, \cdots$ and put

$$\Omega_{\{\alpha_n\}} = \sum_{n=0}^{\infty} \alpha_n f_n \otimes \overline{f_n}.$$

Then $\Omega_{\{\alpha_n\}}$ is a non-singular positive self-adjoint operator in $L^2(\mathbb{R})$. Let π be a self-adjoint representation of the O*-algebra $\mathcal{M}$ defined by

$$\mathcal{D}(\pi) = \mathcal{S}(\mathbb{R}) \otimes \overline{L^2(\mathbb{R})} \text{ and } \pi(X)T = XT, \quad X \in \mathcal{M}, T \in \mathcal{S}(\mathbb{R}) \otimes \overline{L^2(\mathbb{R})}.$$

This has been defined in Section 2.4, D.

Theorem 4.3.13. Let $\mathcal{M}$ be an O*-algebra on $\mathcal{S}(\mathbb{R})$ generated by $\pi_0(\mathcal{A})$ and $f_0 \otimes \overline{f_0}$ and let $\alpha_n > 0$, $n = 0, 1, \cdots$. Then $\lambda_{\Omega_{\{\alpha_n\}}}$ is a quasi-standard generalized vector for $\pi(\mathcal{M})$. Further, $\lambda_{\Omega_{\{\alpha_n\}}}$ is standard if and only if $\alpha_n = e^{\beta n}$, $n \in \mathbb{N} \cup \{0\}$ for some $\beta \in \mathbb{R}$.

Proof. By (4.3.13) it is easily shown that the conditions (i) and (ii) in Theorem 2.4.23 hold. We show only (iii) in Theorem 2.4.23. Let $\mathbb{N}_1 = \{n \in \mathbb{N}; \alpha_n > \frac{1}{n}\} \cup \{0\}$ and $\mathbb{N}_2 = \{n \in \mathbb{N}; \alpha_n \le \frac{1}{n}\}$. For each $n \in \mathbb{N} \cup \{0\}$ we put

$$r'_n = \begin{cases} e^{-n\beta} & , n \in \mathbb{N}_1 \\ 0 & , n \notin \mathbb{N}_1, \end{cases} \qquad r''_n = \begin{cases} e^{-n\beta}\alpha_n & , n \in \mathbb{N}_2 \\ 0 & , n \notin \mathbb{N}_2, \end{cases}$$

$$r_n = r'_n + r''_n,$$

$$A_n = \sum_{k=0}^{n} \frac{r_k}{\alpha_k} f_k \otimes \overline{f_k}.$$

Then, $\{r_n\} \in s_+$, $A_n \in \mathcal{M} \cap (\mathcal{S}(\mathbb{R}) \otimes \overline{L^2(\mathbb{R})})$ and

$$A_n \longrightarrow A \equiv \sum_{k \in \mathbb{N}_1} \frac{1}{\alpha_k} e^{-k} f_k \otimes \overline{f_k} + \sum_{k \in \mathbb{N}_2} e^{-k} f_k \otimes \overline{f_k}$$
$$\in \mathcal{S}(\mathbb{R}) \otimes \overline{L^2(\mathbb{R})}.$$

Further, we have

$$A_n, A_m X A_n \in \mathcal{D}(\lambda_{\Omega_{\{\alpha_n\}}})^\dagger \cap \mathcal{D}(\lambda_{\Omega_{\{\alpha_n\}}}), \qquad n, m \in \mathbb{N} \cup \{0\},$$
$$A_m^2 X A_n \Omega_{\{\alpha_n\}} \xrightarrow[n,m\to\infty]{} A^2 X \Omega_{\{r_n\}}$$

for each $X \in \mathcal{M}$, which implies by the non-singularity of A that $\lambda_{\Omega_{\{\alpha\}}}$ $((\mathcal{D}(\lambda_{\Omega_{\{\alpha_n\}}})^\dagger \cap \mathcal{D}(\lambda_{\Omega_{\{\alpha_n\}}}))^2)$ is total in $L^2(\mathbb{R}) \otimes \overline{L^2(\mathbb{R})}$. Since $\Omega^{it}_{\{\alpha_n\}} \mathcal{S}(\mathbb{R}) \subset \mathcal{S}(\mathbb{R})$ for each $t \in \mathbb{R}$, it follows from Theorem 2.4.23 that $\lambda_{\Omega_{\{\alpha_n\}}}$ is a quasi-standard generalized vector for $\pi(\mathcal{M})$. Suppose $\lambda_{\Omega_{\{\alpha_n\}}}$ is standard. Since

$$\pi_\otimes(a^+)\Omega^{it}_{\{\alpha_n\}} f_n = \alpha_n^{it}\sqrt{n+1} f_{n+1}$$

for each $n \in \mathbb{N} \cup \{0\}$ and $t \in \mathbb{R}$, it follows that α_n/α_{n+1} = constant for $n = 0, 1, 2, \cdots$, which implies $\alpha_n = e^{\beta n}, n = 0, 1, 2, \cdots$, for some $\beta \in \mathbb{R}$. The converse follows from Lemma 4.3.11. This completes the proof.

B. *Dynamics of an interacting Boson model*

We consider standard generalized vectors and standard quasi-weights in a class of interacting Boson models in Fock space.

Let $\mathcal{H}$ be a separable Hilbert space, and let $\mathcal{H}^n$ be the n-fold tensor product of $\mathcal{H}$. We define an operator S_n on $\mathcal{H}^n$ by

$$S_n(f_1 \otimes f_2 \otimes \cdots \otimes f_n) = (n!)^{-1} \sum_\pi f_{\pi_1} \otimes f_{\pi_2} \otimes \cdots \otimes f_{\pi_n},$$

where the sum is over all permutations. We put

$$\mathcal{F}_0(\mathcal{H}) = \mathbb{C},$$
$$\mathcal{F}_n(\mathcal{H}) = S_n \mathcal{H}^n,$$

and

$$\mathcal{F} = \bigoplus_{n=0}^{\infty} \mathcal{F}_n(\mathcal{H}).$$

$\mathcal{F}$ is called the Bose-Fock space. Let A be a self-adjoint operator in $\mathcal{H}$. We put

$$d\Gamma_0(A) = 0,$$
$$d\Gamma_n(A) = A \otimes I \otimes \cdots \otimes I + I \otimes A \otimes I \otimes \cdots \otimes I + \cdots + I \otimes I \otimes \cdots \otimes I \otimes A \quad (n \geq 1),$$

and

$$d\Gamma(A) = \bigoplus_{n=0}^{\infty} d\Gamma_n(A).$$

Then $d\Gamma(A)$ is a self-adjoint operator in $\mathcal{F}$. We denote by $\mathcal{F}_0$ the subspace in $\mathcal{F}$ spanned by vectors $\xi = \{\xi^{(n)}\}_{n=0}^{\infty} \in \mathcal{F}$ such that $\xi^{(n)} = 0$ for all but finitely many n. The subspace $\mathcal{F}_0$ is dense in $\mathcal{F}$. For each $f \in \mathcal{H}$ there exists a closed linear operator $a(f)$ in $\mathcal{F}$ such that

$a(f)\xi = \{0, \cdots, 0, (a(f)\xi)^{(k-1)}, 0, \cdots\}$, where

$$(a(f))^{(k-1)} = \frac{1}{\sqrt{k}} \sum_{j=1}^{k} (f|f_j) \mathcal{S}_{k-1}(f_1 \otimes \cdots \otimes f_{j-1} \otimes f_{j+1} \otimes \cdots \otimes f_k)$$

for $\xi = \{0, \cdots, 0, \xi^{(k)}, 0, \cdots\}$, where

$$\xi^{(k)} = \mathcal{S}_k(f_1 \otimes \cdots \otimes f_k), \qquad (f_1, \cdots, f_k \in \mathcal{H}).$$

The domain of $a(f)$ contains $\mathcal{F}_0$ and $a(f)$ leaves $\mathcal{F}_0$ invariant. For each $f \in \mathcal{H}$ there exists a closed linear operator $a^*(f)$ in $\mathcal{F}$ such that

$$(a^*(f)\xi)^{(n)} = \begin{cases} 0, & n = 0 \\ \sqrt{n}\mathcal{S}_n(f \otimes \xi^{(n-1)}), & n \geq 1 \end{cases}$$

for $\xi = \{\xi^{(n)}\}_{n=0}^{\infty} \in \mathcal{F}_0$. The domain of $a^*(f)$ leaves $\mathcal{F}_0$ invariant. The closed operators $a(f)$ and $a^*(f)$ are called the annihilation operator and the creation operator, respectively. We define a number operator N in $\mathcal{F}$ by

$$\begin{cases} D(N) = \{\xi;\ \xi = \{\xi^{(n)}\}_{n=0}^{\infty},\ \sum_{n=0}^{\infty} n^2 \|\xi^{(n)}\|^2 < +\infty\} \\ N\xi \ = \{n\xi^{(n)}\}_{n=0}^{\infty} \end{cases}$$

and put $\mathcal{D}^{\infty}(N) = \bigcap_{k=0}^{\infty} \mathcal{D}(N^k)$. For each $f \in \mathcal{H}$ the domains of $a(f)$ and $a^*(f)$ contain $\mathcal{D}^{\infty}(N)$, and $a(f)$ and $a^*(f)$ leave $\mathcal{D}^{\infty}(N)$ invariant. Let $\mathfrak{A}_{\mathrm{CCR}}$ be the self-adjoint O*-algebra on $\mathcal{D}^{\infty}(N)$ generated by the identity I and $\{a(f), a^*(f); f \in \mathcal{H}\}$ and it is a CCR-algebra over $\mathcal{H}$, that is,

(a) $a^*(f)$ is linear in f,

(b) $a(f)^{\dagger} = a^*(f), \quad f \in \mathcal{H}$,

(c) $[a(f), a^*(g)] = (f|g)I$,
$[a^*(f), a^*(g)] = [a(f), a(g)] = 0 \qquad$ for $f, g \in \mathcal{H}$.

Let Δ be the Laplacian operator in $L^2(\mathbb{R}^3)$. We put

$$\mathcal{H} = L^2(\mathbb{R}^3),$$

$$h = -\Delta.$$

A two-body potential is a real function Φ over $\mathbb{R}^3 \times \mathbb{R}^3$ whose values $\Phi(x_1, x_2)$ represent the potential energy of interaction between a particle at the point x_1 and a second particle at the point x_2. Thus the total interaction energy of n particles at the points $x_1, x_2, \cdots, x_n$ is given by

$$U^{(n)}(x_1,\cdots,x_n) = \sum_{1\le i<j\le n} \Phi(x_i,x_j).$$

Note that the symmetry of $U^{(n)}$ is reflected by the symmetry property

$$\Phi(x_i,x_j) = \Phi(x_j,x_i).$$

We assume that Φ is bounded, that is, there exists a constant c such that

$$|\Phi(x,y)| < c \text{ for all } x,y \in \mathbb{R}^3.$$

The interaction operator $V^{(n)}$ is defined by

$$(V^{(n)}f)(x_1,\cdots,x_n) = U^{(n)}(x_1,\cdots,x_n)f(x_1,\cdots,x_n) \quad (f \in \mathcal{F}_n(\mathcal{H})).$$

We put

$$H_n = d\Gamma_n(h) + V^{(n)},$$

and

$$H = \bigoplus_{n=0}^{\infty} H_n.$$

Then H is a self-adjoint operator in $\mathcal{F}$. Let $\mathcal{M}$ be the O*-algebra on $\mathcal{D}^\infty(N)$ generated by $\mathfrak{A}_{\mathrm{CCR}}$ and $\{f(N); f$ is a continuous function on $[0,\infty)$ and there exists a polynomial p such that $|f(x)| \le p(x)$ on $[0,\infty)\}$. Then $\mathcal{M}$ is a self-adjoint O*-algebra such that $\mathcal{M}_{\mathrm{w}}' = \mathbb{C}I$ and $\Omega = e^{-\frac{H}{2}}$ is a positive self-adjoint operator in $\mathcal{F}$. We have the following

Theorem 4.3.14. λ_Ω is a quasi-standard generalized vector for $\pi(\mathcal{M})$ on $\mathcal{D}^\infty(N)\otimes\overline{\mathcal{F}}$ in $\mathcal{F}\otimes\overline{\mathcal{F}}$. In particular, if the two-body potential Φ is constant, that is, $\Phi(x,y) = $ constant for all $x,y \in \mathbb{R}^3$, then λ_Ω is a standard generalized vector for $\pi(\mathcal{M})$.

Proof. Ω is a positive self-adjoint operator in $\mathcal{F}$ satisfying Ω^{-1} is densely defined. We put

$$\mathcal{F}_F(\mathcal{H}) = \{\xi = \{\xi^{(n)}\}_{n=0}^{\infty} \in \mathcal{F};\text{ there exists a positive integer } n_0 \text{ s.t. } \xi^{(n)} = 0 \text{ for all } n \ge n_0\}.$$

Since $\mathcal{F}_F(\mathcal{H})\cap\mathcal{D}(\Omega^{-1})$ is included in $\mathcal{D}\cap\mathcal{D}(\Omega^{-1})$, it follows that $\mathcal{D}\cap\mathcal{D}(\Omega^{-1})$ is a core for Ω^{-1}. We put

$$\mathcal{F}_S(\mathcal{H}) = \{\xi = \{\xi^{(n)}\}_{n=0}^{\infty} \in \mathcal{F}_F(\mathcal{H});\ \text{each } \xi^{(n)} \text{ is a finite linear combination of simple tensors}\}.$$

Let $\mathcal{R}$ be the set of all rank one projections constructed from $\mathcal{F}_S(\mathcal{H})$, and let $\mathcal{N}$ be the linear span of $\{R_1\mathcal{M}R_2;\ R_1, R_2 \in \mathcal{R}\}$. Since $\mathcal{R} \subset \mathcal{M}$, we have $\mathcal{N} \subset \mathcal{M}$. And we have

$$\pi(\mathcal{N}) \subset \mathcal{D}(\lambda_\Omega),\ \mathcal{N}^\dagger\mathcal{D} \subset \mathcal{D}(\Omega),$$

and the linear span of $\mathcal{N}^\dagger\mathcal{D}$ is a core for Ω. Since $\mathcal{F}_S(\mathcal{H})$ is a subspace of $\mathcal{D} \cap \mathcal{D}(\Omega)$ such that $\{\xi \otimes \overline{\eta};\ \xi, \eta \in \mathcal{F}_S(\mathcal{H})\} \subset \mathcal{M}$ and $\mathcal{F}_S(\mathcal{H})$ is a core for Ω, by Lemma 2.4.22, we have $\lambda_\Omega((D(\lambda_\Omega) \cap D(\lambda_\Omega)^\dagger)^2)$ is total in $\mathcal{F} \otimes \overline{\mathcal{F}}$. Furthermore, since we have $\Omega^{it}\mathcal{D} \subset \mathcal{D}$ for all $t \in \mathbb{R}$, it follows from Theorem 2.4.23, λ_Ω is a quasi-standard generalized vector for $\pi(\mathcal{M})$.

By the way, if the two-body potential Φ is constant, then we have

$$\Omega^{it}\mathcal{M}\Omega^{-it} = \mathcal{M},\ {}^\forall t \in \mathbb{R}.$$

Hence, by Theorem 2.4.23, λ_Ω is a standard generalized vector for $\pi(\mathcal{M})$. This complets the proof.

Using Theorem 2.2.4 and Proposition 3.2.9, we can construct from λ_Ω the quasi-weight ω_{λ_Ω} on $\mathcal{P}(\pi(\mathcal{M})''_{\text{wc}})$, which satisfies the KMS condition. As in the previous case, we may interpret it as an equilibrium state of the interacting Boson model.

4.4 Standard systems in the BCS-Bogolubov model

Let Λ be a finite region of a lattice and $|\Lambda|$ the number of points in Λ. The local C^*-algebra $\mathfrak{A}_\Lambda$ is generated by the Pauli operators $\sigma_p = (\sigma_p^x, \sigma_p^y, \sigma_p^z)$ at every point $p \in \Lambda$. The σ_p are copies of the Pauli matrices

$$\sigma^x = \begin{pmatrix} 0 & 1 \\ 1 & 0 \end{pmatrix},\quad \sigma^y = \begin{pmatrix} 0 & -i \\ i & 0 \end{pmatrix},\quad \sigma^z = \begin{pmatrix} 1 & 0 \\ 0 & -1 \end{pmatrix}.$$

$\mathfrak{A}_\Lambda$ is isomorphic to the C^*-algebra of all $2^{|\Lambda|} \times 2^{|\Lambda|}$-matrices on the $2^{|\Lambda|}$-dimensional complex Hilbert space $\mathcal{H}_\Lambda = \bigotimes_{p \in \Lambda} C_p^2$, where C_p^2 is the 2-dimensional complex Hilbert space at p. If $\Lambda \subset \Lambda'$ and $A_\Lambda \in \mathfrak{A}_\Lambda$, then $A_\Lambda \to A_{\Lambda'} = A_\Lambda \otimes (\bigotimes_{p \in \Lambda' \setminus \Lambda} I_p)$ defines the natural imbedding of $\mathfrak{A}_\Lambda$ into $\mathfrak{A}_{\Lambda'}$. Let $n = (\ell_x, \ell_y, \ell_z)$ be a unit vector in $\mathbb{R}^3$, and put

$$(\sigma n) = \ell_x\sigma^x + \ell_y\sigma^y + \ell_z\sigma^z.$$

Then, denoting as $Sp(\sigma n)$ the spectrum of σn, we have

$$Sp(\sigma n) = \{1, -1\}.$$

Let $|n\rangle$ be a unit eigenvector associated with 1, and let $\{n\} = \{n_1, n_2, \cdots\}$ be an infinite sequence of unit vectors in $\mathbb{R}^3$. Then $|\{n\}\rangle = \bigotimes_p |n_p\rangle$ is a unit vector in the infinite tensor product $\mathcal{H}_\infty = \bigotimes_p C_p^2$. We put

$$\mathfrak{A} = \bigcup_\Lambda \mathfrak{A}_\Lambda$$

and

$$\mathcal{D}^0_{\{n\}} = \mathfrak{A}|\{n\}\rangle.$$

And we denote the closure of $\mathcal{D}^0_{\{n\}}$ in $\mathcal{H}_\infty$ by $\mathcal{H}_{\{n\}}$. Let (n, n^1, n^2) be an orthonormal basis of $\mathbb{R}^3$. We put

$$n' = \frac{1}{2}(n^1 - in^2)$$

and

$$|m, n\rangle = (\sigma n')^m |n\rangle \quad (m = 0, 1).$$

Then we have

$$(\sigma n)|m, n\rangle = (-1)^m |m, n\rangle \quad (m = 0, 1).$$

Thus $\{|\{m\}, \{n\}\rangle = \bigotimes_p |m_p, n_p\rangle;\ m_p = 0, 1,\ \sum_p m_p < \infty\}$ forms an orthonormal basis in $\mathcal{H}_{\{n\}}$. In this space we define the unbounded self-adjoint operator M by

$$M|\{m\}, \{n\}\rangle = (\sum_p m_p)|\{m\}, \{n\}\rangle.$$

M counts the number of the flipped spins in $|\{m\}, \{n\}\rangle$ with respect to $|\{n\}\rangle$. Now we put

$$\mathcal{D}_{\{n\}} = \bigcap_k \mathcal{D}(M^k),$$

and let $\pi_{\{n\}} : \mathfrak{A} \to \mathcal{L}^\dagger(\mathcal{D}_{\{n\}})$ be the natural realization of $\mathfrak{A}$ on $\mathcal{D}_{\{n\}}$, i.e.

$$\pi_{\{n\}}(\sigma_p^i) \mid \{m\}, \{n\}\rangle = \sigma_p^i \mid m_p, n_p\rangle \otimes (\Pi_{p' \neq p} \otimes \mid m_{p'}, n_{p'}\rangle) \quad (i = x, y, z).$$

The BCS-Hamiltonian in the quasi-spin formulation is given by

$$H_\Lambda = \varepsilon \sum_{p=1}^{|\Lambda|} (1 - \sigma_p^z) - \frac{2g}{|\Lambda|} \sum_{\substack{p,q=1 \\ p \neq q}}^{|\Lambda|} \sigma_p^- \sigma_q^+,$$

where

$$\sigma^- = 2\begin{pmatrix} 0 & 0 \\ 1 & 0 \end{pmatrix}, \quad \sigma^+ = 2\begin{pmatrix} 0 & 1 \\ 0 & 0 \end{pmatrix}.$$

Note that for the limit of the dynamics,

$$\alpha_t(A) = \lim_{\Lambda \to \infty} e^{iH_\Lambda t} A e^{-iH_\Lambda t} \quad (A \in \mathfrak{A})$$

fails to exist on the C^*-algebra $\overline{\mathfrak{A}}$.

Let N be a unit vector in $\mathbb{R}^3$, and let $\{N\}$ be a sequence

$$\{\varepsilon_1 N, \varepsilon_2 N, \varepsilon_3 N, \cdots\} \quad (\varepsilon_p = 1 \text{ or } -1 \text{ for each } p)$$

such that

$$\lim_{k \to \infty} \frac{1}{k} \sum_{p=1}^{k} \varepsilon_p = \eta \neq 0.$$

Then the work of Bogolubov and Haag shows that the total Hamiltonian of the BCS-model is given by a self-adjoint operator H_B in $\mathcal{H}_{\{N\}}$

$$H_B = \alpha \sum_{p=1}^{\infty} \{\varepsilon_p - (\sigma_p N)\}$$

where α is a constant. Then we have the following

Theorem 4.4.1. Let $\mathcal{M}$ be the O*-algebra generated by $\pi_{\{N\}}(\mathfrak{A})$ and $\{f(M);\ f$ is a continuous function on $[0, \infty)$ and there exists a polynomial p such that $|f(x)| \leq p(x)$ on $[0, \infty)\}$ with domain $\mathcal{D}_{\{N\}}$. We put

$$\Omega = e^{-\frac{H_B}{2}}.$$

Then λ_Ω is a standard generalized vector for $\pi_{\{N\}}(\mathcal{M})$ and the quasi-weight ω_{λ_Ω} on $\mathcal{P}(\mathcal{M})$ constructed by λ_Ω satisfies the KMS condition. This quasi-weight may be thought of as an equilibrium state for the BCS-Bogolubov model.

Proof. $\mathcal{M}$ is a self-adjoint O*-algebra such that $\mathcal{M}'_w = \mathbb{C}1$ and Ω is a positive self-adjoint operator in $\mathcal{H}_{\{N\}}$ such that Ω^{-1} is densely defined. And since $\{|\{m\}, \{N\}\rangle;\ \sum_p m_p < \infty\} \subset \mathcal{D}_{\{N\}} \bigcap \mathcal{D}(\Omega^{-1})$, it follows that $\mathcal{D}_{\{N\}} \bigcap \mathcal{D}(\Omega^{-1})$ is a core for Ω^{-1}. Let $\mathcal{R}$ be the set of all rank one projections constructed from $\{|\{m\}, \{N\}\rangle;\ \sum_p m_p < \infty\}$, and let $\mathcal{N}$ be the linear span of $\{R_1 \mathcal{M} R_2;\ R_1, R_2 \in \mathcal{R}\}$. Since $\mathcal{R} \subset \mathcal{M}$, we have $\mathcal{N} \subset \mathcal{M}$. And we have

$$\pi(\mathcal{N}) \subset D(\lambda_\Omega), \ \mathcal{N}^\dagger \mathcal{D}_{\{N\}} \subset \mathcal{D}(\Omega),$$

and the linear span of $\mathcal{N}^\dagger \mathcal{D}_{\{N\}}$ is a core for Ω. Let $\mathcal{E}$ be the linear span of $\{|\{m\}, \{N\}\rangle;\ \sum_p m_p < \infty\}$. Then $\mathcal{E}$ is a subspace of $\mathcal{D}_{\{N\}} \cap \mathcal{D}(\Omega)$ such that $\{\xi \otimes \overline{\eta};\ \xi, \eta \in \mathcal{E}\} \subset \mathcal{M}$ and $\mathcal{E}$ is a core for Ω. Thus, by Lemma 2.4.22, $\lambda_\Omega((D(\lambda_\Omega) \cap D(\lambda_\Omega)^\dagger)^2)$ is total in $\mathcal{H} \otimes \overline{\mathcal{H}}$. Furthermore, since $MH_B = H_B M$ on $\mathcal{D}_{\{N\}}$, we have

$$\Omega^{it}\mathcal{D}_{\{N\}} \subset \mathcal{D}_{\{N\}}, \qquad {}^\forall t \in \mathbb{R}$$

and

$$\Omega^{it}\mathcal{M}\Omega^{-it} = \mathcal{M}, \qquad {}^\forall t \in \mathbb{R}.$$

Therefore, by Theorem 2.4.23, λ_Ω is a standard generalized vector for $\pi_{\{N\}}(\mathcal{M})$.

There is, however, something more: we can find quasi-weights satisfying the KMS condition for the BCS model also in the thermodynamical limit.

Let $\widetilde{\mathfrak{A}}$ be the completion of the spin C*-algebra $\mathfrak{A}$ with respect to the topology ξ_H defined in Lassner [3, 4]. Then $\widetilde{\mathfrak{A}}$ is still a *-algebra and each representation $\pi_{\{N\}}$ is continuous from $\mathfrak{A}$ into $\mathcal{L}^\dagger(\mathcal{D}_{\{N\}})$ endowed with the quasi-uniform topology τ_{qu} which makes of it a complete topological $*$-algebra. Hence $\pi_{\{N\}}$ can be extended by continuity to a $*$-representation $\hat{\pi}_{\{N\}}$ defined on the whole $\widetilde{\mathfrak{A}}$. Indeed, since $\mathcal{L}^\dagger(\mathcal{D}_{\{N\}})$ is complete under τ_{qu}, it follows that $\hat{\pi}_{\{N\}}(\widetilde{\mathfrak{A}}) \subset \mathcal{L}^\dagger(\mathcal{D}_{\{N\}})$. These facts above imply that $\hat{\pi}_{\{N\}}$ is a $*$-representation and, therefore, $\hat{\pi}_{\{N\}}(\widetilde{\mathfrak{A}})$ is an O*-algebra on $\mathcal{D}_{\{N\}}$. Then we can proceed as shown above and get KMS quasi-weights for the BCS model also in the thermodynamical limit.

4.5 Standard systems in the Wightman quantum field theory

The general theory of quantized fields has been developed along two main lines : One is based on the Wightman axioms and makes use of unbounded field operators (Streater-Wightman [1]). The other is the theory of local nets of bounded observables initiated by Haag-Kastler [1] and Araki [1]. Here we characterize the passage from a Wightman field to a local net of von Neumann algebras by the existence of standard systems obtained from the right wedge-region in Minkowski space . The almost results stated here are based in the works of Bisognano-Wichmann [1,2] and Driessler-Summers-Wichmann [1].

Let Φ be one scalar hermitian Wightman field, assumed to be an operator-valued tempered distribution. It is regarded as a linear map of the Schwartz space $S(\mathbb{R}^4)$ into an O*-algebra $\mathcal{L}^\dagger(\mathcal{D})$ such that $\Phi(f)^\dagger = \Phi(f^*)$ for $f \in S(\mathbb{R}^4)$ adhering the assumptions :

(A1) $\mathbb{R}^4$ is the Minkowski space $\mathbb{R}^4$ with the Lorentz metric : $x \circ y = x^4y^4 - x^1y^1 - x^2y^2 - x^3y^3$. The Hilbert space $\mathcal{H}$ obtained by the completion of $\mathcal{D}$ carries a strongly continuous unitary representation $\Lambda \to U(\Lambda)$ of the

Poincaré group P. The subgroup of translations $T(x) = U(I, x)$, $x \in \mathbb{R}^4$ has the common spectral resolution

$$T(x) = \int e^{i(x \circ p)} d\mu(p), \quad x \in \mathbb{R}^4$$

and the spectral measure μ is contained in the closed forward light cone $\overline{V}_+$, where $V_+ = \{x \in \mathbb{R}^4 \ ; \ x \circ x > 0, x^4 > 0\}$.

(A2) The existence of a vacuum state Ω. For $R \subset \mathbb{R}^4$ we denote by $\mathcal{P}(R)$ the O*-algebra on $\mathcal{D}$ generated by $\{\Phi(f) \ ; \ f \in \mathcal{S}(\mathbb{R}^4)$ and $\operatorname{supp} f \subset R\}$. It is assumed that $\Omega \in \mathcal{D}$, $U(\Lambda)\Omega = \Omega$ for each $\Lambda \in P$, $\mathcal{P}(R)\Omega$ is dense in $\mathcal{H}$ for each open subset R of $\mathbb{R}^4$ and $\mathcal{D}_0 \equiv \mathcal{P}(\mathbb{R}^4)\Omega$ is $t_{\mathcal{P}(\mathbb{R}^4)}$-dense in $\mathcal{D}$.

(A3) $U(\Lambda)\mathcal{D} = \mathcal{D}$ for each $\Lambda \in P$ and $U(\Lambda)\Phi(f)U(\Lambda)^{-1} = \Phi(\Lambda f)$ for each $\Lambda \in P$ and $f \in \mathcal{S}(\mathbb{R}^4)$, where $(\Lambda f)(x) = f(\Lambda^{-1}x)$ for $\Lambda \in P$ and $x \in \mathbb{R}^4$.

(A4) $f \to \Phi(f)$ is continuous of $\mathcal{S}(\mathbb{R}^4)$ into $(\mathcal{P}(\mathbb{R}^4), \tau_s)$.

(A5) For each open subsets R and R_1 which are space-like separated , i.e. $R_1 \subset R^c \equiv \{y \in \mathbb{R}^4 \ ; \ (x-y)\circ(x-y) < 0 \text{ for } {}^\forall x \in R\}$, we have $[X, X_1] \equiv XX_1 - X_1X = 0$ for each $X \in \mathcal{P}(R)$ and $X_1 \in \mathcal{P}(R_1)$.

A local net is an assignment $R \to \mathcal{A}(R)$ of regions R of the Minkowski space $\mathbb{R}^4$ with von Neumann algebras $\mathcal{A}(R)$ satisfying the conditions of isotony, i.e. $\mathcal{A}(R_1) \subset \mathcal{A}(R_2)$ if $R_1 \subset R_2$, locality, i.e. $[\mathcal{A}(R_1), \mathcal{A}(R_2)] = 0$ if R_1 and R_2 are space-like separated, and covariance, i.e. $U(\Lambda)\mathcal{A}(R)U(\Lambda)^{-1} = \mathcal{A}(\Lambda R)$ for all $\Lambda \in P$.

Definition 4.5.1. A Wightman field Φ is associated to a local net $\mathcal{A}$ of von Neumann algebras if each field operator $\Phi(f)$ has an extension to a closed operator, $\Phi(f)_e \subset \Phi(f^*)^*$, that is affiliated with $\mathcal{A}(R)$ if the support of the test function f is contained in the interior of R.

In the simplest case the field operators $\Phi(f)$ are essentially self-adjoint for all $f^* = f \in \mathcal{S}(\mathbb{R}^4)$ and $\overline{\Phi(f)}$ and $\overline{\Phi(g)}$ are strongly commuting if $f^* = f$ and $g^* = g$ in $\mathcal{S}(\mathbb{R}^4)$ have space-like supports. In particular, this holds if certain growth conditions on the Wightman functions (Borchers-Zimmermann [1]) or the Schwinger functions (Driessler-Fröhlich [1] and Glimm-Jaffe [1]) are fulfilled. Then Φ is associated to a local net $R \to (\mathcal{P}(R)'_{\mathrm{w}})'$. However, the self-adjointness of the field operators cannot be expected, in general (counter examples are provided by Wick polynomial of free fields). Hence it is natural to look for more general ways of associating a Wightman field with a local net.

Let W_R and W_L be the wedge-regions in Minkowski space $\mathbb{R}^4$ defined by

$$W_R = \{x \in \mathbb{R}^4 \ ; \ x^3 \ > \mid x^4 \mid\}, \quad W_L = \{x \in \mathbb{R}^4 \ ; \ x^3 \ < \ - \mid x^4 \mid\}.$$

We denote by $\mathcal{W}$ the set $\{\Lambda W_R \ ; \ \Lambda \in P\}$ of all wedge-regions Poincaré-equivalent to W_R, and denote by $\mathcal{K}$ the set of all closed double cones K with a non-empty interior. For any $K \in \mathcal{K}$ we have $K = \bigcap\{W \; ; \; W \in \mathcal{W}, \; W \supset K\}$ and $K^c = \bigcup\{W \; ; \; W \in \mathcal{W}, \; W \subset K^c\}$. It is sufficient for most purposes to restrict the choice of regions R to the following types : wedge-regions W, closed double cones K and the causal complements K^c. We investigate when the Wightman field Φ is associated to a local net $W \in \mathcal{W} \to \mathcal{A}(W)\,(K \in \mathcal{K} \to \mathcal{B}(K))$ of von Neumann algebras by the existense of a standard system $(\mathcal{P}(W_R), \Omega, \mathcal{A}(W_R))$.

Theorem 4.5.2. The following statements hold:

(1) Ω is a cyclic and separating vector for $\mathcal{P}(W_R)$ and $\mathcal{P}(W_L)$, and so $X\Omega \to X^\dagger\Omega$, $X \in \mathcal{P}(W_R)$ (resp. $\mathcal{P}(W_L)$) is closable and its closure is denoted by $S_{\mathcal{P}(W_R)\Omega}$ (resp. $S_{\mathcal{P}(W_L)\Omega}$).

(2) Let $S_{\mathcal{P}(W_R)\Omega} = J_{\mathcal{P}(W_R)\Omega}\Delta^{1/2}_{\mathcal{P}(W_R)\Omega}$ be the polar decomposition. $J_{\mathcal{P}(W_R)\Omega}$ equals the antiunitary involution $J = U(\pi_3, 0)\Theta_0$, where π_3 denotes the rotation by angle π about the 3-axis and Θ_0 denotes the canonical TCP-operator , and $\Delta^{1/2}_{\mathcal{P}(W_R)\Omega}$ equals a positive self-adjoint operator $V(i\pi)$ obtained by analytic continuation of a one-parameter unitary group $\{V(t)\}_{t\in\mathbb{R}}$ of velocity transformations in the 3-direction.

(3) $S_{\mathcal{P}(W_L)\Omega} = S^*_{\mathcal{P}(W_R)\Omega} = JV(-i\pi)$.

(4) $J\mathcal{P}(W_R)J = \mathcal{P}(W_L)$, $V(t)\mathcal{P}(W_R)V(t)^{-1} = \mathcal{P}(W_R)$ and $V(t)\,\mathcal{P}(W_L)\,V(t)^{-1} = \mathcal{P}(W_L)$ for all $t \in \mathbb{R}$.

We first consider when there exists a von Neumann algebra $\mathcal{A}(W_R)$ on $\mathcal{H}$ such that $(\mathcal{P}(W_R), \Omega, \mathcal{A}(W_R))$ is a standard system.

Theorem 4.5.3. Let $\mathcal{A}(W_R)$ be a von Neumann algebra on $\mathcal{H}$. Then $(\mathcal{P}(W_R), \Omega, \mathcal{A}(W_R))$ is a standard system if and only if $\mathcal{A}(W_R)' \subset \mathcal{P}(W_R)'_{\mathrm{w}}$ and $\mathcal{A}(W_R) \subset \mathcal{P}(W_L)'_{\mathrm{w}}$.

Proof. Suppose that $(\mathcal{P}(W_R), \Omega, \mathcal{A}(W_R))$ is a standard system. Then it follows from Theorem 2.2.4 that $S_{\mathcal{P}(W_R)\Omega} = S_{\mathcal{A}(W_R)\Omega}$, which implies by Theorem 4.5.2, (4) that

$$\mathcal{A}(W_R) = J\mathcal{A}(W_R)'J \subset J\mathcal{P}(W_R)'_{\mathrm{w}}J = \mathcal{P}(W_L)'_{\mathrm{w}}.$$

Conversely suppose that $\mathcal{A}(W_R)' \subset \mathcal{P}(W_R)'_{\mathrm{w}}$ and $\mathcal{A}(W_R) \subset \mathcal{P}(W_L)'_{\mathrm{w}}$. By Theorem 1.4.2 and Proposition 1.2.3 $\overline{e_{\mathcal{A}(W_R)'}(X)}$ is affiliated with $\mathcal{A}(W_R)$ for each $X \in \mathcal{P}(W_R)$. Hence, since $\mathcal{P}(W_R)\Omega$ is dense in $\mathcal{H}$, it follows that

$$\mathcal{A}(W_R)\Omega \text{ is dense in } \mathcal{H}. \tag{4.5.1}$$

Similarly, we have that $\mathcal{A}(W_R)'\Omega$ is dense in $\mathcal{H}$. Furthermore, we have

$$S_{\mathcal{P}(W_R)\Omega} \subset S_{\mathcal{A}(W_R)\Omega} \quad \text{and} \quad S_{\mathcal{P}(W_L)\Omega} \subset S_{\mathcal{A}(W_R)'\Omega},$$

which implies by Theorem 4.5.2, (3) that

$$S_{\mathcal{A}(W_R)\Omega} = S^*_{\mathcal{A}(W_R)'\Omega} \subset S^*_{\mathcal{P}(W_L)\Omega} = S_{\mathcal{P}(W_R)\Omega} \subset S_{\mathcal{A}(W_R)\Omega}.$$

Hence, $S_{\mathcal{A}(W_R)\Omega} = S_{\mathcal{P}(W_R)\Omega}$. Therefore it follows from Theorem 4.5.2, (4) that $(\mathcal{P}(W_R), \Omega, \mathcal{A}(W_R))$ is a standard system. This completes the proof.

Theorem 4.5.4. Consider the following statements.

(i) Φ is associated to some local net $K \in \mathcal{K} \to \mathcal{B}(K)$ of von Neumann algebras.

(ii) There exists a standard system $(\mathcal{P}(W_R), \Omega, \mathcal{A}(W_R))$ such that

(a) $U(\Lambda)\mathcal{A}(W_R)U(\Lambda)^{-1} = \mathcal{A}(W_R)$ for each $\Lambda \in P$ s.t. $\Lambda W_R = W_R$;

(b) $U(\Lambda)\mathcal{A}(W_R)U(\Lambda)^{-1} \subset \mathcal{A}(W_R)$ for each $\Lambda \in P$ s.t. $\Lambda W_R \subset W_R$;

(c) $U(\Lambda)\mathcal{A}(W_R)U(\Lambda)^{-1} \subset \mathcal{A}(W_R)'$ for each $\Lambda \in P$ s.t. $\Lambda W_R \subset \overline{W_R}^c = W_L$;

(d) $\{U(\Lambda)\mathcal{A}(W_R)U(\Lambda)^{-1} \ ; \ \Lambda W_R \subset K^c\}'' \subset \mathcal{P}(K)'_{\mathrm{w}}, \ K \in \mathcal{K}$.

(iii) Φ is associated to some local net $W \in \mathcal{W} \to \mathcal{A}(W)$ of von Neumann algebras.

(iv) There exists a standard system $(\mathcal{P}(W_R), \Omega, \mathcal{A}(W_R))$ satisfying the conditions (a), (b) and (c) in (ii).

Then the following implications hold :

$$\begin{array}{c} \text{(i)} \iff \text{(ii)} \\ \Downarrow \\ \text{(iii)} \iff \text{(iv)}. \end{array}$$

Proof. (iii) $\Rightarrow$ (iv) Since Φ is associated to a local net $W \in \mathcal{W} \to \mathcal{A}(W)$, it follows that $\mathcal{A}(W_R)' \subset \mathcal{P}(W_R)'_{\mathrm{w}}$ and $\mathcal{A}(W_R) \subset \mathcal{A}(W_L)' \subset \mathcal{P}(W_L)'_{\mathrm{w}}$, which implies by Theorem 4.5.3 that $(\mathcal{P}(W_R), \Omega, \mathcal{A}(W_R))$ is a standard system. It is clear that $(\mathcal{P}(W_R), \Omega, \mathcal{A}(W_R))$ satisfies the conditions (a), (b) and (c). Furthermore, since $\mathcal{A}(W_R) \subset \mathcal{A}(W_L)'$, $\mathcal{A}(W_L)\Omega$ is dense in $\mathcal{H}$ and $V(t)\mathcal{A}(W_L)V(t)^{-1} = \mathcal{A}(W_L)$ for all $t \in \mathbb{R}$ by the covariance of $W \to \mathcal{A}(W)$, it follows from Theorem 2 in Bisognano-Wichmann [1] that $\mathcal{A}(W_L) = \mathcal{A}(W_R)'$, which implies

$$\mathcal{A}(\overline{W}^c) = \mathcal{A}(W)', \quad W \in \mathcal{W}. \tag{4.5.2}$$

(iv) $\Rightarrow$ (iii) We put

$$\mathcal{A}(W) = U(\Lambda)\mathcal{A}(W_R)U(\Lambda)^{-1}, \quad W = \Lambda W_R \in \mathcal{W}.$$

Then we see by (a) that $\mathcal{A}(W)$ is well-defined for each $W \in \mathcal{W}$, and $W \to \mathcal{A}(W)$ has the covariance property. The condition (b) implies that $W \to \mathcal{A}(W)$ has the isotony property. Furthermore, by (c) we have $\mathcal{A}(W_L) \subset \mathcal{A}(W_R)'$, which implies by the covariance of $W \to \mathcal{A}(W)$ that $\mathcal{A}(\overline{W}^c) \subset \mathcal{A}(W)'$ for each $W \in \mathcal{W}$. Hence it follows that $W \to \mathcal{A}(W)$ is local. Since

$(\mathcal{P}(W_R), \Omega, \mathcal{A}(W_R))$ is standard and $W \to \mathcal{A}(W)$ has the covariance property, it follows that $\mathcal{A}(W)' \subset \mathcal{P}(W)'_{\mathrm{w}}$ for each $W \in \mathcal{W}$. Thus we see that Φ is associated to a local net $W \to \mathcal{A}(W)$.

(i) $\Rightarrow$ (ii) We put

$$\mathcal{A}(W) = \{\mathcal{B}(K)\,;\, K \subset W\}'', \; W \in \mathcal{W}.$$

Then it follows from the isotony and covariance of $K \to \mathcal{B}(K)$ that $W \to \mathcal{A}(W)$ has the same properties. Take arbitrary $W, W_1 \in \mathcal{W}$ s.t. $W_1 \subset \overline{W}^c$. Then it follows from the locality of $K \to \mathcal{B}(K)$ that $\mathcal{B}(K_1) \subset \mathcal{B}(K)'$ for each K, $K_1 \in \mathcal{K}$ s.t. $K \subset W$ and $K_1 \subset \overline{W}^c$, which implies

$$\mathcal{A}(\overline{W}^c) = \{\mathcal{B}(K_1); K_1 \subset \overline{W}^c\}'' \subset \mathcal{A}(W)' = \bigcap\{\mathcal{B}(K)'; K \subset W\}.$$

Hence, $W \to \mathcal{A}(W)$ is local. Since $\mathcal{B}(K)' \subset \mathcal{P}(K)'_{\mathrm{w}}$ for each $K \in \mathcal{K}$, we have

$$\begin{aligned}\mathcal{A}(W)' = \bigcap\{\mathcal{B}(K)'; K \subset W\} &\subset \bigcap\{\mathcal{P}(K)'_{\mathrm{w}}; K \subset W\}\\ &= \mathcal{P}(W)'_{\mathrm{w}}\end{aligned}$$

for each $W \in \mathcal{W}$, which implies that Φ is associated to the local net $W \to \mathcal{A}(W)$. By the equivalence of (iii) and (iv) $(\mathcal{P}(W_R), \Omega, \mathcal{A}(W_R))$ is a standard system satisfying the conditions (a), (b) and (c) in (ii). The condition (d) follows from

$$\{\mathcal{A}(W); W \subset K^c\}'' = \{\mathcal{B}(K_1); K_1 \subset K^c\}'' \subset \mathcal{B}(K)' \subset \mathcal{P}(K)'_{\mathrm{w}}$$

for each $K \in \mathcal{K}$.

(ii) $\Rightarrow$ (i) We put

$$\mathcal{A}(W) = U(\Lambda)\mathcal{A}(W_R)U(\Lambda)^{-1}, \; W = \Lambda W_R \in \mathcal{W}.$$

As showed in the proof of (iv) $\Rightarrow$ (iii), Φ is associated to the local net $W \to \mathcal{A}(W)$ of von Neumann algebras. We now put

$$\mathcal{B}(K) = \bigcap\{\mathcal{A}(W); K \subset W\}, \; K \in \mathcal{K}.$$

Then it follows from the isotony and covariance of $W \to \mathcal{A}(W)$ that $K \to \mathcal{B}(K)$ has the same properties. By (4.5.2) we have

$$\begin{aligned}\mathcal{B}(K)' = \{\mathcal{A}(W)'; K \subset W\}'' &= \{\mathcal{A}(\overline{W}^c)\,;\, \overline{W}^c \subset K^c\}''\\ &\subset \mathcal{P}(K)'_{\mathrm{w}}, \;\; K \in \mathcal{K},\end{aligned}$$

and so it is proved as in (4.5.1) that

$$\mathcal{B}(K)\Omega \text{ is dence in } \mathcal{H}. \tag{4.5.3}$$

Hence, $\mathcal{A} \equiv \{\mathcal{B}(K); K \subset W_R\}'' \subset \mathcal{A}(W_R)$ and $\mathcal{A}\Omega$ is dence in $\mathcal{H}$. Furthermore, it follows from the covariance of $K \to \mathcal{B}(K)$ that $V(t)\mathcal{A}V(t)^{-1} = \mathcal{A}$ for each $t \in \mathbb{R}$, and so $\mathcal{A} = \mathcal{A}(W_R)$ by Theorem 2 in Bisognano-Wichmann [1], which implies that $K \to \mathcal{B}(K)$ is local. In fact, we have by (4.5.2)

$$\begin{aligned}\mathcal{B}(K)' &= \{\mathcal{A}(W)'; K \subset W\}'' = \{\mathcal{A}(\overline{W}^c); \overline{W}^c \subset K^c\}'' \\ &= \{\mathcal{B}(K_1); K_1 \subset K^c\}'' \\ &\supset \mathcal{B}(K_0)\end{aligned}$$

for each K, $K_0 \in \mathcal{K}$ s.t. $K_0 \subset K^c$. Thus we see that Φ is associated to the local net $K \to \mathcal{B}(K)$. This completes the proof.

Remark 4.5.5. (1) Suppose that Φ is associated to a local net $K \in \mathcal{K} \to \mathcal{B}(K)$. Then Ω is a cyclic and separating vector for $\mathcal{B}(K)$, and $\mathcal{B}(K) \subset \mathcal{P}(K^c)'_{\mathrm{w}}$ for each $K \in \mathcal{K}$. In fact, by (4.5.3) Ω is cyclic for $\mathcal{B}(K)$. It follows from the association of $\mathcal{B}$ with Φ and the locality of $\mathcal{B}$ that

$$\begin{aligned}\mathcal{P}(K^c)'_{\mathrm{w}} &= \bigcap\{\mathcal{P}(K_1)'_{\mathrm{w}}; K_1 \subset K^c\} \supset \bigcap\{\mathcal{B}(K_1)'; K_1 \subset K^c\} \\ &\supset \mathcal{B}(K),\end{aligned}$$

which implies as in (4.5.1) that $\mathcal{B}(K)'\Omega$ is dense in $\mathcal{H}$.

(2) Suppose that Φ is associated to a local net $W \in \mathcal{W} \to \mathcal{A}(W)$. Then, $\mathcal{A}(\overline{W}^c) = \mathcal{A}(W)'$ for each $W \in \mathcal{W}$. This was already proved in the proof of Theorem 4.5.4 (in equation (4.5.2)).

Corollary 4.5.6. Suppose that $\mathcal{P}(W_R)'_{\mathrm{w}}$ is a von Neumann algebra. Then the following statements are equivalent.

(i) $(\mathcal{P}(W_R), \Omega, (\mathcal{P}(W_R)'_{\mathrm{w}})')$ is standard.

(ii) $(\mathcal{P}(W_R)'_{\mathrm{w}})' = \mathcal{P}(W_L)'_{\mathrm{w}}$.

(iii) Φ is associated to a local net $W \to (\mathcal{P}(W)'_{\mathrm{w}})'$.

Proof. (iii) ⇒ (ii) This follows from Remark 4.5.5, (2).

(ii) ⇒ (i) This follows from Theorem 4.5.2.

(i) ⇒ (iii) It is clear that the conditions (a) and (b) in Theorem 4.5.4 hold. By Theorem 4.5.1 and the standardness of $(\mathcal{P}(W_R), \Omega, (\mathcal{P}(W_R)'_{\mathrm{w}})')$ we have $\mathcal{P}(W_L)'_{\mathrm{w}} = J\mathcal{P}(W_R)'_{\mathrm{w}}J = (\mathcal{P}(W_R)'_{\mathrm{w}})'$, which implies that

$$\begin{aligned}U(\Lambda)(\mathcal{P}(W_R)'_{\mathrm{w}})'U(\Lambda)^{-1} &= (\mathcal{P}(\Lambda W_R)'_{\mathrm{w}})' \subset (\mathcal{P}(W_L)'_{\mathrm{w}})' \\ &= \mathcal{P}(W_R)'_{\mathrm{w}}\end{aligned}$$

for each $\Lambda \in P$ s.t. $\Lambda W_R \subset W_L$. Therefore, it follows from Theorem 4.5.4 that Φ is associated to a local net $W \to (\mathcal{P}(W)'_{\mathrm{w}})'$.

Corollary 4.5.7. Suppose that $\mathcal{P}(K)'_{\mathrm{w}}$ is a von Neumann algebra for all $K \in \mathcal{K}$. Then the following statements are equivalent.

(i) $(\mathcal{P}(W_R), \Omega, (\mathcal{P}(W_R)'_{\mathrm{w}})')$ is a standard system.

(ii) Φ is associated to a local net $W \in \mathcal{W} \to (\mathcal{P}(W)'_{\mathrm{w}})'$ of von Neumann algebras.

(iii) Φ is associated to some local net $K \in \mathcal{K} \to \mathcal{B}(K)$ of von Neumann algebras.

(iv) Φ is associated to a local net $K \in \mathcal{K} \to (\mathcal{P}(K)'_{\mathrm{w}})'$ of von Neumann algebras.

(v) Φ is associated to a local net $K \in \mathcal{K} \to \mathcal{P}(K^c)'_{\mathrm{w}}$ of von Neumann algebras.

Proof. Since $\mathcal{P}(W_R)'_{\mathrm{w}} = \bigcap\{\mathcal{P}(K)'_{\mathrm{w}}; K \subset W_R\}$ and $\mathcal{P}(K)'_{\mathrm{w}}$ is a von Neumann algebra for all $K \in \mathcal{K}$, it follows that $\mathcal{P}(W_R)'_{\mathrm{w}}$ is a von Neumann algebra.

(i) $\Rightarrow$ (v) By Corollary 4.5.6 the standard system $(\mathcal{P}(W_R), \Omega, (\mathcal{P}(W_R)'_{\mathrm{w}})')$ satisfies the conditions (a), (b) and (c) in Theorem 4.5.4 and $(\mathcal{P}(W_R)'_{\mathrm{w}})' = \mathcal{P}(W_L)'_{\mathrm{w}}$. Furthermore, we have

$$U(\Lambda)(\mathcal{P}(W_R)'_{\mathrm{w}})'U(\Lambda)^{-1} = \mathcal{P}(\Lambda W_L)'_{\mathrm{w}} \subset \mathcal{P}(K)'_{\mathrm{w}}$$

for each $\Lambda \in P$ s.t. $\Lambda W_R \subset K^c$, and so it follows since $\mathcal{P}(K)'_{\mathrm{w}}$ is a von Neumann algebra that $\{U(\Lambda)(\mathcal{P}(W_R)'_{\mathrm{w}})'U(\Lambda)^{-1}; \Lambda W_R \subset K^c\}'' \subset \mathcal{P}(K)'_{\mathrm{w}}$. Therefore, it follows from Theorem 4.5.4 that Φ is associated to a local net $K \to \mathcal{B}(K)$, where $\mathcal{B}(K) = \bigcap\{(\mathcal{P}(W)'_{\mathrm{w}})'; K \subset W\}$, $K \in \mathcal{K}$. By Remark 4.5.5 we have

$$\mathcal{B}(K) \subset \mathcal{P}(K^c)'_{\mathrm{w}} \subset \mathcal{P}(\overline{W}^c)'_{\mathrm{w}} = (\mathcal{P}(W)'_{\mathrm{w}})'$$

for each $W \in \mathcal{W}$ s.t. $K \subset W$, and so $\mathcal{B}(K) = \mathcal{P}(K^c)'_{\mathrm{w}}$. Therefore Φ is associated to a local net $K \to \mathcal{P}(K^c)'_{\mathrm{w}}$.

(v) $\Rightarrow$ (iii) This is trivial.

(iii) $\Rightarrow$ (iv) By the association of $\mathcal{B}$ with Φ we have

$$(\mathcal{P}(K_1)'_{\mathrm{w}})' \subset \mathcal{B}(K_1) \subset \mathcal{B}(K)' \subset \mathcal{P}(K)'_{\mathrm{w}}$$

for each $K, K_1 \in \mathcal{K}$ s.t. $K_1 \subset K^c$, and so $K \to (\mathcal{P}(K)'_{\mathrm{w}})'$ is local. It is clear that $K \to (\mathcal{P}(K)'_{\mathrm{w}})'$ is isotony and covariant. Furthermore, since $\mathcal{P}(K)'_{\mathrm{w}}$ is a von Neumann algebra, it follows that Φ is associated to a local net $K \to (\mathcal{P}(K)'_{\mathrm{w}})'$.

(i) $\Leftrightarrow$ (ii) This follows from Corollary 4.5.6.

(iv) $\Rightarrow$ (i) It follows from Theorem 4.5.4 that $(\mathcal{P}(W_R), \Omega, \mathcal{A}(W_R))$ is a standard system for some von Neumann algebra $\mathcal{A}(W_R)$, which implies by Theorem 2 in Bisognano-Wichmann [1] that $\mathcal{A}(W_R) = (\mathcal{P}(W_R)'_{\mathrm{w}})'$.

Remark 4.5.8. (1) A Wightman field Φ is said to be satisfy a *generalized H-bounded* if there exists a nonnegative number $\alpha < 1$ such that $\overline{\Phi(f)}e^{-H^{\alpha}}$

is a bounded operator for all f, where H denotes the Hamiltonian of Φ. Driessler-Summer-Wichmann [1] showed that if Φ satisfies a generalized H-bounded then $\mathcal{P}(K)'_{\mathrm{w}}$ is a von Neumann algebra for every $K \in \mathcal{K}$.

(2) The Borchers tensor algebra $\underline{S}$ is defined by

$$\underline{S} = \{\underline{f} = (f_0, f_1, \cdots); f_0 \in \mathbb{C}, f_n \in \mathcal{S}(\mathbb{R}^{4n}) \text{ for } n \geq 1\}$$

equipped with the natural algebraic operations. The sequence of Wightman distributions W_n of the field Φ defines a positive linear functional $\mathcal{W}$ (Wightman functional) by $\mathcal{W}(\underline{f}) = \sum_n W_n(f_n), \underline{f} \in \underline{S}$ (Borchers [2]). Borchers-Yngvason [2] defined the notion of centrally positivity of a positive linear functional ω on a $*$-algebra $\mathcal{A}$ with respect to a hermitian element a_0 of $\mathcal{A}$ as follows : ω is *centrally positive with respect to* a_0 if ω is positive on all elements of the form $\sum_n a_0^n a_n, a_n \in \mathcal{A}$ such that $\sum_n t^n a_n \in \mathcal{P}(\mathcal{A})$ for all $t \in \mathbb{R}$. This is an adaptation of a centrally positive operator defined by Powers [2]. Borchers-Yngvason [2] showed that ω is centrally positive with respect to a_0 if and only if $\pi_\omega(a_0)$ has a self-adjoint extension $\widehat{\pi_\omega}(a_0)$, eventually in an extended Hilbert space $\widehat{\mathcal{H}_\omega}$, such that the restrictions to $\mathcal{D}(\pi_\omega)$ of all bounded functions of $\widehat{\pi_\omega}(a_0)$ are contained in $\pi_\omega(\mathcal{A})'_{\mathrm{w}}$, and using this, they obtained the following result: Let Φ be a Wightman field such that $\mathcal{P}(K)'_{\mathrm{w}}$ is a von Neumann algebra for every $K \in \mathcal{K}$. The following statements are equivalent:

(i) Φ is associated to a local net $K \in \mathcal{K} \longrightarrow (\mathcal{P}(K)'_{\mathrm{w}})'$.

(ii) Let f be any element of $\mathcal{S}(\mathbb{R}^4)$ such that support of f is contained in the interior of a double cone K, and $\underline{S}(f, K^C)$ the subalgebra of $\underline{S}$ generated by f and $\{g \in \mathcal{S}(\mathbb{R}^4);$ support of $g \subset K^C\}$. Then $\mathcal{W}\lceil_{\underline{S}(f,K^C)}$ is centrally positive with respect to f.

We investigate the connection between a standard system and the association of a local net of von Neumann algebras with Φ considering a commutant $\tilde{\mathcal{P}}(W_R)'_{ss}$ which is stronger than the weak commutant $\mathcal{P}(W_R)'_{\mathrm{w}}$ and is always a von Neumann algebra.

For $K \in \mathcal{K}$ we denote by $\mathcal{F}(K)$ the set of all real test functions f such that $\Phi(f) \in \mathcal{P}(K)$ and the Fourier transform $\tilde{f}$ of f satisfies the condition $\tilde{f}(p) \neq 0$ for all p. For $f \in \mathcal{F}(K)$ we denote by $\mathcal{R}(f)$ the von Neumann algebra, generated by $\overline{\Phi(f)}$, i.e. $\mathcal{R}(f) = \{C \in \mathcal{B}(\mathcal{H}); \Phi(f)C \supset C\overline{\Phi(f)}\}'$, and denote by $\mathcal{F}_\ell(K)$ the set of all elements f of $\mathcal{F}(K)$ such that $U(\Lambda)\mathcal{R}(f)U(\Lambda)^{-1} \subset \mathcal{R}(f)'$ for each $\Lambda \in P$ s.t. $\Lambda K \subset K^c$. For $R \subset M$ we define a commutant $\tilde{\mathcal{P}}(R)'_{ss}$ of $\mathcal{P}(R)$ which is always a von Neumann algebra as follows :

$$\tilde{\mathcal{P}}(R)'_{ss} = \{C \in \mathcal{P}(R)'_{\mathrm{w}} \,;\, C\mathcal{D} \subset \tilde{\mathcal{D}}(\mathcal{P}(R)) \equiv \bigcap_{X \in \mathcal{P}(R)} \mathcal{D}(\overline{X}) \text{ and } C^*\mathcal{D} \subset \tilde{\mathcal{D}}(\mathcal{P}(R))\}.$$

Theorem 4.5.9. Consider the following statements.
(i) $(\mathcal{P}(W_R), \Omega, (\tilde{\mathcal{P}}(W_R)'_{ss})')$ is standard.
(ii) $\tilde{\mathcal{P}}(W_R)'_{ss} = \mathcal{P}(W_R)'_{\mathrm{w}}$ and $(\mathcal{P}(W_R)'_{\mathrm{w}})' = \mathcal{P}(W_L)'_{\mathrm{w}}$.
(iii) Φ is associated to a local net $W \in \mathcal{W} \to (\tilde{\mathcal{P}}(W)'_{ss})'$.
(iv) Φ is associated to a local net $K \in \mathcal{K} \to \mathcal{B}(K)$ such that

$$(\tilde{\mathcal{P}}(W_R)'_{ss})' = \{\mathcal{B}(K); K \subset W_R\}''.$$

(v) $\mathcal{F}_\ell(K) = \mathcal{F}(K)$ for each $K \in \mathcal{K}$.
Then the statements (i) ~ (iv) are equivalent, and they imply the statement (v).

Proof. It is easily shown by the definition of $\tilde{\mathcal{P}}(W_R)'_{ss}$ and Theorem 4.5.2 that

$$\begin{aligned} \tilde{\mathcal{D}}(\mathcal{P}(W)) &= U(\Lambda)\tilde{\mathcal{D}}(\mathcal{P}(W_R)), \\ \tilde{\mathcal{P}}(W)'_{ss} &= U(\Lambda)\tilde{\mathcal{P}}(W_R)'_{ss}U(\Lambda)^{-1}, \ W = \Lambda W_R \ ; \end{aligned} \tag{4.5.4}$$

$$\begin{aligned} J\tilde{\mathcal{D}}(\mathcal{P}(W_R)) &= \tilde{\mathcal{D}}(\mathcal{P}(W_L)), \\ J\tilde{\mathcal{P}}(W_R)'_{ss}J &= \tilde{\mathcal{P}}(W_L)'_{ss} \ ; \end{aligned} \tag{4.5.5}$$

$$\tilde{\mathcal{P}}(R_2)'_{ss} \subset \tilde{\mathcal{P}}(R_1)'_{ss} \ \text{ if } R_1 \subset R_2. \tag{4.5.6}$$

(i) ⇒ (iv) By (4.5.4) the conditions (a) and (b) in Theorem 4.5.4 hold. By (4.5.5) and the standardness of $(\mathcal{P}(W_R), \Omega, (\tilde{\mathcal{P}}(W_R)'_{ss})')$ we have

$$(\tilde{\mathcal{P}}(W_R)'_{ss})' = J\tilde{\mathcal{P}}(W_R)'_{ss}J = \tilde{\mathcal{P}}(W_L)'_{ss}, \tag{4.5.7}$$

which implies that the condition (c) in Theorem 4.5.4 holds. Furthermore, we have by (4.5.4), (4.5.6) and (4.5.7)

$$\{(\tilde{\mathcal{P}}(W)'_{ss})'; W \subset K^c\}'' \subset \tilde{\mathcal{P}}(K)'_{ss} \subset \mathcal{P}(K)'_{\mathrm{w}}$$

for each $K \in \mathcal{K}$, and so the condition (d) holds.

(iv) ⇒ (iii) This follows from the proof of (i) ⇒ (ii) in Theorem 4.5.4.

(ii) ⇒ (i) This follows from Theorem 4.5.4. Thus we see that the statements (i), (iii) and (iv) are equivalent.

(i) ⇒ (v) Let $f \in \mathcal{F}(K)$. We put

$$\mathcal{B}(K) = \bigcap\{(\tilde{\mathcal{P}}(W)'_{ss})' \ ; \ K \subset W\}, \ K \in \mathcal{K}.$$

Then it is clear that $\mathcal{R}(f) \subset \mathcal{B}(K)$, and so it follows from the locality of $K \in \mathcal{K} \to \mathcal{B}(K)$ that

$$U(\Lambda)\mathcal{R}(f)U(\Lambda)^{-1} \subset \mathcal{B}(\Lambda K) \subset \mathcal{B}(K)' \subset \mathcal{R}(f)'$$

for each $\Lambda \in P$ s.t. $\Lambda K \subset K^c$. Hence, $f \in \mathcal{F}_\ell(K)$.

(ii) ⇒ (i) This follows from Corollary 4.5.6.

(i) ⇒ (ii) Let $f \in \mathcal{F}(K)$. Then, $f \in \mathcal{F}_\ell(K)$ as seen above. We put

$$\mathcal{A}(W) = \{U(\Lambda)\mathcal{R}(f)U(\Lambda)^{-1} \ ; \ \Lambda K \subset W\}, \ W \in \mathcal{W}.$$

Then it follows from Theorem 4.6, Lemma 4.7 and Theorem 4.8 in Driessler-Summers-Wichmann [1] that $(\mathcal{P}(W_R), \Omega, \mathcal{A}(W_R))$ is a standard system such that $\mathcal{A}(W) = \mathcal{P}(\overline{W}^c, f)'_{\mathrm{w}}$ for all $W \in \mathcal{W}$, where $\mathcal{P}(R, f)$ denotes the O*-algebra on $\mathcal{D}$ generated by $\{U(\Lambda)\Phi(f)\, U(\Lambda)^{-1} \ ; \ \Lambda K \subset R\}$. Hence we have

$$\mathcal{P}(W_R)'_{\mathrm{w}} \subset \mathcal{P}(W_R, f)'_{\mathrm{w}} = \mathcal{A}(W_L) = \mathcal{A}(W_R)' \subset \mathcal{P}(W_R)'_{\mathrm{w}},$$

and so $\mathcal{P}(W_R)'_{\mathrm{w}} = \mathcal{A}(W_R)'$ and it is a von Neumann algebra. Hence, $(\mathcal{P}(W_R), \Omega, (\mathcal{P}(W_R)'_{\mathrm{w}})')$ is a standard system, and so we have by (4.5.10)

$$\tilde{\mathcal{P}}(W_R)'_{ss} \subset \mathcal{P}(W_R)'_{\mathrm{w}}$$

and

$$\begin{aligned}(\tilde{\mathcal{P}}(W_R)'_{ss})' = \tilde{\mathcal{P}}(W_L)'_{ss} &= J\tilde{\mathcal{P}}(W_R)'_{ss}J \subset J\mathcal{P}(W_R)'_{\mathrm{w}}J \\ &= \mathcal{P}(W_L)'_{\mathrm{w}} \\ &= (\mathcal{P}(W_R)'_{\mathrm{w}})'.\end{aligned}$$

Hence, $\tilde{\mathcal{P}}(W_R)'_{ss} = \mathcal{P}(W_R)'_{\mathrm{w}}$. This completes the proof.

Remark 4.5.10. Suppose $\mathcal{F}_\ell(K) \neq \phi$ for some $K \in \mathcal{K}$. Then, by the proof of (i) ⇒ (ii) in Theorem 4.5.9 Φ is associated to a local net $W \to (\mathcal{P}(W)'_{\mathrm{w}})'$ of von Neumann algebras. Driessler-Summers-Wichmann [1] have showed that Φ is not necessarily associated to a local net $K \to \mathcal{P}(K^c)'_{\mathrm{w}}$, but it is weakly associated to a local net $K \to \mathcal{P}(K^c)'_{\mathrm{w}}$ in the following sense :

$$(\mathcal{P}(K^c)'_{\mathrm{w}})' \subset \mathcal{P}(K, f)'_{\mathrm{w}}, \ K \in \mathcal{K}.$$

Corollary 4.5.11. Suppose that $\mathcal{P}(W_R)$ is essentially self-adjoint. Then the following statements are equivalent.

(i) $(\mathcal{P}(W_R), \Omega, (\mathcal{P}(W_R)'_{\mathrm{w}})')$ is standard.

(ii) Φ is associated to a local net $W \in \mathcal{W} \to (\mathcal{P}(W)'_{\mathrm{w}})'$ of von Neumann algebras.

(iii) Φ is associated to some local net $K \in \mathcal{K} \to \mathcal{B}(K)$ of von Neumann algebras.

(iv) Φ is associated to a local net $K \in \mathcal{K} \to \mathcal{P}(K^c)'_{\mathrm{w}}$ of von Neumann algebras.

Proof. Since $\mathcal{P}(W_R)$ is essentially self-adjoint, it follows that $\mathcal{P}(W_R)'_{\mathrm{w}} = \tilde{\mathcal{P}}(W_R)'_{ss}$ and it is a von Neumann algebra, so that by Corollary 4.5.6 and Theorem 4.5.9 the statements (i) and (ii) are equivalent.

(iv) ⇒ (iii) This is trivial.

(iii) ⇒ (i) This follows from Theorem 4.5.4 and Theorem 2 in Bisognano-Wichmann [1]. (i) ⇒ (iv) This is proved as in the proof of (i) ⇒ (iii) in Corollary 4.5.7.

We finally give some examples of modular systems and quasi-standard systems for a Wightman field.

Example 4.5.12. Let Φ be a Wightman field. We could give some examples of standard systems $(\mathcal{P}(W), \Omega, \mathcal{A}(W))$ for wedge-regions W. But, it is difficult to give examples of standard systems for domains except wedge-regions. Suppose that $(\mathcal{P}(W_R), \Omega, \mathcal{A}(W_R))$ is a standard system. Then, for each open subset R of W_R $(\mathcal{P}(R), \Omega, \mathcal{A}(W_R))$ is a quasi-standard. But, we don't know whether $(\mathcal{P}(R), \Omega, \mathcal{A}(R))$ is a standard (or modular) system for some local net $R \to \mathcal{A}(R)$ of von Neumann algebras. There is only one example known in case of a massless free field where the domain in the double cone. By Theorem 2 in Hislop-Longo [1] we see that for a massless free field Φ $(\mathcal{P}(O), \Omega, (\mathcal{P}(O)'_{\mathrm{w}})')$ is a modular system for each open double cone O in $\mathbb{R}^4$.

Notes

4.1. The quantum moment problem was first studied by Sherman [1] who gave an affirmative answer for a countably generated O*-algebra $\mathcal{M}$ which contains the restriction to $\mathcal{D}$ of the inverse of some compact operator. Woronowicz [1], [2] proved that the O*-algebra $\mathcal{L}^\dagger(\mathcal{S}(\mathbb{R}))$ and the O*-algebra generated by the position and momentum operators on $\mathcal{S}(\mathbb{R})$ are QMP-solvable. Schmüdgen [2], [3], [21] generalized these results to more general O*-algebras. Theorem 4.1.6 extends the corresponding results of Sherman [1], Woronowicz [1], [2] and Lassner-Timmermann [1], and the proof is according to that of Theorem 2 in Schmüdgen [2]. We here use the algebraic conjugate dual of $\mathcal{D}^\dagger$ of a dense subspace of $\mathcal{D}$ in a Hilbert space $\mathcal{H}$ and the ordered $*$-vector space $\mathcal{L}(\mathcal{D}, \mathcal{D}^\dagger)$. These have also appeared in Inoue [6, 16] and Tomita [2]. Let $\mathcal{M}$ be a closed O*-algebra on $\mathcal{D}$ in $\mathcal{H}$ and $\mathcal{D}^\dagger_{\mathcal{M}}$ a topological conjugate dual of the locally conver space $\mathcal{D}_{\mathcal{M}} \equiv \mathcal{D}[t_{\mathcal{M}}]$. The rigged Hilbert space $\mathcal{D}[t_{\mathcal{M}}] \subset \mathcal{H} \subset \mathcal{D}^\dagger_{\mathcal{M}}$ and the space of all continuous linear operators $C(\mathcal{D}_{\mathcal{M}}, \mathcal{D}^\dagger_{\mathcal{M}})$ from $\mathcal{D}_{\mathcal{M}}$ to the locally convex space $\mathcal{D}^\dagger_{\mathcal{M}}$ equipped with the strong topology β whose subspace $\mathcal{L}(\mathcal{D}_{\mathcal{M}}, \mathcal{D}^\dagger_{\mathcal{M}})$ of all continuous sesquilinear forms on $\mathcal{D}_{\mathcal{M}} \times \mathcal{D}^\dagger_{\mathcal{M}}$ have been studied in Kürsten [1, 3], Trapani [1] and Schmüdgen [21]. Theorem 4.1.7 and Theorem 4.1.10 are due to Schmüdgen [2], [21]. Theorem 4.1.13 is due to Lemma 5.2 in Inoue-Ueda-Yamauchi [1]. Theorem 4.1.14 is due to Lemma 5.2 and Theorem 5.3 in Inoue-Ueda-Yamauchi [1]. Quantum moment problem for partial O*-algebras has been studied in Antoine-Inoue [1] and Kürsten [3].

4.2. This work is due to Inoue-Kürsten [1].

4.3. Lemma 4.3.6 and Theorem 4.3.7 are due to Lemma 8.2 and Theorem 8.3 in Powers [1], respectively. Guder and Hudson [1] have considered the Tomita-Takesaki theory in the CCR-algebra $\pi_0(\mathcal{A})$, and obtained the same results as those in Theorem 4.1.3. Theorem 4.3.14 is due to Example 5.2 in Antoine-Inoue-Ogi-Trapani [1].

4.4 This work is due to Antoine-Inoue-Ogi-Trapani [1].

4.5 This work is due to Inoue [14].

References

J. ALCANTARA and D. A. DUBIN

[1] I*-algebras and their applications. *Publ. RIMS Kyoto Univ.* **17**(1981), 179-199.

[2] States on the current algebra. *Rep. Math. Phys.* **19**(1984), 13-26.

J.P. ANTOINE and A. INOUE

[1] Normal forms on partial O*-algebras. *J. Math. Soc.* **32**(1991), 2074-2081.

J.P. ANTOINE, A. INOUE and H. OGI

[1] Standard genelarized vectors for partial O*-algebras. *Ann. Inst. Henri Poincaré* **67**(1997), 223-258.

J.P. ANTOINE, A. INOUE, H. OGI and C. TRAPANI

[1] Standard generalized vectors in the space of Hilbert-Schmidt operators. *Ann. Inst. H. Poincaré* **63**(1995), 177-210.

J.P. ANTOINE, A. INOUE, and C. TRAPANI

[1] Partial ∗-algebras of closable operators I. The basic theory and the abelian case. *Publ. RIMS, Kyoto Univ.* **26**(1990), 359-395.

[2] Partial ∗-algebras of closable operators II. States and representations of partial ∗-algebras. *Publ. RIMS, Kyoto Univ.* **27**(1991), 399-430.

[3] Partial ∗-algebras of closable operators: A review. *Reviews Math. Phys.* **8**(1996), 1-42.

J.P. ANTOINE and W. KARWOWSKI

[1] Partial ∗-algebras of closable operators in Hilbert space. *Publ. RIMS, Kyoto Univ.* **21**(1985), 205-236.

H. ARAKI

[1] Einführung in die axiomatische Quantenfeldtheorie (Lecture notes, ETH Zürich 1962).

[2] Von Neumann algebras of local observables for free scalar field. *J. Math. Phys.* **5**(1964), 1-13.

[3] Some properties of modular conjugation operator of von Neumann algebras and a non-commutative Radon-Nikodym theorem with a chain rule. *Pacific J. Math.* **50**(1974), 309-354.

H. ARAKI and J. P. JURZAK

[1] On a certain class of $*$-algebras of unbounded operators. *Publ. RIMS, Kyoto Univ.* **18**(1982), 1013-1044.

D. ARNAL and J. P. JURZAK

[1] Topological aspects of algebras of unbounded operators. *J. Functional Anal.* **24**(1977), 397-425.

R. ASCOLI, G. EPIFANIO and A. RESTIVO

[1] On the mathematical description of quantized fields. *Commun. Math. Phys.* **18**(1970), 291-300.

F. BAGARELLO and C. TRAPANI

[1] "Almost" mean-field Ising model: an algebraic approach. *J. Stat. Phys.* **65**(1991), 469-482.
[2] States and representations of CQ*-algebras. *Ann. Inst. Henri Poincaré* **61**(1994), 103-133.
[3] CQ*-algebras: structure properties. *Publ. RIMS, Kyoto Univ.* **32** (1996), 85-116.

S. J. BHATT

[1] Representability of positive functionals on abstract star algebras without identity with applications to locally convex $*$-algebras. *Yokohama Math. J.* **29**(1981), 7-16.
[2] Structure of normal homomorphisms on a class of unbounded operator algebras. *Indian J. Pure Appl. Math.* **21**(1990), 150-154.
[3] An irreducible representation of a symmetric star algebra is bounded. *Trans. Amer. Math. Soc.* **292**(1985), 645-652.

S. J. BHATT, A. INOUE and H. OGI

[1] Unbounded C*-seminorms and unbounded C^*-spectral algebras. *Preprint, Fukuoka University* (1998).
[2] Admissibility of weights on non-normed $*$-algebras. *to appear in Trans. Amer. Math. Soc.*

J. J. BISOGNANO and E. H. WICHMANN

[1] On the duality condition for a Hermitian scalar field. *J. Math. Phys.* **16**(1975), 985-1007.
[2] On the duality condition for quantum fields. *J. Math. Phys.* **17**(1976), 303-321.

H.J. BORCHERS

[1] On the structure of the algebra of field operators. *Nuovo Cimento* **24**(1962), 214-236.
[2] Algebraic aspects of Wightman field theory, In Statistical Mechanics and Field Theory. *Lecture, 1971 ; Haifa Summer School, New York-Jerusalem-London* (1972), 31-79.

H.J. BORCHERS and J. YNGVASON

[1] On the algebra of field operators. The weak commutant and integral decomposition of states. *Commun. Math. Phys.* **42**(1975), 231-252.
[2] Positivity of Wightman functionals and the existence of local nets. *Commun. Math. Phys.* **127**(1990), 607-615.

H.J. BORCHERS and W. ZIMMERMANN

[1] On the self-adjointness of field operators. *Neuvo Cim.* **31**(1963), 1047-1059.

O. BRATTELI and D.W. ROBINSON

[1] Operator Algebras and Quantum Statistical Mechanics I. *Springer Verlag, New York, Heidelberg, Berlin* (1979).
[2] Operator Algebras and Quantum Statistical Mechanics II. *Springer Verlag, New York, Heidelberg, Berlin* (1981).

R. M.BROOKS

[1] Some algebras of unbounded operators. *Math. Nachr.* **56**(1973), 47-62.

A. CONNES

[1] Une classification de facteurs de type III. *Ann. École Norm. Sup., 4-ieme Sér.* **6**(1973), 133-252.
[2] Caractérisation das espaces vectoriels ordonnés sous-jacents aux algebres de von Neumann. *Ann. Inst. Fourier, Grenoble.* **24**(1974), 121-155.

F. DEBACKER-MATHOT

[1] Integral decomposition of unbounded operator families. *Commun. Math. Phys.* **71**(1980), 47-58.

J. DIXMIER

[1] Sur la relation $i(PQ - QP) = I$. *Composito Math.* **13**(1958), 263-269.
[2] Sur les algebres de Weyl I. *Bull. Soc. Math. France* **96**(1968), 209-242.
[3] C*-Algebras. *North-Holland Publ. Comp., Amsterdam,* (1977).
[4] Von Neumann Algebras. *North-Holland Publ. Comp., Amsterdam,* (1981).

P. G. DIXON

[1] Generalized B*-algebras. *Proc. London Math. Soc.* **21**(1970), 693-715.
[2] Unbounded operator algebras. *Proc. London Math. Soc.* **23**(1971), 53-69.

W. DRIESSLER and J. FRÖHLICH

[1] The reconstruction of local observable algebras from the Euclidean Green's functions of relativistic quantum field theory. *Ann. Inst. Henri Poincaré* **27**(1977), 221-236.

W. DRIESSLER, S. J. SUMMERS and E. H. WICHMANN

[1] On the connection between quantum fields and von Neumann algebras of local operators. *Commun. Math. Phys.* **105**(1986), 49-84.

D. A. DUBIN and M. A. HENNINGS

[1] Quantum Mechanics, Algebras and Distributions. *Pitman Research Notes in Math. Series, Longman, Harlow* (1990).

D. A. DUBIN and J. SOTELO-CAMPOS

[1] A theory of quantum measurement based on the CCR algebra $L^\dagger(W)$. *Z. Anal. Anw.* **5**(1986), 1-26.

G. O. S. EKHAGUERE

[1] Unbounded partial Hilbert algebras. *J. Math. Phys.* **30**(1989), 1964-1975.

G. EPIFANIO, T. TODOROV and C. TRAPANI

[1] Complete sets of compatible non self-adjoint observables: an unbounded aproach. *J. Math. Phys.* **37**(1996), 1148-1160.

G. Epifanio and C. Trapani

[1] V*-algebras: an extension of the concept of von Neumann algebras to unbounded operators. *J. Math. Phys.* **25**(1984), 2633-2637.

[2] Quasi $*$-algebras valued quantized fields. *Ann. Inst. H. Poincaré* **46**(1987), 175-185.

W. G. FARIS

[1] Self-adjoint operators. *Lecture Notes in Mathematics,433 Springer-Verlag,* (1975).

J. FRIEDRICH and K. SCHMÜDGEN

[1] n-Positivity of unbounded $*$-representations. *Math. Nachr.* **141**(1989), 233-250.

I. M. GELFAND and N. YA. VILENKIN

[1] Generalized functions. Vol.4, *Academic Press, New York,* 1972.

B. FUGLEDE

[1] On the relation $PQ - QP = -iI$. *Math. Scand.* **20**(1967), 79-88.

[2] Conditions for two self-adjoint operators to commute or to satisfy the Weyl relation. *Math. Scand.* **51**(1982), 163-178.

J. GLIMM and A. JAFFE

[1] Quantum Physics. *Berlin, Heidelberg, New York: Springer.* (1981).

S.P. GUDDER

[1] A Radon-Nikodym theorem for ∗-algebras. *Pacific J. Math.* **70**(1979), 141-149.

S.P. Gudder and R.L. Hudson

[1] A noncommutative propability theory. *Trans. Amer. Math. Soc.* **245**(1978), 1-41.

S.P. GUDDER and W. SCRUGGS

[1] Unbounded representations of ∗-algebras. *Pacific J. Math.* **70**(1977), 369-382.

R. HAAG and D. KASTLER

[1] An algebraic approach to quantum field theorym. *J. Math. Phys.* **5**(1964), 848-861.

U. HAAGERUP

[1] Normal weights on W*-algebras *J. Functional Analysis.* **19**(1975), 305-317.

G. C. HEGERFELDT

[1] Extremal decomposition of Wightman functions and states on nuclear ∗-algebras by Choquet theory. *Commun. Math. Phys.* **45**(1975), 133-135.

W. D. HEINRICHS

[1] The density property in Fréchet-domains of unbounded operator ∗-algebras. *Math. Nachr.* **165**(1994), 49-60.

[2] On unbounded positive ∗-representations on Fréchet-domain. *Publ. RIMS, Kyoto Univ.* **30**(1994), 1123-1138.

[3] Topological tensor products of unbounded operator algebras on Fréchet domains. *Publ. RIMS, Kyoto Univ.* **33**(1997), 241-255.

P. D. HISLOP and R. LONGO

[1] Modular structure of the local algebras associated with the free massless scalar field theory. *Commun. Math. Phys.* **84**(1982), 71-85.

I. IKEDA and A. INOUE

[1] Invariant subspaces of closed ∗-representations. *Proc. Amer. Math. Soc.* **116**(1992), 737-745.

[2] On types of positive linear functionals of ∗-algebras. *J. Math. Anal. Appl.* **173**(1993), 276-288.

I. IKEDA, A. INOUE and M. TAKAKURA

[1] Unitary equivalence of unbounded $*$-representations of $*$-algebras. *Math. Proc. Camb. Phil. Soc.* **122**(1997), 269-279.

A. Inoue

[1] On a class of unbounded operator algebras. *Pacific J. Math.* **65** (1976), 77-95.
[2] On a class of unbounded operator algebras II. *Pacific J. Math.* **66** (1976), 411-431.
[3] On a class of unbounded operator algebras III. *Pacific J. Math.* 69(1977), 105-115.
[4] Unbounded generalization of left Hilbert algebras. *J. Functional Analysis* **34**(1979), 339-362
[5] Unbounded generalization of left Hilbert algebras II. *J. Functional Analysis* **35**(1980), 230-250.
[6] Positive linear functionals on dynamical systems. *Fukuoka Univ. Sci. Reports* **12**(1982), 9-16.
[7] On regularity of positive linear functionals. *Japanese J. Math.* **19**(1983), 247-275.
[8] A Radon-Nikodym theorem for positive linear functionals on $*$-algebras. *J. Operator Theory* **10**(1983), 77-86.
[9] An unbounded generalization of the Tomita-Takesaki theory. *Publ. RIMS, Kyoto Univ.* **22**(1986), 725-765.
[10] An unbounded generalization of the Tomita-Takesaki theory II. *Publ. RIMS, Kyoto Univ.* **23**(1987), 673-726.
[11] Self-adjointness of the $*$-representation generalized by the sum of two positive linear functionals. *Proc. Amer. Math. Soc.* **107**(1989), 665-674.
[12] Standard O_p^*-algebras. *J. Math. Anal. Appl.* **161**(1991), 555-565.
[13] Extension of unbounded left Hilbert algebras to partial $*$-algebras. *J. Math. Phys.* **32**(1991), 323-331.
[14] Modular structure of algebras of unbounded operators. *Math. Proc. Camb. Phil. Soc.* **111**(1992), 369-386.
[15] Modular systems induced by trace functionals on algebras of unbounded operators. *J. Math. Phys.* **35**(1994), 435-442.
[16] Weak regularity of positive linear functionals on $*$-algebras without identity. *Bull. Inst. Adv. Res. Fukuoka* **163**(1994), 37-58.
[17] Standard generalized vectors for algebras of unbounded operators. *J. Math. Soc. Japan* **47**(1995), 329-347.
[18] O^*-algebras in standard system. *Math. Nachr.* **172**(1995), 171-190.
[19] Standard systems for semifinite O^*-algebras. *Proc. Amer. Math. Soc.* **125**(1997), 3303-3312.

A. Inoue and W. Karwowski

[1] Cyclic generalized vectors for algebras of unbounded operators. *Publ. RIMS, Kyoto Univ.* **30**(1994), 577-601.

A. Inoue, W. Karwowski and H. Ogi

[1] Standard weights on algebras of unbounded operators. *to appear in J. Math. Soc. Japan.*

A. Inoue, K. Kuriyama and S. Ôta

[1] Topologies on unbounded operator algebras. *Mem. Fac. Sci. Kyushu Univ.* **33**(1979), 355-375.

A. Inoue, H. Kurose and S. Ôta

[1] Extensions of unbouded representations. *Math. Nachr.* **155**(1992), 257-268.

A. Inoue and K. D. Kürsten

[1] Trace representation of weights on algebras of unbounded operators. Reprint, Fukuoka University (1998).

A. Inoue and H. Ogi

[1] Regular weights on algebras of unbounded operators. *J. Math. Soc. Japan* **50**(1998), 227-252.

A. Inoue and S. Ôta

[1] Derivations on algebras of unbounded operators. *Trans. Amer. Math. Soc.* **261**(1980), 567-577.

A. Inoue, S. Ôta and J. Tomiyama

[1] Derivations of operator algebras into spaces of unbounded operators. *Pacific J. Math.* **96**(1981), 389-404.

A. Inoue and K. Takesue

[1] Self-adjoint representations of polynomial algebras. *Trans. Amer. Math. Soc.* **280**(1983), 393-400.

[2] Spatial theory for algebras of unbounded operators II. *Proc. Amer. Math. Soc.* **87**(1983), 295-300.

A. Inoue, H. Ueda and T. Yamauchi

[1] Commutants and bicommutants of algebras of unbounded operators. *J. Math. Phys.* **28**(1987), 1-7.

P.E.T. Jorgensen

[1] Selfadjoint extension of operators commuting with an algebra. *Math. Z.* **169**(1979), 41-62.

[2] Operators and Representation Theory. *North-Holland, Amsterdam,* 1988.

P.E.T. Jorgensen and R.T. Moore

[1] Operator Commutation Relations. *D. Reidel Publ. Comp. Dordrecht* (1984).

R. JOST

[1] The general theory of quantized fiedls. *Providence RI: Am. Math. Soc.* (1963).

H. JUNEK

[1] Maximal O_p^*-algebras on DF-domains. *Z. Anal. Anw.* **9**(1990), 403-414.

J. P. JURZAK

[1] Simple facts about algebras of unbounded operators. *J. Functional Analysis.* **21**(1976), 469-482.
[2] Unbounded operator algebras and DF-spaces. *Publ. RIMS, Kyoto Univ.* **17**(1981), 755-776.
[3] Unbounded Non-commutative Integration. *D. Reidel Publ. Comp., Dordrect,* 1986.

R.V. KADISON

[1] Algebras of unbounded functions and operators. *Expo. Math.* **4**(1986), 3-33.

R.V. KADISON and J.R. RINGROSE

[1] Fundamentals of the Theory of Operator Algebras. Vol. I, *Academic Press, New York,* 1983.
[2] Fundamentals of the Theory of Operator Algebras. Vol. II, *Academic Press. New York,* 1986.

A. KASPAREK and VAN DAELE

[1] On the strong unbounded commutant of an O*-algebra. *Proc. Amer. Math. Soc.* **105**(1989), 111-116.

K. KATAVOLOS and I. KOCH

[1] Extension of Tomita-Takesaki theory to the unbounded algebra of the canonical commutation relations. *Rep. Math. Phys.* **16**(1979), 335-352.

T. KATO

[1] Perturbation Theory for Linear Operators. *Springer-Verlag, Berlin,* (1966).

J. L. KELLY

[1] General topology. *D. Van Nstrand Co., Toronto-New York-London* (1955).

H. KOSAKI

[1] Lebesgue decomposition of states on a von Neumann algebra. *Amer J. Math.* **107**(1985), 679-735.

G. KÖTHE

[1] Topological vector spaces. Vol. II. *Springer-Verlag, Berlin* (1979).

H. KUROSE and H. NAKAZATO

[1] Geometric construction of $*$-representations of the Weyl algebra with degree 2. *Publ. RIMS, Kyoto Univ.* **32**(1996), 555-579.

H. KUROSE and H. OGI

[1] On a generalization of the Tomita-Takesaki theorem for a quasifree state on a self-dual CCR-algebra. *Nihonkai Math. J.* 1(1990), 19-42.

K. D. KÜRSTEN

[1] The completion of the maximal O_p^*-algebra on a Frechet domain. *Publ. RIMS Kyoto Univ.* **22**(1986), 151-175.

[2] Two-sided closed ideals of certain algebras of unbounded operators. *Math. Nachr.* **129**(1986), 157-166.

[3] Duality for maximal O_p^*-algebras on Frechet domains. *Publ. RIMS Kyoto Univ.* **24** (1988), 585-620.

[4] On topological linear spaces of operators with unitary domain of definition. *Wiss. Z. Univ. Leipzig, Math.-Nat.-wiss. Reihe* **39**(1990), 623-655.

[5] On commutatively dominated O_p^*-algebras with Fréchet domains. *J. Math. Anal. Appl.* **157**(1991), 506-526.

K. D. KÜRSTEN and M. MILDE

[1] Calkin representations of unbounded operator algebras acting on non-separable doamins. *Math. Nachr.* **154**(1991), 285-300.

G. LASSNER

[1] Topological algebras of operators. *Rep. Math. Phys.* **3**(1972), 279-293.

[2] Topologien auf O_p^*-algebren. *Wiss. Z. KMU Leipzig, Math.-Naturw. R.* **24**(1975), 465-471.

[3] Topological algebras and their applications in quantum statistics. *Wiss. Z. KMU Leipzig, Math.-Naturw. R.* **30**(1981), 572-595.

[4] Alegbras of unbounded operators and quantum dynamics. *Physica* **124 A**(1984), 471-480.

G. LASSNER and G.A. LASSNER

[1] On the continuity of entropy. *Rep. Math. Phys.* **15**(1980), 41-46.

[2] Qu*-algebrass and twisted product. *Publ. RIMS, Kyoto Univ.* **25**(1989), 279-299.

G. LASSNER and W. TIMMERMANN

[1] Normal states on algebras of unbounded operators. *Rep. Math. Phys.* **3**(1972), 295-305.

[2] Classification of domains of operator algebras. *Rep. Math. Phys.* **9**(1976), 205-217.

G. Lassner and A. Uhlmann

[1] On positive functionals on algebras of test functions for quantum fields. *Commun. Math. Phys.* **7**(1968), 152-159.

F. Löffler and W. Timmermann

[1] The Calkin representation for a certain class of algebras of unbounded operators. *Rev. Roum. Pur. et Appl.* **31**(1986), 891-903.
[2] Singular states on maximal Op^*-algebras. *Publ. RIMS, Kyoto Univ.* **22**(1988), 671-687.
[3] On the structure of the state space of maximal Op^*-algebras. *Publ. RIMS, Kyoto Univ.* **22**(1986), 1063-1078.
[4] On the irreducibility of generalized Calkin representations. *Rev. Roum. Math. Pur. et Appl.* **33**(1988), 413-421.

F. Mathot

[1] Topological properties of unbounded bicommutants. *J. Math. Phys.* **26**(1985), 1118-1124.

E. Nelson

[1] Analytic vectors. *Ann. Math.* **70**(1959), 572-615.

A.E. Nussbaum

[1] Reduction theory for unbounded closed operators in Hilbert space. *Duke Math. J.* **31**(1964), 33-44.
[2] On the integral representation of positive linear functionals. *Trans. Amer. Math. Soc.* **128**(1967), 460-473.
[3] A commutativity theorem for unbounded operators. *Trans. Amer. Math. Soc.* **140**(1969), 485-493.
[4] Quasi-analytic vectors. *Ark. Math.* **6**(1967), 179-191.

H. Ogi

[1] On KMS states for self-dual CCR algebras and Bogoliubov automorphism groups. *Math. Proc. Camb. Phil. Soc.* **110**(1991), 191-197.
[2] The normality of strongly positive states on the CCR algebra. *Rep. Math. Phys.* **31**(1992), 139-146.
[3] On ground states for CCR-algebras and Bogoliubov automorphism groups. *Math. Proc. Camb. Phil. Soc.* **119**(1996), 419-424.

H. Ogi and A. Inoue

[1] Regular quasi-weights on algebras of unbounded operators. *GROUP 21, Physical Applications and Mathematical Aspects of Geometry, Group, and Algebras Vol.1*(1996), 339-343. eds. H-D Doebner et al. World Scientific.

[2] On an interacting Boson model in O*-algebra frame work. *to appear in 5th International Wigner Symposium Proceedings, World Scientific.*

S. ÔTA

[1] Unbounded representations of a $*$-algebra on indefinite metric space. *Ann. Inst. Henre Poincaré* **48**(1988), 333-353.

G. PEDERSEN and M. TAKESAKI

[1] The Radon-Nikodym theorem for von Neumann algebras. *Acta Math.* **130**(1973), 53-88.

R.T. POWERS

[1] Self-adjoint algebras of unbounded operators. *Commun. Math. Phys.* **21**(1971), 85-124.

[2] Self-adjoint algebras of unbounded operators II. *Trans. Amer. Math. Soc.* **187**(1974), 261-293.

[3] Algebras of unbounded operators. *Proc. Sym. Pure Math.* **38**(1982), 389-406.

M. REED and B. SIMON

[1] Methods of Modern Mathematical Physics. Vol. I. *Academic Press, New York* (1972).

[2] Methods of Modern Mathematical Physics. Vol. II. *Academic Press, New York* (1975).

M.A. RIEFFEL and A. VAN DAELE

[1] A bounded operator approach to Tomita-Takesaki theory. *Pacific J. Math.* **69**(1977), 187-221.

R. ROUSSEAU, A. VAN DAELE and L. VANHEESWIJCK

[1] A necessary and sufficient condition for a von Neumann algebra to be in standard form. *J. London Math. Soc.* **15**(1977), 147-154.

D.RUELLE

[1] On the asymptotic condition in quantum field theory. *Helv. Phys. Acta.* **35**(19629, 147-163.

S.SAKAI

[1] C*-algebras and W*-algebras. *Springer-Verlag*, 1971.

[2] Operators Algebras in Dynamical Systems. *Cambridge Univ. Press*, 1991.

K. SCHMÜDGEN

[1] The order structure of topological $*$-algebras of unbounded operators I. *Rep. Math. Phys.* **7**(1975), 215-227.

[2] On trace representation of linear functionals on unbounded operator algebras. *Comm. Math. Phys.* **63**(1978), 113-130.
[3] An example of a positive polynomial which is not a sum of squares of polynomials. A positive, but not strongly positive functional. *Math. Nachr.* **88**(1979), 385-390.
[4] A proof of a theorem on trace representation of strongly positive linear functionals on O_p^*-algebras. *J. Operator Theory* **2**(1979), 39-47.
[5] Uniform topologies on enveloping algebras. *J. Funct. Analysis* **39**(1980), 57-66.
[6] On topologization of unbounded operator algebras. *Rep. Math. Phys.* **17**(1980), 359-371.
[7] Two theorems about topologies on countably generated O_p^*-algebras. *Acta. Math. Acad. Sci. Hungar.* **35**(1980), 139-150.
[8] Graded and filtrated $*$-algebras I. Graded normal topologies. *Rep. Math. Phys.* **18**(1980), 211-229.
[9] On the Heisenberg commutation relation II. *Publ. RIMS, Kyoto Univ.* **19**(1983), 601-671.
[10] On domains of powers of closed symmetric operators. *J. Operator Theory* **9**(1983), 53-75. Correction : ibid, **12**(1984), 199.
[11] On restrictions of unbounded symmetric operators. *J. Operator Theory* **11**(1984), 379-393.
[12] On commuting unbounded self-adjoint operators. *Acta Sci. Math. Szeged* **47**(1984), 131-146.
[13] On commuting unbounded self-adjoint operators. III. *Manuscripta Math.* **54**(1985), 221-247.
[14] On commuting unbounded self-adjoint operators. IV. *Math. Nachr.* **125** (1986), 83-102.
[15] A note on commuting unbounded self-adjoint operators affiliated to properly infinite von Neumann algebras. *Bull. London Math. Soc.* **16**(1986), 287-292.
[16] Topological realizations of Calkin algebras on Frechet domains of unbounded operator algebras. *Z. Anal. Anw.* **5**(1986), 481-490.
[17] Unbounded commutants and intertwining spaces of unbounded symmetric operators and $*$-representations. *J. Functional Analysis* **71**(1987), 47-68.
[18] Strongly commuting self-adjoint operators and commutants of unbounded operator algebras. *Proc. Amer. Math. Soc.* **102**(1988), 365-372.
[19] Spaces of continuous sesquilinear forms associated with unbounded operator algebras. *Z. Anal. Anw.* **7**(1988), 309-319.
[20] A note on the strong operator topology of countably generated O^*-vector spaces. *Z. Anal. Anw.* **8**(1989), 425-430.
[21] Unbounded operator Algebras and Representation Theory. *Akademie-Verlag Berlin* (1990).
[22] Non-commutative moment problems. *Math. Z.* **206**(1991), 623-650.

K. SCHMÜDGEN and J. FRIEDRICH

[1] On commuting unbounded self-adjoint operators. II. *J. Integral Equ. and Operator Theory* **7**(1984), 815-867.

M. SCHRÖDER and W. TIMMERMANN

[1] Invariance of domains and automorphisms in algebras of unbounded operators. *Proc. Int. Conf. on Operator algebras and Group Representations, Romania* (1980), 134-139.

I.E. SEGAL

[1] A noncommutative extension of abstract integration. *Ann. Math.* **57**(1953), 401-457.

T. SHERMAN

[1] Positive linear functionals on $*$-algebras of unbounded operators. *J. Math. Anal. Appl.* **22**(1968), 285-318.

S. STRATILA and L. ZSIDO

[1] Lectures on von Neumann algebras. *Abacus Press, Tunbridge Wells* (1979).

S. STRATILA

[1] Modular Theory in Operator Algebras. *Abacus Press, Tunbridge Wells* (1981).

R. F. STREATER and A. S. WIGHTMAN

[1] P. C. T. spin and statistics and all that. *New York : Benjamin* (1964).

M. TAKESAKI

[1] Tomita's theory of modular Hilbert algebras and its applications. *Lecture Notes in Mathematics*, 128 *Springer*, (1970).

[2] Theory of Operator Algebras I. *Springer-Verlag, New York, 1979.*

K. TAKESUE

[1] Spatial theory for algebras of unbounded operators. *Rep. Math. Phys.* **21**(1985), 347-355.

W. THIRRING

[1] On the mathematical structure of the B.C.S.-model. II. *Commun. Math. Phys.* **7**(1968), 181-189.

W. THIRRING and A. WEHRL

[1] On the mathematical structure of the B.C.S.-model. *Commun. Math. Phys.* **4**(1967), 303-324.

W. Timmermann

[1] On an ideal in algebras of unbounded operators. *Math. Nachr.* **91**(1979), 347-355.
[2] Ideals of algebras of unbounded operators. *Math. Nachr.* **92**(1979), 99-100.
[3] On commutators in algebras of unbounded operators. *Z. Anal. Anw.* **7**(1988), 1-14.

M. Tomita

[1] Standard forms of von Neumann algebras. *The Vth functional analysis symposium of the Math. Soc. of Japan, Sendai* (1967).
[2] Foundations of noncommutative Fourier analysis. Japan-US Seminar on C*-algebras and Applications to Physics, Kyoto 1974.

C. Trapani

[1] Quasi ∗-algebras of operators and their applications. *Rev. Math. Phys.* **7**(1995), 1303-1332.

A. Uhlmann

[1] Über die Definition der Quantenfelder nach Wightman and Haag. *Wizz. Z. KMU Leipzig, Math. Naturw. R.* **11**(1962), 213-217

A. Van Daele

[1] A new approach to the Tomita-Takesaki Theory of generalized Hilbert algebras. *J. Functional Analysis* **15**(1974), 378-393.

A. N. Vasiliev

[1] Theory of representations of a topological (non-Banach) involutory algebras. *Theor. Math. Phys.* **2**(1970), 113-123.

J. von Neumann

[1] Die Eindentigkeit der Schrödingerschen operatoren. *Math. Ann.* **104**(1931), 570-578.

S.L. Woronowicz

[1] The quantum moment problem I. *Rep. Math. Phys.* **1**(1970), 135-145.
[2] The quantum moment problem II. *Rep. Math. Phys.* 1(1971), 175-183.

J. Yngvanson

[1] On the algebra of test functions for field operators. *Commun. Math. Phys.* **34**(1973), 315-333.

Index

Springer and the environment

At Springer we firmly believe that an international science publisher has a special obligation to the environment, and our corporate policies consistently reflect this conviction.

We also expect our business partners – paper mills, printers, packaging manufacturers, etc. – to commit themselves to using materials and production processes that do not harm the environment. The paper in this book is made from low- or no-chlorine pulp and is acid free, in conformance with international standards for paper permanency.

Springer

Printing: Weihert-Druck GmbH, Darmstadt
Binding: Buchbinderei Schäffer, Grünstadt